Gut–Brain Connection, Myth or Reality?

Role of The Microbiome in Health and Disease

Gut–Brain Connection, Myth or Reality?

Role of The Microbiome in Health and Disease

Editor

Adrien A. Eshraghi *MD, MSc, FACS*

Professor of Otolaryngology,
Neurological Surgery, Pediatrics and Biomedical Engineering
Chief of Jackson Memorial Otology and Neurotology Clinics
Director of Hearing Research and Communication Disorders Laboratory
Co-Director of University of Miami Ear Institute
Vice-Speaker of Medical Faculty Council
University of Miami Miller School of Medicine

World Scientific

NEW JERSEY • LONDON • SINGAPORE • BEIJING • SHANGHAI • HONG KONG • TAIPEI • CHENNAI • TOKYO

Published by

World Scientific Publishing Co. Pte. Ltd.
5 Toh Tuck Link, Singapore 596224
USA office: 27 Warren Street, Suite 401-402, Hackensack, NJ 07601
UK office: 57 Shelton Street, Covent Garden, London WC2H 9HE

British Library Cataloguing-in-Publication Data
A catalogue record for this book is available from the British Library.

GUT–BRAIN CONNECTION, MYTH OR REALITY?
Role of The Microbiome in Health and Diseases

ISBN 978-981-122-114-9 (hardcover)
ISBN 978-981-122-115-6 (ebook for institutions)
ISBN 978-981-122-116-3 (ebook for individuals)

For any available supplementary material, please visit
https://www.worldscientific.com/worldscibooks/10.1142/11853#t=suppl

Dedicated To My Inspirations,
Melody and Nicolas

Grateful For My Mentors,
Dr. Fred F. Telischi, Dr. Thomas J. Balkany
and Dr. Thomas R. Van de Water

Thankful For My Collaborators,
Dr. Rahul Mittal, Dr. Jenna Bergman, Camron Davies
and Dr. Rebecca Sherry Eshraghi

Foreword by Professor Luc Montagnier

The ever-advancing knowledge and discovery of the gut–brain connection is one of the most intriguing developments in medicine in recent years. Indeed, the bidirectional influence of the gut microbiome on the brain and vice versa is not to be underestimated. In fact, the gut's influence on the brain is so dynamic that it is referred to as the "Second Brain".

This book entitled ***Gut–Brain Connection: Myth or Reality?*** comprises comprehensive chapters that will significantly enhance our understanding on the gut–brain axis in our well-being and **the role of the microbiome in health and disease**. While the chapters have been written in a scientific manner using evidence-based research, it is also presented in an inclusive style to make it accessible to all readers. The audience of this book includes physicians and other health care providers, researchers, medical students, and enthusiastic patients who are interested in delving deeper into the field of the gut–brain axis and the role of the microbiome in our well-being.

The gut microbiome plays a crucial role in the regulation of the gut–brain axis. Our gut contains trillions of microbes that determine predisposition to a number of disorders. Under normal conditions, the gut microbiome contributes to tissue homeostasis and produces many factors such as neurotransmitters to facilitate bidirectional communication between gut–brain axis. Gut microbes produce numerous short-chain fatty acids (SCFA) such as butyrate, propionate and acetate, which have profound effects on the physiology and brain functions such as appetite, metabolism, digestion, causing neuro-inflammation and much more.

Alteration in the gut microbiome can deleteriously disrupt the gut–brain axis and can have implications in various chronic gastrointestinal and neurological disorders. Recent studies have shown that manipulation of gut microbiome through diet, prebiotics, probiotics and microbial transplant therapy has promising results in reducing clinical manifestations of a wide variety of disorders. All these exciting findings have been presented in a comprehensively in this book.

I congratulate Professor Adrien Eshraghi and his team of experts on this outstanding and thought-provoking compilation of comprehensive independent chapters in this book. This book discusses the role of the microbiome in relation to gut–brain axis in various disorders such as autism spectrum disorders, migraines, Alzheimer's disease, multiple sclerosis, anxiety and depression, accelerated aging, obesity, type two diabetes, HIV and chronic GI disease. I encourage every curious and inquisitive mind to learn more about cutting edge advances in how the microbiome and gut brain connection influences health and disease as outlined in this book!

Professor Luc Montagnier
Nobel Prize Laureate in Medicine

Preface

The gut–brain connection is one of the most promising physiological targets in a rapidly increasing number of pathological health conditions. Recent evidence suggests that the gastrointestinal (GI) microbiome may play a significant role in many disease states. In this book, *Gut–Brain Connection: Myth or Reality*, expert researchers and clinicians from diverse fields share their perspectives and knowledge of the human microbiome, the pathophysiology underlying the gut–brain axis, the clinical implications of this bidirectional relationship, and the future perspectives within the field. Timely topics such as genomic sequencing, transcriptomics, proteomics, and metabolomics are also discussed.

Every chapter is independent and begins with a summary outlining the key points, presents in a comprehensive manner the subject, and concludes with a collection of relevant citations. Figures and tables are also included to enrich the content. In this manner, this book may be considered a reference for researchers, clinicians, medical students, other healthcare providers, and enthusiastic patients who are interested in exploring whether the gut–brain connection is indeed, a myth or reality.

The first chapter expands on the historical aspects about our understanding of the gut microbiome to highlight the evolution and innovation in the field of microbiology that led to today's technological advancements. With the progress in the field of gene sequencing in the 21st century came access to proteomics and metabolomics, which allowed for better identification of the microbiome, its metabolites, and a greater understanding of the role these interactions play in epigenetic regulation. As more knowledge is gleaned about the host–microbiota interaction, the apparent impact of the microbiome in health and disease can be more deeply explored.

In the second chapter, the crucial role of the microbiome from the onset of human development is explored. An infant's gut microbiome is influenced by the mode of pregnancy delivery, diet, and antigen exposure. Any disruption in the development of the GI microbiome can affect cognitive function and may lead to the development of neurological disorders later in life.

Chapter 3 investigates the direct influence that the microbiome can have on the central nervous system by regulating levels of neurotransmitters such as serotonin, dopamine, and gamma-aminobutyric acid (GABA). Uncovering the molecular mechanisms underlying how the gut microbiota influence neurological functioning may provide novel opportunities for targeted interventions in various neurological and GI disorders.

The connection between the gut microbiome and neuroinflammation is discussed in Chapter 4. The gut microbiome regulates the immune system function through various physiological mechanisms and serves a vital role in protecting against pathogens. Conversely, dysbiosis in the gut can lead to immune dysfunction, systemic inflammation, and chronic disease. Research suggests that the intestinal microbiota can affect the functioning of the neuroimmune system and the blood–brain barrier.

Besides neuroinflammation, our symbiotic gut microbiome is also inextricably linked with host neurophysiology, blurring the lines between host and tenant. Bacteria not only produce many biologically active molecules such as short-chain fatty acids and serotonin, but they are also able to modify their function and by-products based on their microbial neighbors and environment. Chapter 5 details these interactions and explores the concept of quorum sensing; the mechanism by which the gut microbiome can function as a more cohesive unit and adapt to a fluctuating intestinal environment.

Chapter 6 explores the current tools used to investigate the anatomical and functional anatomy of the brain concerning the gut microbiome, such as functional and diffusion weight magnetic resonance imaging (MRI). After discussing the limitations of current and novel technologies such as tractography, we highlight future implications and directions needed to provide precise, non-invasive imaging of the gut–brain axis. We also discuss the non-invasive gut imaging modality of electrogastrography (EGG), and the improvements needed to better map the gut–brain axis.

We then look at evidence that demonstrates how microbial products are an essential component of host metabolic pathways in Chapter 7. These metabolites circulate throughout the entire human body and cross the blood–brain barrier to the central nervous system, serving as a bidirectional communication network between the gut and brain. Further understanding of the inner workings of this network is the key to managing the pathological processes that result from dysbiosis.

There has been an increased interest in modulating the gut–brain axis through prebiotics and psychobiotics. Prebiotics are naturally occurring non-digestible food components, whereas psychobiotics include live bacteria that are capable of affecting the gut–brain axis. A combination of prebiotics and psychobiotics, known as synbiotics, have a synergistic effect and can be beneficial in ameliorating inflammatory disorders. In Chapter 8, we discuss the possible therapeutic applications of these findings.

Chapter 9 is clinically oriented and discusses the mounting research that implicates the gut microbiome in a wide range of pathological states due to its ability to alter the body's inflammatory response. We discuss how a dysbiotic gut microbiome can induce systemic inflammation leading to neuroinflammation, while a eubiotic microbiome can reduce inflammation and induce immunotolerance. In this comprehensive chapter, we will examine the role that the microbiome and diet plays in a wide range of human disorders, including its role in accelerated aging, obesity, type 2 diabetes, Alzheimer's disease, multiple sclerosis, autism spectrum disorder, central processing disorders, anxiety, and depression.

In Chapter 10, we explore the connection between gut dysbiosis and human immunodeficiency virus (HIV). Gut dysbiosis that occurs with HIV-1 infection may disrupt normal intestinal homeostasis and contribute to HIV-associated immune activation and inflammation. The gut microbiome can play a crucial role in the neurological manifestations of HIV infections, such as dementia. Restoring healthy gut microbiome through supplementation with prebiotics, probiotics, and microbial transplantation has a great potential for designing novel treatment paradigms for HIV-1-associated gut and neurological comorbidities.

We further delve into microbiome-related therapeutic interventions in Chapter 11 with fecal microbiome transplant (FMT) therapy. FMT consists of transplanting healthy donor fecal microbes into the recipient patient, either orally, through an enema, or colonoscopy. Several recent studies have demonstrated the beneficial effects of FMT therapy for specific neurological disorders. Further studies aimed at better defining the composition of a healthy donor sample as well as determining the safety and efficacy of FMT therapy are necessary before it can serve as an effective therapeutic modality.

Zebrafish is an emerging animal model in translational research for understanding many physiological conditions, including the gut–brain axis. In Chapter 12, we discuss this model that offers several advantages such as transparent larvae, genetic and physiological homology to humans, external fertilization, the rapid development of organ systems, genetic tractability, and a fully sequenced genome. Further studies employing gene-editing tools such as CRISPR/Cas9 will facilitate in identifying genes involved in the bidirectional communication between the gut–brain axis using zebrafish models.

Chapter 13 discusses the beneficial role that occupational therapists play in modulating the gut–brain axis. Occupational therapy can help in promoting a healthy gut microbiome in various ways. They can improve the effects of chronic disease by providing targeted interventions such as physical exercises, dietary changes, and sleep modifications aimed at improving the overall functioning of the gut–brain axis.

In the final chapter of this book, we discuss the tightly regulated interactions between the GI tract and enteric nervous system (ENS). The ENS sends neurochemical signals to modulate immune function within the GI tract, which contains trillions of microbes that collectively make up the microbiome. To understand the changes in gut–brain axis function in these disorders, preclinical animal models are an essential tool to elucidate the complex interactions between host genetics and influences from the external environment. Exciting advances in this field are focused on deciphering mechanisms of the gut–brain axis to identify potential targets for personalized medicine and improve health.

Finally, I would like to express my deep gratitude to all the contributors and experts that accepted the opportunity to share their outstanding research in these very exciting chapters. I would also like to recognize and give a special thank you to Dr. Rahul Mittal, Dr. Jenna Bergman, Camron Davies, and my wife, Dr. Rebecca Sherry Eshraghi, for their invaluable input and help in putting together this comprehensive textbook. Without their tireless dedication and countless hours spent during the weekends, we would not have been able to accomplish this collection of beautiful articles that touch this novel area of scientific advances. Finally, I hope you, dear reader, enjoy being part of this adventure.

Adrien A. Eshraghi MD, MSc, FACS
Professor of Otolaryngology, Neurological Surgery,
Pediatrics and Biomedical Engineering
University of Miami Miller School of Medicine

Contents

List of Corresponding Authors of the Chapters

Adrien A. Eshraghi, MD, MSc, FACS, Professor of Otolaryngology, Neurological Surgery, Pediatrics and Biomedical Engineering, Chief, Jackson Memorial Otology and Neurotology Clinics, Director, University of Miami Hearing Research Laboratory, Co-Director, University of Miami Ear Institute, University of Miami-Miller School of Medicine, Miami, USA.

Melanie Gareau, PhD, Associate Professor, Department of Anatomy, Physiology and Cell Biology, School of Veterinary Medicine, University of California Davis, Davis, USA.

Mehdi Mirsaeidi, MD, MPH, Professor and Chief, Division of Pulmonary and Critical Care, College of Medicine-Jacksonville, University of Florida, Jacksonville, FL, USA.

Rebecca S. Eshraghi, PhD, DNM, BBA, Hearing Research and Communication Disorders Laboratory, University of Miami-Miller School of Medicine, Miami, USA; Director postgraduate program and Dissertation Chair, Quantum University for Integrative Medicine, Honolulu, USA.

Sylvia Daunert, PharmD, MS, PhD, Professor and Lucille P. Markey Chair, Director, Dr. JT Macdonald Foundation Biomedical Nanotechnology Institute, Department of Biochemistry and Molecular Biology, University of Miami-Miller School of Medicine, Miami, USA.

Santanu Banerjee, PhD, Assistant Professor, Department of Surgery, University of Miami-Miller School of Medicine, Miami, USA.

Sudhir Aggarwal, MD, PhD, Medical Director, Lower Merion Neurology, Paoli, PA, USA; Assistant Professor, Department of Neurology, Lewis Katz School of Medicine, Temple University, Philadelphia, PA, USA.

Rahul Dev Jayant, PhD, Assistant Professor, Department of Pharmaceutical Sciences, School of Pharmacy, Texas Tech University Health Sciences Center, Amarillo, USA.

Rahul Mittal, PhD, Associate Scientist, Hearing Research and Communication Disorders Laboratory, Division of Neurotology, Department of Otolaryngology, University of Miami-Miller School of Medicine, Miami, USA.

Nelson Gonzalez, OTR/L, RFS, CHHC, CCT, CPMT, CLT, Executive Director for Rehab Without Walls NeuroSolutions, Louisville, Kentucky, USA.

Hamid R. Djalilian, MD, Professor of Otolaryngology and Biomedical Engineering, Director of Otology, Neurotology and Skull Base Surgery, Medical Director of Cochlear Implant Program, Department of Otolaryngology, Head and Neck Surgery, University of California, Irvine, USA.

Elisa Hill-Yardin, PhD, Associate Professor, School of Health and Biomedical Sciences, RMIT University, Bundoora, VIC 3083, Australia.

Chapter 1

The Role of Microbiome in Health and Neurosensory Disorders: Historical and Clinical Perspectives

Adrien A. Eshraghi*, Jenna Bergman and Camron Davies

Abstract

The assemblage of bacteria, viruses, and fungi within the human body, collectively known as the microbiome, has recently emerged as a critical player in human physiology and disease. In particular, the gut is a biological niche, home to millions of microbes that possibly impact every organ system in the human body. Research into the gut microbiota has come a long way since the ancient identification of microorganisms in the 16th century. In recent history, the development of next-generation sequencing has allowed for the identification of various microbes in the gut, while proteomics and metabolomics has further characterized the effect of microbial–host interactions. Perhaps the most intriguing topic of research in this field is the connection between the gut microbiome and the brain; a bidirectional relationship typically referred to as the gut–brain axis. Investigation

**Hearing Research and Communication Disorders Laboratory, Department of Otolaryngology, University of Miami-Miller School of Medicine, Miami, FL, USA.*
Corresponding Author: Adrien A. Eshraghi, MD, MSc, FACS, Hearing Research and Communication Disorders Laboratory, Department of Otolaryngology, Neurological Surgery, Pediatrics and Biomedical Engineering, University of Miami-Miller School of Medicine, 1600 NW 10th Avenue, Miami, FL 33136, USA, Telephone number 305-243-8078, E-mail: aeshraghi@med.miami.edu.

into gut–brain axis has revealed that the microbiome not only interacts locally with gastrointestinal tract, but also directly with the central nervous system (CNS) through neuroendocrine and metabolic pathways. For example, microbial metabolites demonstrate potent epigenetic regulatory activity, which can impact the immune system and in turn, effect the CNS. Furthermore, the vagus nerve serves as a physical link between the gut and brain, potentially modulating microbiota composition and intestinal inflammation. Accordingly, dysbiosis in the gut has been implicated in several neuropathologic states including autism spectrum disorder, Parkinson's disease, and depression. Further investigation into microbial phenotyping and metabolic profiling of the gut microbiome is necessary to explain the link between dysbiosis and disease and develop novel therapeutic interventions to treat these conditions.

INTRODUCTION

While it may be surprising to many, over 99% of the genetic information in our body is not ours. Indeed, most of the cells in our body are not human. The human body contains almost 10 times more bacteria than human cells. Combined, these bacteria have an estimated 3.3 million genes compared to a human's 23,000. While the vast majority of bacteria are found in the gut, there are unique microbiomes in almost every part of the body, each playing an integral role in our physiology (Fig. 1.1).

Most of the bacterial species that reside in humans come from four phyla: Actinobacteria, Firmicutes, Proteobacteria and Bacteroidetes.[1] The composition and function of each microbiome is dependent on its location within the human body while the distribution and ratio of microbial species can vary tremendously between individuals depending on age, health status, and even geographic location. Though the gut microbiome receives most of the attention, the skin, oral, nasal, pulmonary, and vaginal microbiomes are also areas of active research. For instance, dermatologic

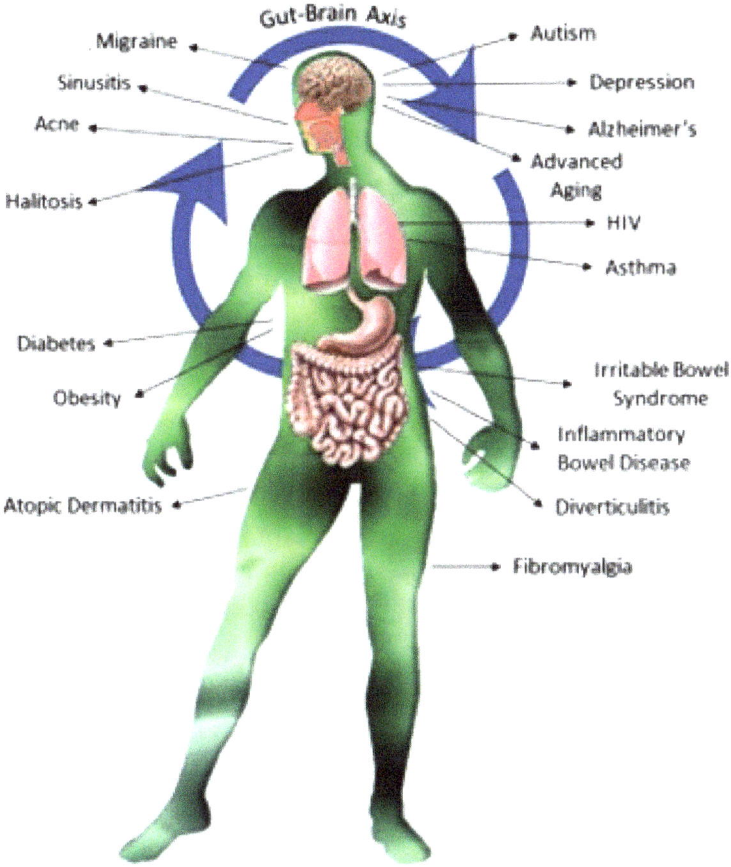

Fig. 1.1: The gut–brain connection. Dysbiosis of the gut microbiota is associated with the pathogenesis of disorders in many different parts of the human body. Conversely, the central nervous system may influence the gut and microbiota in different organ systems. The bidirectional relationship between the human microbiota, their metabolic products, and the host immune system may play a significant role in the development and progression of several disease processes.

conditions such as otitis externa are beginning to be associated with dysbiosis on the skin, while maternal vaginal flora may play a role in neonatal risk for infection. Irrespective of location in the human body, perturbations to the microbiome such as a decrease in microbial diversity or an excessive bacterial load can lead to inflammation and pathologic disease.

In particular, a growing body of evidence over the last 10 years has revealed that disturbances in the gut microbiome have a profound effect on disease susceptibility, immune responses, and neurologic functioning. The observation that the gut microbiome can affect metabolic processes and functioning in multiple organ systems is well demonstrated in experiments using germ-free (GF) mice that are raised in sterile environments and devoid of all microorganisms. For example, a number of studies have found differences in neurotransmitters, receptors, and associated metabolites in specific brain regions of GF mice compared to wild-type mice.[2] Another study found that GF mice have an exaggerated hypothalamic–pituitary–adrenal axis response to stress that is able to be reversed after colonization with bacteria.[3,4] There is also compelling evidence that the microbial composition in the gut is directly responsible for triggering epigenetic changes that affect the immune system, suggesting that the microbiome may play a role in the development of chronic inflammatory conditions.[5]

Today, the gut–brain connection is an increasingly important area of research with a significant focus on the bidirectional signaling between the brain, gut, and intestinal microbiome. Numerous studies have examined the physiologic mechanisms underlying this axis, which in turn has led to an unprecedented shift in our understanding of the development of many chronic diseases such as autism spectrum disorder (ASD), Parkinson's disease, and depression. A deeper understanding about the gut–brain axis and its role in the underlying pathogenesis of disease will reveal opportunities for novel therapeutic strategies and prophylactic interventions.

Historical Aspects of the Gut Microbiome

Gut–Brain Connection in Ancient Times

Specific references to the gut–brain connection can be found throughout the history in many societies and cultures that traditionally put emphasis on the importance of intestinal health. In the 2nd century

BCE, the concept of "Qi" was established and became a fundamental part of traditional Chinese medicine (TCM). "Qi" refers to the body's vital energy and recognizes the abdomen as the location for the human soul.[6] Recent research has suggested that many TCMs can affect the microbiome. For example, red ginseng has been shown to suppress *Staphylococcus aureus*, Salmonella, and *Escherichia coli* and has various anti-inflammatory properties.[7]

An emphasis on a healthy gut can also be seen with the traditional ancient Indian approach to food; Thali. Thali has been used in India for thousands of years to promote a nutritionally balanced meals with a variety of flavors, textures, and colors. Many plant-based foods are used in Thali, which provide an abundant source of fiber and phytochemicals that exert anti-inflammatory effects by promoting bacterial diversity. Thali has been used for centuries to promote health and well-being and plays a major role in Ayurvedic medicine as a therapy for gastrointestinal (GI) illness and cancer.[8]

In Traditional Persian Medicine (TPM), particular attention is paid to the function of the GI tract for many ailments due the long-standing belief in the connection between the gut and other organ systems.[9] In a medical text from 980 A.D., the father of modern medicine, Ibn Sina (Avicenna), explained gastric dysfunction as the most common cause for headache and described a neurovascular communication between the stomach and CNS. Many TPM therapies that target the gut are still used today to treat a variety of diseases that affect different parts of the body.[10]

Discovery of the Human Microbiota

Research into the composition of the human microbiome began with Antonie van Leewenhoek, who, as early as the 1680s, had studied his own oral and fecal microbiota. He noted differences in organisms found in each habitat and also between samples from healthy and sick patients.[11] Almost 200 years later in 1878, German scientist Henrich Anton de Mary developed the concept of symbiosis after he observed that lichen algae could live and thrive independently of fungi, but lichen fungi were entirely dependent on the algae for survival.[12] His theory was further developed a century later when evolutionary biologist Lynn Margulis described endosymbiosis, in which the evolution of mitochondria and chloroplasts were the result of a symbiotic merging of archaeal and prokaryotic organisms. Margulis was the first to declare humans as holobionts, a group of organisms that share space and resources symbiotically with microbial counterparts, and together make a whole.[13]

The term "microbiome" was previously thought to originate in 2001 by Nobel laureate-microbiologist Joshua Lederberg.[14] However, a closer look at the literature revealed that in reality, the term was created long before that and had been used in multiple scientific articles throughout the 20th century. Indeed, in the 1960s, mention of a "microbiota compatible with health" was written in the papers of experiments analyzing the health and behavior of GF mice.[14] Despite controversy over initial use of the term microbiome, it is well established that the study of microorganisms within the human body began much earlier in history after the invention of the microscope.

The Emergence of the Gut–Brain Connection in Western Medicine

The gut–brain connection to health emerged in Western medicine around the 18th century. Decades following the discovery of bacteria, Scottish physician Robert Whytt developed the theory of "nervous sympathy" in 1765 to describe the connection between organs within the human body. Whytt observed that the gut possessed an abundant supply of nerve endings and from this realization deduced that "nervous energy" could be disseminated throughout the body. English surgeon John Abernethy was intrigued by Whytt's idea of "nervous energy" and discussed in his 1829 book, *The Abernethian Code of Health and Longevity*, phenomena such as excessive worry causing reduced appetite, and conversely, stomach pain resulting in emotional distress.[15] In 1868, the Prussian physician Hermann Senator hypothesized that "self-infection" through bacteria in the intestines could lead to disease. Senator's claims ultimately formed the basis of French doctor Charles Bouchard's theory of "auto-intoxication," which was a breakthrough in the scientific community's understanding of intestinal microbes in health. Bouchard stated in his lectures that "if these [digestive] organisms were not properly eliminated, or if too many of them were produced, they could cause internal 'poisoning', leading to disease." These theories were explored further in the 20th century by American physician and nutritionist John Harvey Kellogg who treated many patients that "required change of the intestinal flora" by modifying their diet.[16]

The momentum of clinical research relating to the gut–brain connection had significantly slowed by 1930 for many reasons. The dramatic decline was in part due to physicians who used the theory of auto-intoxication to justify extreme and unnecessary treatments. For example, Scottish Physician Sir William Arbuthnot Lane performed countless colon resections as a way of treating intestinal stasis. Others criticized the model for failing to explain conditions such as fecal impaction and intestinal malrotation. By the mid-20th century, most physicians no longer looked to the gut–brain relationship as an explanation for various diseases.[16]

A resurgence of widespread interest in the human microbiome came in the early 2000s due to a number of pioneering studies that used gene sequencing techniques to analyze the microbiota. This research laid the foundation for the 2007 launch of the $215 million Human Microbiome Project developed by the National Institutes of Health to conduct a survey of the microbial communities in and on the human body.

Advancements in Genomic Sequencing, Transcriptomics, and Metabolomics

Next-Generation Sequencing and Database Development

The development of gene sequencing methods in the 1990s allowed for a better understanding of various types of bacteria and their function. Next-generation sequencing (NGS) techniques massively increased the throughput of genomic sequencing and genetic analysis by enabling billions of reads per run. The development of NGS has allowed for an incredible increase in breadth and depth of genomic studies. Not only are researchers able to more accurately sample the gut microbiome with more reads, but they can also identify specific genes that are enriched or depleted. Alongside the sequencing techniques came the development of bioinformatic databases such as the Kyoto Encyclopedia of Genes and Genomes (KEGG), which is used to identify the biological function of specific genes. These advancements have given researchers unprecedented insight into the structure and function of the gut microbiome, however, critical information about the genome is still unknown. For example, NGS does not provide direct information about the behavior of DNA in different environments.[17] In a sense, sequencing the genome is equivalent to knowing someone's name. While it may provide a good idea of who they are, where they come from, and who their family may be, it offers very little information on their work, how they do it, or how they are affected by different environments. To understand these aspects of a person's or bacteria's life in the context of the microbiome, we need more information.

Functional Genomics

The current forefront of microbiome research seeks to provide this missing information. Functional genomics seeks to understand and integrate the vast amounts of data generated from high-throughput experiments in not only metagenomics, but also transcriptomics, proteomics, and metabolomics. Transcriptomics gives insight on how genes are expressed, and if they are down-regulated, up-regulated, or entirely silenced. Proteomics expands upon this technique by identifying the structure and function of genes after they are translated and is able to determine where they operate and how they work. Finally, metabolomics examines the products of metabolism and what role these metabolites play in physiology. With this flow of information,

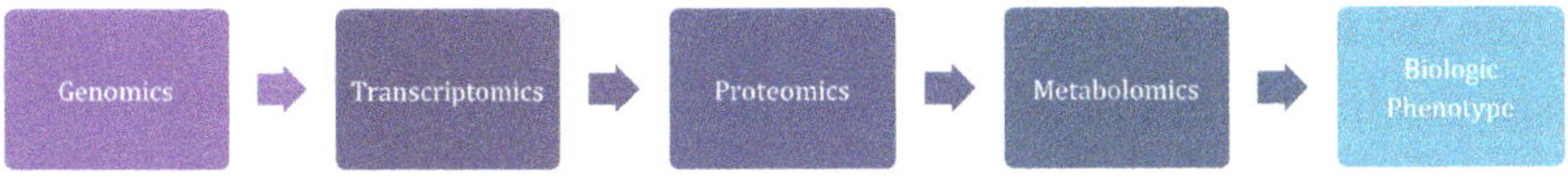

Fig. 1.2: Information flow from genomics to biological phenotype. Genomics looks at all of an individual's or organism's genes and how they interact with each other and the environment. Transcriptomics gives insight on how genes are expressed while proteomics expands upon this technique by identifying the structure and function of genes after they are translated into proteins. Finally, metabolomics examines the products of metabolism and what role these metabolites play in physiology. All of this information helps to characterize the observable characteristics we see in an organism; their biologic phenotype.

researchers can paint a more vivid picture of the gut microbiome and its interactions with human body (Fig. 1.2). Transcriptomics, proteomics, and metabolomics are relatively underdeveloped and underutilized compared to genomics due to the increased difficulty of their experimental techniques.[18] Nevertheless, these fields, especially metabolomics, are likely to be the future of microbiome and gut–brain research. The metabolomic profile is a combination of genomics, transcriptomics, and proteomics. Indeed, it is the metabolites that bacteria produce, which are often more important to the physiologic functions of the GI tract, than the specific species of bacteria themselves. By providing information on the precise bioactive molecules that microbiota produce, researchers can understand how microorganisms ultimately interface with the rest of the body, including the brain. Metabolomic analysis, combined with other "omic" techniques, will continue to increase our understanding of the gut-brain axis.

Epigenetics

It is important to note that the role of the microbiome goes beyond just the metabolism of nutrients and extends to affect our DNA through epigenetic regulation. While genetics are the relatively immutable, physical DNA bases that compose our genome, epigenetics are chemical modifications that can be superimposed upon these bases. Literally, epi — above — genetics, these chemical modifications regulate the expression of DNA. Importantly, like DNA bases themselves, epigenetic modifications can remain in place as cells divide. Epigenetics are engrafted based on several factors having to do with one's parental epigenome and environmental influences.[19] These modifications selectively turn on or off gene expression, which allows for greater control over cellular function than genetics alone provides.

There are a number of ways that epigenetics can control gene expression. Two of the most common processes are histone modification and RNA silencing. In order to arrange the estimated 2 m of human DNA into the nucleus of a cell, it must be compressed into an appropriate size.[20] This is accomplished by winding the double helix DNA strands twice around tightly packed proteins called histones. Histones play a major role in the expression of genes and their associated proteins. For example, when DNA is wrapped around histones, it is unreadable by RNA polymerase and unable to be expressed by the cell. In order to regulate how tightly DNA is bound to these proteins, enzymes within the cells can methylate or acetylate histones. Methylation further compresses the histones, while acetylation relaxes them, allowing DNA to be expressed more easily (Fig. 1.3). Microbial metabolites are also understood to promote acetylation of histones in key regulatory immune genes. Specifically, studies have shown that byproducts of bacterial metabolism, such as short-chain fatty acid, have wide-ranging effects outside of the gut by inhibiting histone deacetylase (HDAC) activity in immune cells.[21] Homeostasis of the gut microbiome can produce the necessary metabolites that allow for the expression of genes that foster a protective immune system.

The gut microbiome is also thought to regulate gene expression through RNA silencing.[23] Bacteria influences the ability of cells to produce short RNA strands (micro-RNA [miRNA]) that are complementary to certain target messenger RNA (mRNA) strands, which serve as templates for protein synthesis. When

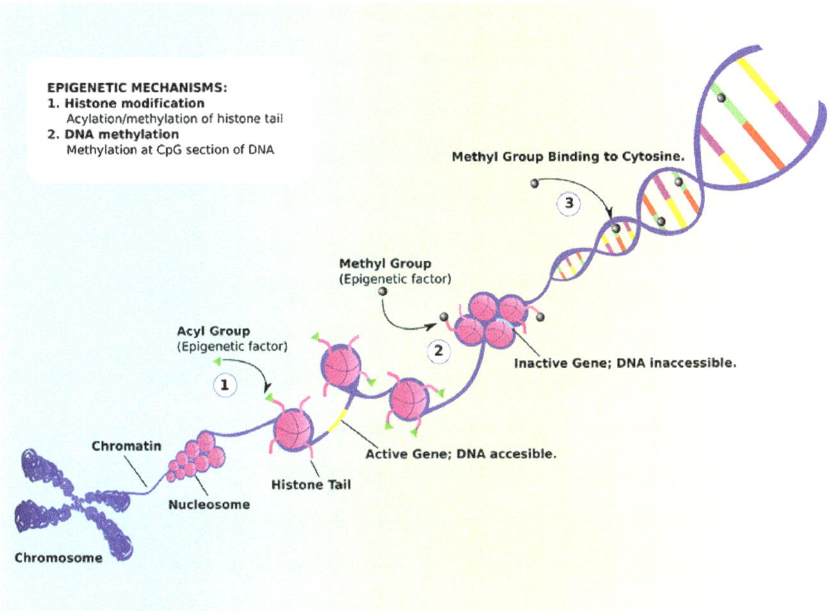

Fig. 1.3: Epigenetic mechanisms. A histone is a protein that helps to comprise the structure of chromatin and is composed of DNA-wrapped protein octamers. The modification of histones plays a fundamental role in most biological processes that are involved in the manipulation and expression of DNA. Histone modifications, and the enzymes implementing them, can contribute to chromatin compaction, nucleosome dynamics, and transcription. Microbial metabolites are understood to promote acetylation of histones in key regulatory immune genes. (Taken from Eshraghi *et al.*[22] which is an open-access article distributed under the terms of the Creative Commons Attribution License (CC BY). The use, distribution or reproduction in other forums is permitted, provided the original author(s) and the copyright owner are credited and that the original publication in this journal is cited, in accordance with accepted academic practice.)

miRNAs bind their complementary section, they inhibit translation of the mRNA into a functional protein. Numerous studies have demonstrated the gut microbiome's ability to affect miRNA production in both the gut and areas of the brain, including the prefrontal cortex, amygdala, and hippocampus.[24,25] Conversely, host production of miRNA has been shown to alter the gut microbiome's composition and function as well.[26] Together both our inherited genetics and epigenetics, as shaped by our environment and gut microbiome, influence our genetic phenotype (Fig. 1.4.).

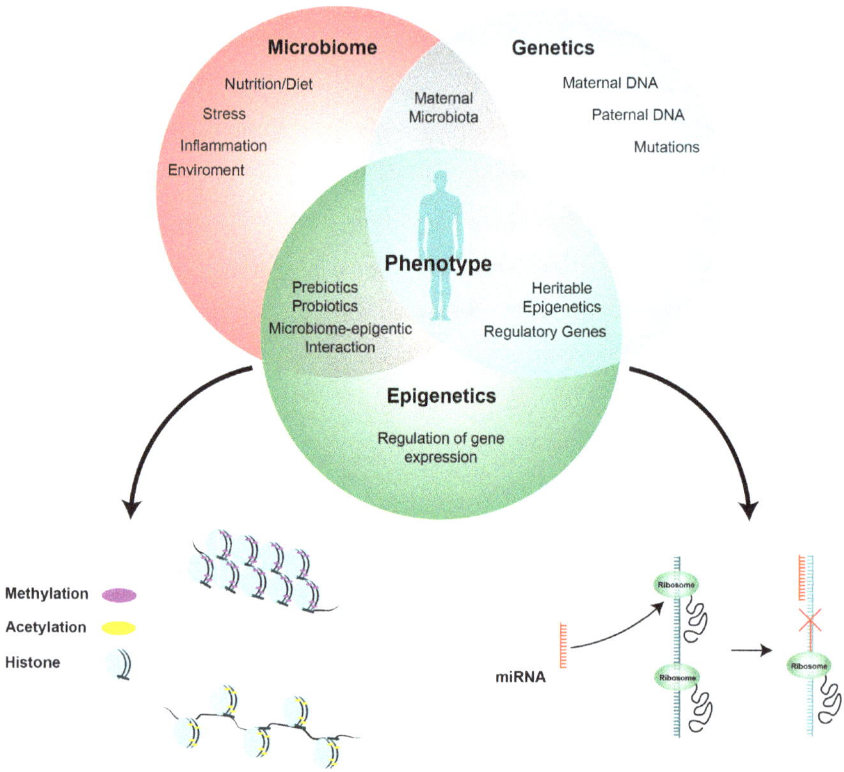

Fig. 1.4: The role of the microbiome, genetics, and epigenetics in regulating the phenotype. While genetics are relatively stable, certain factors like the maternal microbiota help shape early phenotypic development. The microbiome may further be altered throughout one's life via nutrition, stress, inflammation, the environment, and via antibiotics, probiotics and prebiotics. In turn, the microbiome can induce epigenetic changes that regulate gene expression. Two of the main mechanisms of epigenetics regulation altered by the microbiome include histone modification and RNA silencing. Bacterial by-products like SCFAs can modify the acetylation and methylation of histones, which alters the cell's ability to express DNA as proteins. Histone acetylation expands DNA strands, allowing cellular machinery to transcribe DNA to RNA. Histone methylation condenses DNA preventing its transcription. Although RNA silencing has many mechanisms, the gut microbiome is implicated in micro-RNA (miRNA) production. miRNA's are short strands of RNA that bind to complementary messenger RNA strands; this complementary binding prevents ribosomes from translating the mRNA into proteins, thereby silencing the gene.

The Role of the Microbiome in Health Disease

Head and Neck Disorders

As the field of research into the human microbiome has expanded and progressed, the microbiota in different parts of the body have been implicated in a growing list of disease processes. For example, dysbiosis in the polymicrobial communities that reside within the paranasal sinus cavities have been well characterized in patients with sinus disease. Microbial analysis of the sinuses in patients with chronic sinusitis demonstrates a significantly less diverse microbiome than in healthy controls. Certain microbial species, such as *Pseudomonas aeruginosa*, exert a pathogenic effect on mucosal tissue and often correlate with disease severity.[27] There is even research describing microbiomes in locations previously believed to be sterile. Studies describe a diverse microbiome in the middle ear and mastoids of healthy individuals. Research has revealed that pathogenic bacteria implicated in chronic otitis media, such as *Pseudomonas* and *Hemophilus*, are also present in healthy individuals, which may suggest that the source of some infections come from within.[28]

Investigation into the pathophysiology of Meniere's disease (MD) has revealed an interesting connection to the gut. During symptomatic episodes of MD, patients were observed to have increased intestinal permeability as evidenced by increased intestinal absorption of lactulose and mannose and increased levels of fecal calprotectin.[29] In addition, several studies have reported a significant improvement of vestibular symptoms associated with MD after dietary changes.[30] Further research into the intestinal inflammation seen in patients with MD may reveal that dysbiosis of the gut microbiome influences the characteristic vestibular symptoms.

In the oral cavity, several unique tissues such as the tongue, oral mucosa, and enamel provide an opportunity for microorganisms to flourish. In fact, the formation of biofilms, a bacterial community encased within a polysaccharide material, on the surface of teeth, forms what is commonly called plaque. Studies suggest that differences in the bacterial composition of plaque is responsible for the increased susceptibility of dental caries among certain patients.[31]

In 2020, Polster *et al.* revealed that patients with cavernous hemangiomas (CA), a common neurovascular abnormality, have distinctive microbiomes compared to healthy individuals. Furthermore, patients with CA have distinct disease characteristics depending on the microbial composition of their gut. The activity of lipopolysaccharide-dependent pathways was found to be significantly more active in patients with CA, which is consistent with a prior study that found a positive correlation between gram-negative bacteria and CA lesions in mice.[32]

Migraines are linked to several GI disorders including irritable bowel syndrome, ulcerative colitis, Crohn's disease, and celiac disease.[33] These inflammatory conditions increase intestinal permeability and allow for the opportunity of bacteria and bacterial components that serve as endotoxins to pass into the blood. Several studies have implicated endotoxins such as lipopolysaccharide in the pathogenesis of migraines.[33] Interestingly, diet has been shown to have a significant impact on migraines although the exact mechanism and physiologic process behind this is not well known.

The role that the microbiome plays in the development and risk of cancer is also of increasing interest to the medical field. For example, research has revealed that the viriome of human papillomavirus-associated oropharyngeal squamous cell carcinoma may impact tumor behavior, oncologic outcomes, and treatment response.[34] The microbiome is also believed to play a role in carcinogenesis by regulating inflammation, metabolizing hormones, and generating tumor promoting metabolites.[35]

Neurosensory Disorders

As the research community has gleaned more knowledge on the gut–brain axis, the microbiota has been implicated in a growing number of neurologic disease processes. The gut microbiome is thought to play a significant role in normal cognitive development and functioning. Alterations to the gut microbiota have been linked to dysfunctional central processing reflected by decreased attention and changes to neural activity in various regions of the brain.[36] A deeper look into the clinical implications of this connection is explored in Chapter 9.

Several researchers have investigated the frequent association between many neurological and psychiatric conditions and GI distress. For example, ASD is a neurodevelopmental disorder that has traditionally been thought of as only a neuropathologic process, however, recent research has suggested that the gut microbiome may also play a major role in the development or some subtype of this disorder. Approximately 70% of children with ASD report comorbid GI distress such as bloating, constipation, and diarrhea, suggesting that gut physiology is altered.[33] A study by Kang *et al.* demonstrated that children with autism and GI issues lack a diverse microbiome and that the severity of autistic characteristics correlated with a decreased prevalence of certain bacterial species.[37]

Alterations in gut microbiota composition have also been observed in Alzheimer's (AD) and Parkinson's disease. One study found that the changes in microbiota correlated with a pathological load of amyloid-β plaques and phosphorylated tau proteins in patients who were diagnosed with AD.[33] In Parkinson's patients, research supports a correlation between an altered gut microbiome and severity of symptoms. This finding supports the frequently observed occurrence of GI dysfunction such as constipation years before motor symptoms appear.[33]

It is now widely recognized that the gut microbiota plays a critical role in the maintenance of health and development of disease. Despite significant evidence of the gut–brain connection, many questions remain regarding the underlying mechanism of various diseases. This book will further discuss how the microbiome influences the gut–brain axis and the disease processes implicated in dysbiosis.

FUTURE DIRECTIONS

The increasing number of scientific research publications over the last 20 years indicates that the microbiome–gut–brain axis has once again become an important area of research in the medical community. This explosion of metagenetic research is due in large part to the advances in technology that made high-throughput genetic sequencing possible and cost effective. Further development of transcriptomics, proteomics, and metabolomics will enhance the concept of precision medicine.

While much of the current scientific focus has been on the gut microbiome, advancement in sequencing techniques and availability has led to numerous discoveries of unique microbiomes in nearly every part of the human body. Future research will be useful in confirming the existence and characterization of various microbiota and their role in health and disease. Moreover, while bacteria are the predominate microorganisms in the microbiome, increasing attention is being paid to viruses and their ability to insert DNA into human and bacterial genomes. Given this possibility, increased research into the virome and mycobiome demands further attention.

A greater understanding of how the microbiome influences both disease and normal physiology is needed. While many studies have found associations between disease and altered microbiomes, in many cases the temporality of this association is uncertain. Many researchers are looking to use the microbiota as a possible avenue for therapeutic intervention. Modulating the neuroimmune axis through the microbiome could play a key role in the management of several diseases. Prebiotic and probiotic supplementation, as well as fecal microbial transplants (FMT), continues to be examined for the treatment of chronic inflammatory and neurologic disorders.

The converse effect of how the CNS affects the gut also warrants investigation as this link may also lend itself to be a mode of intervention for diseases associated with dysbiosis. Many areas of the body are innervated by the vagus nerve and may be influenced by its autonomic input, presenting novel opportunities for therapeutic intervention.[38] CNS stimulation modalities via the vagus nerve such as direct vagus nerve stimulation (VNS), baroreceptor stimulation, and low-level tragus stimulation (LLTS) have demonstrated a beneficial effect on diseases associated with neuroimmune dysfunction.[39] Further discussion of the connection between the autonomic nervous system and the gut or microbiome can be found in Chapter 3. While these novel therapies appear to be very promising, it is still necessary to wait for further research and appropriate randomized clinical trials before implementing them in clinical practices.

This volume of research will address what is currently known about the composition and metabolites of our gut microbiome. We will explore the bidirectional interaction between gut–brain axis, further investigate the development of various disease processes, and discuss the future therapeutic possibilities associated with dysbiosis of the microbiome.

REFERENCES

1. Pflughoeft KJ, Versalovic J. (2012) Human microbiome in health and disease. *Annu Rev Pathol* **7**: 99–122. doi: 10.1146/annurev-pathol-011811-132421. Epub 2011 Sep 9. PMID: 21910623.
2. Diaz Heijtz R, *et al.* (2011) Normal gut microbiota modulates brain development and behavior. *Proc Natl Acad Sci USA* **108**(7): 3047–3052, doi:10.1073/pnas.1010529108. https://www.jstor.org/stable/41002259.
3. Mayer EA, Knight R, Mazmanian SK, Cryan JF, Tillisch K. (2014) Gut microbes and the brain: Paradigm shift in neuroscience. *J Neurosci* **34**(46): 15490–15496, doi:10.1523/JNEUROSCI.3299-14.2014. https://www.ncbi.nlm.nih.gov/pubmed/25392516.
4. Sudo N, *et al.* (2004) Postnatal microbial colonization programs the hypothalamic–pituitary–adrenal system for stress response in mice. *J Physiol* **558**(1): 263–275, doi:10.1113/jphysiol.2004.063388. https://onlinelibrary.wiley.com/doi/abs/10.1113/jphysiol.2004.063388.

5. Eshraghi RS, *et al.* (2018) Early disruption of the microbiome leading to decreased antioxidant capacity and epigenetic changes: Implications for the rise in autism. *Front Cell Neurosci* **12**: 256, doi:10.3389/fncel.2018.00256. https://www.ncbi.nlm.nih.gov/pubmed/30158857.
6. Umeda N. (2019) Gut flora "the second brain" connects Eastern and Western medicine: Intestinal hyper-permeability or Qi deficiency can affect brain, mind, and whole body. *Longhua Chin Med* **2**: 6, doi:10.21037/lcm.2019.04.01.
7. Xu J, Chen H, Li S. (2017) Understanding the molecular mechanisms of the interplay between herbal medicines and gut microbiota. *Med Res Rev* **37**(5): 1140–1185, doi:10.1002/med.21431. https://onlinelibrary.wiley.com/doi/abs/10.1002/med.21431.
8. Shondelmyer K, Knight R, Sanivarapu A, Ogino S, Vanamala JKP. (2018) Ancient Thali diet: Gut microbiota, immunity, and health. *Yale J Biol Med* **91**(2): 177–184. https://www.ncbi.nlm.nih.gov/pubmed/29955222.
9. Shirzad M, Mosaddegh M, Minaii B, Nikbakht Nasrabadi A, Ahmadian-Attari MM. (2013) The relationship between heart and stomach in Iranian traditional medicine: A new concept in cardiovascular disease management. *Int J Cardiol* **165**(3): 556–557, doi:10.1016/j.ijcard.2012.09.006.
10. Naeimi M, Gorji N, Memariani Z, Moeini R, Kamalinejad M, Kolangi F. (2020) Gastroprotective herbs for headache management in Persian medicine: A comprehensive review. *J Integr Med* **18**(1): 1–13.
11. Ursell LK, Metcalf JL, Parfrey LW, Knight R. (2012) Defining the human microbiome. *Nutr Rev* **70**(8): S38–S44, doi:10.1111/j.1753-4887.2012.00493.x. https://onlinelibrary.wiley.com/doi/abs/10.1111/j.1753-4887.2012.00493.x.
12. Oulhen N, Schulz BJ, Carrier TJ. (2016) English translation of Heinrich Anton de Bary's 1878 speech, 'Die Erscheinung der Symbiose' ('De la symbiose'). *Symbiosis* **69**(3): 131, doi:10.1007/s13199-016-0409-8.
13. Margulis L, Fester R. (1991). Symbiosis as a source of evolutionary innovation: Speciation and morphogenesis. Cambridge (Mass.): MIT Press.
14. Prescott SL. (2017) History of medicine: Origin of the term microbiome and why it matters. *Human Microbiome J* **4**: 24–25, doi:10.1016/j.humic.2017.05.004. https://www.clinicalkey.es/playcontent/1-s2.0-S245223171730012X.
15. Miller I. (2018) The gut-brain axis: Historical reflections. *Microb Ecol Health Dis* **29**(2): 1542921–1542929, doi:10.1080/16512235.2018.1542921. http://www.tandfonline.com/doi/abs/10.1080/16512235.2018.1542921.

16. Mathias M. (2018) Autointoxication and historical precursors of the microbiome-gut-brain axis. *Microb Ecol Health Dis* **29**(2): 1548249, doi:10.1080/16512235.2018.1548249. http://www.tandfonline.com/doi/abs/10.1080/16512235.2018.1548249.
17. Wang WL, Xu SY, Ren ZG, Tao L, Jiang JW, Zheng SS. (2015) Application of metagenomics in the human gut microbiome. *世界胃肠病学杂志：英文版* **21**(3): 803–814, doi:10.3748/wjg.v21.i3.803. http://lib.cqvip.com/qk/84123X/201503/90888889504849534851484855.html.
18. Marx V. (2020) Boost that metabolomic confidence. *Nat Methods* **17**(1): 33–36, doi:10.1038/s41592-019-0694-2. https://www.ncbi.nlm.nih.gov/pubmed/31863087.
19. Simmons D. (2008) Epigenetic influence and disease. *Nat Ed* **1**(1): 6. https://www.nature.com/scitable/topicpage/epigenetic-influences-and-disease-895/.
20. Piovesan A, Pelleri MC, Antonaros F, Strippoli P, Caracausi M, Vitale L. (2019) On the length, weight and GC content of the human genome. *BMC Res Notes* **12**(1): 106, doi:10.1186/s13104-019-4137-z. https://www.ncbi.nlm.nih.gov/pubmed/30813969.
21. Paul B, *et al.* (2015) Influences of diet and the gut microbiome on epigenetic modulation in cancer and other diseases. *Clin Epigenetics* **7**(1): 112, doi:10.1186/s13148-015-0144-. https://www.ncbi.nlm.nih.gov/pubmed/26478753.
22. Eshraghi AA, *et al.* (2018) Epigenetics and autism spectrum disorder: Is there a correlation? *Front Cell Neurosci* **12**: 78, doi:10.3389/fncel.2018.00078. Published 2018 Mar 27.
23. Bhaskaran N, Quigley C, Paw C, Butala S, Schneider E, Pandiyan P. (2018) Role of short chain fatty acids in controlling T_{regs} and immunopathology during mucosal infection. *Front Microbiol* **9**: 1995, doi:10.3389/fmicb.2018.01995. https://search.proquest.com/docview/2101920920.
24. Hoban AE, *et al.* (2017) Microbial regulation of microRNA expression in the amygdala and prefrontal cortex. *Microbiome* **5**(1): 102, doi:10.1186/s40168-017-0321-3. https://www.ncbi.nlm.nih.gov/pubmed/28838324.
25. Yuan C, Burns MB, Subramanian S, Blekhman R. (2018) Interaction between host MicroRNAs and the gut microbiota in colorectal cancer. *mSystems* **3**(3): 205, doi:10.1128/mSystems.00205-17. https://search.proquest.com/docview/2048072421.

26. Liu S, *et al.* (2016) The host shapes the gut microbiota via fecal MicroRNA. *Cell Host Microbe* **19**(1): 32–43, doi:10.1016/j.chom.2015.12.005.
27. Cho D, Hunter RC, Ramakrishnan VR. (2020) The microbiome and chronic rhinosinusitis. *Immunol Allergy Clin North Am* **40**(2): 251–263, doi:10.1016/j.iac.2019.12.009.
28. Xu Q, Gill S, Xu L, Gonzalez E, Pichichero ME. (2019) Comparative analysis of microbiome in nasopharynx and middle ear in young children with acute otitis media. *Front Genetics* **10**: 1176, doi:10.3389/fgene.2019.01176. https://www.ncbi.nlm.nih.gov/pubmed/31803245.
29. Di Berardino F, *et al.* (2018) Intestinal permeability and Ménière's disease. *Am J Otolaryngol* **39**(2): 153–156, doi:10.1016/j.amjoto.2017.12.002.1016/j.joim.2019.11.004
30. Luxford E, Berliner KI, Lee J, Luxford WM. (2013) Dietary modification as adjunct treatment in Ménière's disease: Patient willingness and ability to comply. *Otol Neurotol* **34**(8): 1438–1443, doi:10.1097/MAO.0b013e3182942261.
31. Pflughoeft KJ, Versalovic J. (2012) Human microbiome in health and disease. *Annu Rev Pathol* **7**(1): 99–122, doi:10.1146/annurev-pathol-011811-132421. https://www.ncbi.nlm.nih.gov/pubmed/21910623.
32. Polster SP, *et al.* (2020) Permissive microbiome characterizes human subjects with a neurovascular disease cavernous angioma. *Nat Commun* **11**(1): 1–11, doi:10.1038/s41467-020-16436-w. https://search.proquest.com/docview/2407308336.
33. Cryan JF, O'Riordan KJ, Cowan C, Sandhu KV, Bastiaanssen T, Boehme M, Codagnone MG, Cussotto S, Fulling C, Golubeva AV, Guzzetta KE, Jaggar M, Long-Smith CM, Lyte JM, Martin JA, Molinero-Perez A, Moloney G, Morelli E, Morillas E, O'Connor R, … Dinan TG. (2019). The microbiota-gut-brain axis. *Physiol Rev*, **99**(4): 1877–2013. https://doi.org/10.1152/physrev.00018.2018.
34. Carey RM, *et al.* (2020) The virome of HPV-positive tonsil squamous cell carcinoma and neck metastasis. *Oncotarget* **11**(3): 282–293, doi:10.18632/oncotarget.27436. Published 2020 Jan 21.
35. Schwabe RF, Jobin C. (2013). The microbiome and cancer. *Nat Rev Cancer*, **13**(11): 800–812. https://doi.org/10.1038/nrc3610.

36. Cryan JF, Dinan TG. (2012) Mind-altering microorganisms: The impact of the gut microbiota on brain and behaviour. *Nat Rev Neurosci* **13**(10): 701–712, doi:10.1038/nrn3346.
37. Kang D, *et al*. (2013) Reduced incidence of prevotella and other fermenters in intestinal microflora of autistic children. *PLoS One* **8**(7): e68322, doi:10.1371/journal.pone.0068322. https://www.ncbi.nlm.nih.gov/pubmed/23844187.
38. Eshraghi AA, Buchman CA, Telischi FF. (2002). Sensory auricular branch of the facial nerve. *Otol Neurotol*, **23**(3): 393–396. https://doi.org/10.1097/00129492-200205000-00028.
39. Jiang Y, Po SS, Amil F, Dasari TW. (2020) Non-invasive low-level tragus stimulation in cardiovascular diseases. *Arrhythm Electrophysiol Rev* **9**(1): 40–46, doi:10.15420/aer.2020.01.
40. Holscher HD. (2017) Dietary fiber and prebiotics and the gastrointestinal microbiota. *Gut Microbes* **8**(2): 172–184, doi:10.1080/19490976.2017.1290756. http://www.tandfonline.com/doi/abs/10.1080/19490976.2017.1290756.
41. Ulubil SA, Eshraghi AA, Telischi FF. (2009) Sectioning the sensory auricular branch of the facial nerve to treat recalcitrant otalgia. *Otol Neurotol* **30**(4): 522–524, doi:10.1097/MAO.0b013e31819e9084.

Chapter 2

Development of the Microbiota–Gut–Brain Axis in Early Life

Danielle (Hyun Jung) Kim, Kavi M. Rude and Mélanie G. Gareau*

Abstract

The microbiota–gut–brain (MGB) axis is important for health regulation. It is established in early life, beginning with the colonization of microbiota at birth, and subsequent maturation of the gastrointestinal tract following diet changes, antigen exposure, and coupled with significant neurodevelopment during the first years of life. Given this concurrent development, insult to any of the compartments in the MGB axis may impact proper communication, leading to impairment in signaling. In this chapter, we highlight the key steps in establishing the MGB axis in neonatal life, the known pathways involved in this communication signaling, and examples of disease where impairments in signaling play a role, including inflammatory bowel disease and autism spectrum disorders.

INTRODUCTION

It is now well-known that microorganisms colonize the human body, with approximately equal numbers of bacterial cells to human

*Department of Anatomy, Physiology and Cell Biology, School of Veterinary Medicine, University of California Davis, Davis, CA, USA.
Corresponding Author: Melanie Gareau, University of California Davis, Davis, CA 95616, USA, E-mail: mgareau@ucdavis.edu.

cells,[1] and around 100 times more genes that function to maintain beneficial host–microbe interactions.[2,3] Consequently, it is important to understand how these interactions develop, particularly during colonization and development in early life. For example, gut microbes help to train the immune system by communicating with specific mucosal leukocytes and epithelial cells lining the gut. This is done through fermented metabolites and synthesized vitamins produced by the microbes, regulating gut–immune homeostasis.[2,4] Because the diversity of gut microbiome usually reflects the gut health status, it is critical to establish a diverse, functional, and healthy gut microbiome that will help maintain health.

Initial colonization of the gut microbiota during early neonatal life is a critical step in setting the baseline for overall health. The neonatal gut microbiome is less diverse than that of adults,[5] increasing the impact of gut dysbiosis. Impaired colonization of the gut microbiome in infants predisposes them to various health conditions including sepsis, necrotizing enterocolitis, and neurodevelopmental disabilities.[6] The first 3 years of life is a crucial window for the development of the gut microbiome and hinges on the pioneer microbiome transferred from the maternal microbiome, a process that can be influenced by mode of delivery and early feeding patterns.[6] Rapid development of the nervous system coincides with the developmental period of the infant gut microbiota, and increasing evidence has suggested the presence of a tight association between the gut microbiome and brain function.[6–8] While there exists more pre-clinical research on rodents compared to human trials,[6] murine models have highlighted the impact of the gut microbiome in modulating brain development, including neurotransmitter systems, resident immune cells, the hypothalamic–pituitary–adrenal (HPA) axis, vagus nerve, and ultimately regulation of behavior including cognition, mood, sociability, and anxiety.[8,9] Herein we discuss the development of the microbiota–gut–brain (MGB) axis in early life and how bacterial–host interactions shape physiology of the host into adulthood (Fig. 2.1).

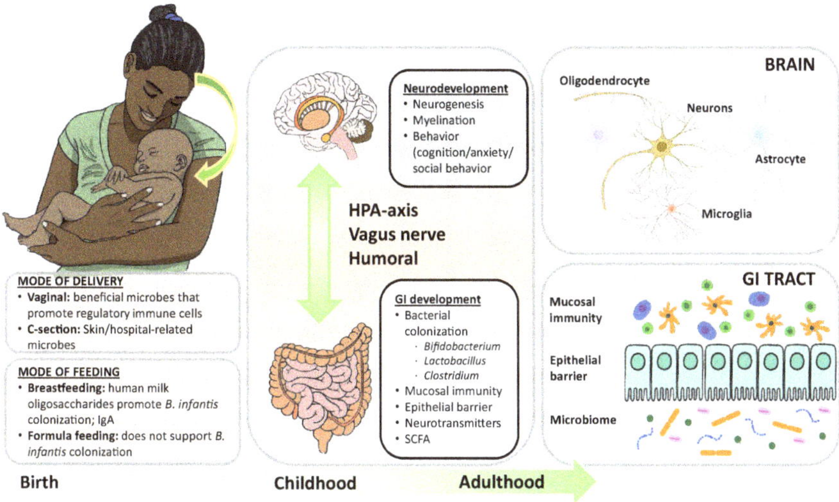

Fig. 2.1: Establishment of the microbiota–gut–brain axis in early life. Colonization of the microbiome occurs at birth and depends on mode of delivery and early feeding patterns. Development of the gastrointestinal tract and neurodevelopment occur simultaneously, with barrier function and mucosal immunity in the gut, and neurogenesis and myelination in the brain. Maintenance of proper host–microbe interactions in the gut can help maintain mood and behavior.

Early Colonization of the Infant Gut Microbiome

The development and maturation of bodily systems that occurs during the first year of life is crucial to the fate of an individual's future health.[10] One key feature is the introduction of bacterial species into the gut, which begins shortly after birth, although there is evidence to support that newborns may encounter microbial components far earlier in placenta, fetal membranes, and umbilical cord blood.[11] Thus, the mode of delivery, whether vaginal or cesarean (C-section), is one of the major factors that "seeds" the gut microbiome of the newborns[12] and initiates the development of the gut microbiome.[13]

Impacts on Mode of Delivery in Infant Microbiome

Vaginal delivery allows the newborn to come into direct contact with the maternal vaginal and skin microbiomes, which serves as the basis for initial colonization of the infant.[2,14] Microbes transferred to the infant during vaginal delivery are typically beneficial and protective, namely strains of *Lactobacilli*, *Bacteroides*, and *Bifidobacterium*, which can decrease the risk of disease such as obesity, later in life.[14] While vaginal delivery is the most common method of birth, up to one out of three women give birth via C-section in the United States.[15] Infants delivered via C-section do not encounter the same microbes as vaginally-delivered infants, but instead have a distinct hospital-acquired gut microbiome that lack many of these beneficial bacteria seen in vaginally-delivered infants. Babies born by C-section have lower levels of intestinal *Lactobacilli* and notably less *Bacteroides* spp. and *Bifidobacterium* spp., as seen in the gastrointestinal (GI) tract of vaginally-delivered infants.[12,16] Comparatively, these infants also have high levels of species that are commonly found in the hospital environment such as *Enterobacter* species, *Haemophilus* species, *Staphylococcus* species,[12] and opportunistic bacteria such as *Enterococcus* and *Klebsiella*,[17] which vaginally-delivered babies lack. As emerging evidence shows that the composition of the initial colonizing gut microbiota could have long-lasting effects on the infant's developing gut,[2] it becomes more obvious that babies born via C-section tend to have higher correlation with incidents to immune-related disorders (as described in detail below) as they lack important beneficial bacteria for their early development.

Other factors, like breastfeeding, can help with establishment of beneficial bacteria (e.g., *Bifidobacterium*) by providing them with human milk oligosaccharides that promote their growth[18,19] and help restore the infant gut microbiome. In contrast, usage of antibiotics during delivery to decrease the risk of post-operative infection in moms[20] and delayed breastfeeding[2] can further impair colonization of the microbiota following birth via C-section in the infant. In order to promote a healthy infant gut microbiome, it is suggested that babies are delivered via vaginal route followed by breastfeeding until the baby reaches 6 months of age with minimal exposure to antibiotics.[21] Long-term studies on these impacts into adulthood remain to be performed, however, many allergy studies suggest that the impacts of colonization decrease over time with age.[22]

Immune Development

The gut microbiome has a well-established role in the development of immune function.[23] Not only do commensal microbes serve as a physical first-line defense mechanism for the host, microbes also induce production of anti-microbial peptides to induce protective immunity against gut pathogens. Importantly, microbes can regulate immune cells by promoting their pro- and anti-inflammatory responses. In addition, digestion of dietary foods leads to the production of microbial metabolites that can subsequently benefit host immunity. Thus, the type and the abundance of bacterial species colonized in the gut can serve as a marker for immune health and may reflect an altered defense system following dysbiosis.

The immune system develops concurrently with the colonization of the gut in early life. Germ-free (GF) mouse studies have confirmed the importance of microbes for development of the immune system, with significant decreases in microglia in the brain and decreased mucosal immunity in the gut compared to conventionally colonized controls.[24] Breastfeeding also greatly impacts the infant's immune system,[25] with breast milk containing nutrients such as proteins, fats, and carbohydrates, as well as oligosaccharides that *Bifidobacterium* species (e.g., *Bifidobacterium infantis*) depend on for colonization.[25] Breast milk also contains over 600 different bacterial species,[26] including lactic acid bacteria such as *Lactobacillus, Lactococcus, Enterococcus,* and *Bifidobacterium* species.[25,27] Multiple immunomodulatory factors such as immunoglobulins (e.g., IgA, IgG), anti-microbial compounds (e.g., lactoferrin, lysozyme), and immune regulatory cytokines (e.g., interleukin [IL]-10, transforming growth factor [TGF]-β) present in breast milk also induce tolerance to intestinal bacteria by inducing T regulatory cells that in turn promote IL-10 production.[25,28] *Bifidobacterium* species are known to be directly correlated with anti-inflammatory effects and secretion of IgA.[25,29,30] Together these factors help promote the development of the immune system, including responses against pathogens, while preventing inappropriate immune responses to innocuous antigens.

Neurodevelopment

Early life is a critical time for rapid neurodevelopment. Studies in GF mice have identified a broad spectrum of neurodevelopmental and behavioral changes that can be associated with alterations of the gut microbiome. GF mice display anxiolytic behavior,[9,31] cognitive deficits,[32] and increased locomotor activity, with elevated levels of noradrenaline, dopamine, and serotonin turnover found in the striatum.[33] Coupled with these changes, GF mice also show decreased expression of synaptic plasticity-related genes (e.g., brain-derived neurotrophic factor [BDNF]) in the hippocampus, amygdala, and cingulate cortex,[33] suggesting microbes can play a role in maintaining neuron function, which may contribute to the observed behavioral deficits. BDNF is also important in maintaining neuronal growth and survival,[8,31,34] however decreased BDNF levels in GF mice were coupled with increased neurogenesis, which could not be restored by colonization of mice at weaning.[35] These findings suggest a complex interplay between the microbiome, neurogenesis, and neurodevelopment.

Microglia are the resident immune cell in the brain, which play a critical role in neuroinflammation. In addition, they are important for maintaining neuronal function, by pruning dendrites, and, consequently, regulating cognitive function. In GF mice, the number of microglia was significantly decreased throughout the brain, including the prefrontal cortex (PFC), hippocampus, cerebellum, and olfactory bulb compared to conventionally colonized controls.[34] Impaired microglia numbers were coupled with reduced branching and thinner spines in the hippocampus.

Maturation of the GI Tract

The GI tract is a critical site for coordination of a variety of cell types including bacteria, immune cells, neurons, and enteroendocrine cells (EECs).[36,37] Establishment of intestinal physiology, including secretory state and permeability, occurs in neonatal life,[38] with maturation occurring longitudinally along the gut.[39] The GI tract regulates interactions between luminal constituents and the host.[36,37] Nutrients from the diet, along with beneficial bacterial metabolites need to be absorbed, while limiting access to noxious bacteria and antigens, to maintain host gut physiology.[36] Dysbiosis of the gut can thus impair the development of the GI tract and its ability to maintain homeostasis.

Intestinal epithelial cells (IEC), including enterocytes and EECs, can be regulated by the gut microbiome.[36,37] EECs are capable of sensing luminal nutrients and respond to the dietary compounds that help shape host physiology.[36] Serotonin, glucagon-like peptide 1 (GLP-1), peptide tyrosine-tyrosine (PYY), and insulin are some of the hormones that are secreted by EECs that regulate host metabolism such as glucose, insulin, and fat metabolisms.[36] Here, gut microbes can play a role in inducing production of gut hormones, as well as, influencing other IEC types.

Bacterial metabolites, namely short-chain fatty acids (SCFA), are involved in host bile acid metabolism and influence the development of the GI tract.[36] SCFAs bind to G protein-coupled receptors (GPCRs), free fatty acid receptors (FFARs), monocarboxylate transporters (MCTs), or sodium-dependent monocarboxylate transporters (SMCTs) to maintain their effects.[40] SCFA receptors on EECs enable cells to sense circulating SCFAs, such as butyrate and propionate, in the human intestinal lumen, which can both serve as energy source for colonocytes.[36,41] Immune cells in the GI tract also have these SCFA receptors that enable SCFA signaling and regulate immune responses.[42] SCFA can also influence PYY secretion, which has been implicated in the development of obesity and metabolic disease.[36] However, the effect of bacterial metabolomes in the host GI tract is region- and bacterial-specific.[36] Bacteroidetes species are enriched with encoding carbohydrate metabolism pathway, whereas, bile acid-tolerant Firmicutes are enriched with genes related to bile acid metabolism pathways.[43]

The MGB Signaling

The bidirectional communication between the gut and the brain has long been established, with stress response being the best characterized. A role for the gut microbiome in mediating this bidirectional communication has been established recently, with elegant studies in GF mice identifying a wide range of brain and behavioral impairments found in the absence of a gut microbiome.[8] The gut and brain communicate via multiple pathways including bacterial SCFAs, neurotransmitters, hormones, HPA signaling, and enteric nervous system (ENS)–central nervous system (CNS) signaling. These pathways enable interdependent signaling that regulate the MGB axis.

MGB Axis: SCFA

SCFA are organic monocarboxylic acids usually with a short carbon chains up to six carbons in length that serve as the main beneficial bacterial metabolites produced in the colon via bacterial fermentation of indigestible polysaccharides, such as fiber and resistant starch.[41,44] Some of the most abundant SCFA are acetate, propionate, and butyrate, which can be beneficial to gut physiology by providing nutrition for colonocytes and regulating immune cells.[41] While SCFA can act locally in the colon and peripheral tissues, emerging evidence shows that SCFAs can cross the blood–brain barrier (BBB) into the CNS, binding to GPCRs[40] and induced beneficial effects such as maintaining BBB integrity. The BBB serves to limit circulation of molecules into the CNS while promoting nutrient transport into the brain, a process that is closely associated with brain development and CNS homeostasis.

GF mice were found to have evidence of reductions of the tight junction proteins claudin and occludin that regulate BBB permeability, resulting in a leaky BBB compared to colonized controls. This effect was rescued when these mice were recolonized with SCFA-producing microbes.[45] Since SCFAs can inhibit histone deacetylases (HDACs) and affect gene expression,[40] they could regulate gut–brain signaling epigenetically. For example, sodium butyrate, which is a potent HDAC inhibitor, can improve learning and memory in mice with brain atrophy[46] and enhance BDNF expression that is important in neurogenesis and cognition.[8] However, the dose of SCFA is important in its function, since various studies show counter effects of butyrate in different doses.[47,48]

SCFA also have neuroactive properties that can help shape complex, neural circuitry responsible for maintaining cognition and behavior. SCFA, particularly acetate, can modulate the level of neurotransmitters such as glutamate, glutamine, and gamma-aminobutyric acid (GABA) in hypothalamus of mice.[49] Propionate and butyrate can alter gene transcription of tyrosine hydroxylase,[50] which is important in dopamine, noradrenaline, and adrenaline synthesis. In addition, SCFA can impact the microglia, which upon activation can induce release of neurotransmitters, hormones, and immunomodulatory cytokines that are important in CNS homeostasis. Compromised innate immune responses and microglial defects in GF mice were partially restored following oral supplementation with SCFA.[24] These defects in microglia seen in GF mice could be recapitulated in mice deficient in the SCFA receptor, FFAR2 (*Ffar2*$^{-/-}$), confirming the importance of SCFA in maintaining microglia function.[24] While most SCFA–MGB axis studies were performed in adult mice, a clear role for neonatal life in establishing their function is suggested, with only partial recovery of microglia function, with SCFA supplementation after weaning. Given that beneficial SCFAs are mostly produced from bacterial species that are highly correlated with species seen in the neonatal gut, bacterial colonization during early life is crucial for SCFA function and maintaining the MGB axis into adulthood.

MGB Axis: Neurotransmitters

Neurotransmitters are critical for maintaining communication of neural networks. Given that neural development occurs predominantly during early life, alterations in neuron–neurotransmitter signaling can gradually lead to behavioral and mood disorders later in life. Recent studies have found that many gut microbes have the capability to produce various neurotransmitters that can subsequently impact mood, behavior, and cognition in the host.

Dopamine is important for reward-motivated behavior and is a major transmitter that is induced by microbes. It serves as a precursor for epinephrine and norepinephrine synthesis, which plays a role in arousal and alertness and is involved in learning memory and attention.[51,52] Dopamine can also benefit bacteria, with several pathogenic bacteria able to utilize dopamine or norepinephrine to expand their growth and increase in motility, biofilm formation, and virulence.[53,54] In line with this, GF mice were found to have low levels of norepinephrine.[55] In addition, GF mice also were found to have elevated activity of dopamine- and glutamate-related transcripts that were associated with increased sensitivity to the behavioral effects of cocaine, which could be normalized with oral supplementation of SCFA.[56]

Serotonin also plays a critical role in gut–brain signaling, regulating both mood and GI motility.[51,57] Given that the gut is the main site of serotonin, producing 95% of the total body serotonin, GF mice display significantly reduced levels of serotonin in the serum and colon compared to conventionally colonized controls.[58] Serotonin is dependent on the availability of tryptophan, an essential amino acid that has to be supplied by diet. Spore-forming bacteria can produce serotonin, highlighting the role of the microbiome in serotonin biosynthesis.[59] In contrast to gut levels, adult GF mice have increased plasma tryptophan concentrations and increased serotonin in the hippocampus.[9] The altered levels of tryptophan and serotonin were normalized with bacterial colonization post-weaning.[60] Thus, the gut microbiome plays a crucial role in regulation of tryptophan and its availability in the maintenance of serotonin levels. In addition, serotonin produced within the ENS[57] may also facilitate gut–brain signaling, although evidence regarding the ENS–CNS route is still limited.

Finally, GABA is also produced by gut bacteria, for example, by *Bifidobacterium* and *Lactobacillus* genera, playing a critical role in maintaining neural excitation[61] and impacting neuronal circuit in the CNS.[51] GABA is an inhibitory neurotransmitter that is important for the maintenance of mental health, with deficits in GABA signaling associated with anxiety-like behavior. In mouse studies, supplementation with *Lactobacillus rhamnosus* strain JB-1 reduced anxiety- and depressive-like behaviors with changes in cerebral GABAergic activity, which was dependent on the vagus nerve.[62] In humans, many people have reported alleviation of anxiety or sleep problems[63] or reduced psychological stress[64] by consuming GABA-supplemented foods.

MGB Axis: HPA Axis Signaling

The HPA axis represents the best characterized pathway for gut–brain signaling, with early life being particularly sensitive to exposure to stress. The HPA axis is a central integrative system, which plays a crucial role in regulating physiological adaptation to stress.[31] Briefly, stress signals induce production of corticotrophin-releasing factor (CRF) and arginine vasopressin from the hypothalamus. CRF stimulates secretion of adrenocorticotropin hormone (ACTH) from the pituitary gland, which in turn induces secretion of glucocorticoids (GCs) in the adrenal cortex. GCs regulate the responses to physical and psychological stress, including blood pressure, GI motility, and so on. HPA axis signaling is prone to environmental stimuli, particularly during early life.[55,57] Exposure to stress in neonatal rats can impair intestinal physiology and behavior at weaning and in adulthood.[38,65] These stress responses are associated with the gut microbiota, with administration of probiotics restoring deficits in the gut and brain.[66] GF mice were found to have an exaggerated HPA axis response to stress, such that the level of plasma ACTH and corticosterone was substantially higher than conventionally colonized controls.[31] Monocolonization of *Bifidobacterium infantis*, an important beneficial microbe in neonates, restored the altered HPA stress response.[31] Therefore, the gut microbiome is critical in regulation of the HPA axis, with early neonatal colonization likely to be involved in establishment of the axis.

MGB Axis: Neural Signaling

The ENS is an intrinsic neural network in the GI tract that is composed of sensory, motor, and interneurons, which are important for maintaining GI physiology and function. The diversity of neurons, glial cells, and neurotransmitters and their receptors are as complex as that seen within the CNS. The ENS independently regulates basic GI functions such as motility, epithelial secretion, and blood flow; however, the main control of the gut functions from the CNS is provided by vagal innervation. The vagus nerve is the main afferent nerve that connects the abdominal cavity to the brain, transmitting signals from the gut up to the brain to regulate food intake, satiety, and so on, and can regulate behaviors. Sickness behaviors, such as depression, anxiety, sleepiness, that helps the host cope with infection is regulated by the vagus nerve.[67] Intact vagal nerve signaling is necessary for transmission of microbiota signals to the brain, as shown in studies demonstrating the absence of *Lactobacillus* species-derived beneficial effects, such as central GABA receptor expression, corticosterone responses, and behaviors, following vagotomy.[62,68] Similarly, the anxiolytic effect of *Bifidobacterium* species was not shown in vagotomized mice in a mouse model of colitis mouse.[69] In contrast, the transient changes in gut microbiota, exploratory behavior, and hippocampal BDNF expression due to antibiotic administration in mice were independent of vagal nerve integrity.[70] Together these findings indicate a complex role for the vagal pathway in maintaining MGB axis signaling.

Another factor recently shown to regulate the MGB axis is myelination. Myelin is a lipid membrane that facilitates efficient signal transduction and connects axons readily upon novel environmental stimuli that can shape behavioral responses.[8,71,72] Myelination is initiated by oligodendrocytes, a glial cell within the CNS.[72] Myelination predominantly occurs during early life, concurrently with the establishment of the gut microbiota, although formation of myelin continues into adulthood.[71,72] Significant myelination is observed in the PFC region of the brain, which typically exhibits late myelination.[8,71] The primary function of the PFC resides in processing memory, decision-making, and social behavior;[71] therefore, any impact on PFC myelination can detrimentally affect behavior and cognitive function. Abnormal myelination in the PFC has been reported in various behavioral and neurological diseases, including anxiety, depression, autism, and schizophrenia.[73,74] Increasing evidence has identified a role for the gut microbiome in regulating behavior via changes in myelination. Gacias *et al.* showed that social avoidance and despair-like behaviors observed in non-obese diabetic mice could be prevented by daily gavage of antibiotics.[75] Impaired social behavioral was associated with impaired myelination in the PFC and changes in the gut microbial metabolome, including p-cresol, which can inhibit myelination via inhibiting oligodendrocyte differentiation.[75] In GF mice, Hoban *et al.* showed evidence of hypermyelination in the PFC with upregulated transcripts of myelin-associated genes and transcription factors compared to conventionally colonized controls.[76] SCFAs could also impact myelin, with butyrate able to regulate myelination via microglial activity.[77] Butyrate is also known to directly affect maturation and differentiation of oligodendrocyte progenitor cells (OPCs), and consequently affect myelination, due to its inhibitory HDAC activity.[78] Indeed, Chen *et al.* demonstrated that butyrate can ameliorate cuprizone-induced demyelination in mice.[77] It is interesting to note that the ENS, particularly enteric glial cells, are capable of expressing myelin-associated transcription factors, such as proteolipid protein 1 (Plp1), with unknown function.[79] More studies are needed to further elucidate the function of myelin proteins within the gut that could regulate the MGB axis.

The Role of Gut Microbiome Disruption in Abnormal Physiological and Cognitive Function

The link between dysbiosis of the gut microbiome and disease has been well established, highlighting its critical role in maintaining health. Etiologies of many inflammatory, metabolic, and neurologic disorders are in part attributed to gut microbiome dysbiosis and the subsequent pathophysiology. Despite the wide range of microbial profiles identified within the human microbiome,[80] certain phylum have been specifically associated with disease states, including obesity,[81] and inflammatory bowel diseases (IBDs).[82] While the mechanism of dysbiosis in IBD pathogenesis has yet to be fully elucidated, reduction in inflammation following antibiotic treatment,[83] beneficial effects of probiotics,[84] and the effectiveness of fecal microbiota transplants in subsets of patients is consistent with the role of dysbiosis in disease.[85] Compared to healthy individuals, the microbiota in patients with IBD have well-documented decreases in overall microbial diversity and a proportional abundance of some pro-inflammatory species relative to others.[86,87] Given that these differences in the microbiota have been identified within family members and twin pair studies suggest that dysbiosis, over genetics or environment, may have a stronger association with disease state.[88] However, it should be noted that whether dysbiosis in IBD reflects the host microbial community's response to intestinal inflammation, or whether dysbiosis has a more causative effect on intestinal inflammation, or both, has not been shown. Patients with IBD have a high risk of developing comorbid behavioral impairments, including anxiety, depression and mild cognitive deficits, compared to other chronic diseases.[89] This is particularly true for pediatric IBD patients[90] where concurrent neurodevelopment, dysbiosis and maturation of the GI tract may converge into psychological impairments.

Neurodegenerative disease, for example, Parkinson's disease (PD), have recently been found to be associated with gut dysbiosis. The degeneration of dopamine-generating brain cells in the basal ganglia present as bradykinesia, resting tremor, rigidity, and late postural instability. In addition to these symptoms, many patients experience GI symptoms, including impaired motility, potentially resulting from the autonomic nervous system and ENS dysfunctions, well before motor impairments.[91] Excessive stimulation of the innate immune system produces systemic inflammation, leading to gut dysbiosis, ileal bacterial overgrowth, and increased intestinal permeability.[92] There is also evidence supporting the cross-reaction between bacterial proteins and human antigens, disturbing the adaptive immune system in patients with PD.[92] Early manifestations of PD frequently include constipation, bloating, early satiety and nausea, and are thought to arise from reduced stomach contractions, impaired gastric[93] and colonic motility,[94] and delayed gastric emptying.[95] As PD is frequently diagnosed later in the adulthood, there is a potential that immature gut–brain signaling during early life can accelerate the onset of PD during adult life; however, this remains to be studied further.

Autism spectrum disorder (ASD) is a neurological and developmental disorder associated with social impairments, communication deficits and learning disabilities that typically manifests in early childhood with a higher incidence in boys.[96] A significant subset of patients suffers from comorbid GI impairments, including altered bowel habits coupled with dysbiosis and mucosal immune dysregulation.[97] Up to 70% of children with ASD have comorbid GI symptoms, which have been speculated to contribute to the severity of the disorder.[98] Unsurprisingly, these GI disturbances have long been associated with compositional changes and metabolic activity of the gut microbiota as well as increased intestinal permeability.[99] Treatment of ASD patients with probiotics has identified an improvement in GI function as well as behaviors, in subgroups of children with a GI impairment.[100] ASD behaviors start to manifest as the microbiome is still plastic during early colonization; it is hypothesized to play a role in disease etiology. Mouse models of ASD have identified behavioral impairments coupled with increased intestinal permeability, which could be ameliorated by treatment with a bacterial metabolite.[101] Better understanding of the MGB axis in children with ASD could provide better treatment options for children in the future.

CONCLUSIONS

Development of the MGB axis in early life is critical for regulating physiology and maintaining homeostasis. Disruption of the MGB axis in early life can have detrimental impacts in composition of the microbiome, GI function and behavior well into adulthood, often resulting in development of disease. Additional studies focusing on the unique development seen in early life is crucial to fully understand the processes that shape the MBG axis, what aspects are reversible after insult, and which are permanent after development is complete.

ACKNOWLEDGMENT

MGG is funded by the NIH (1R01AT009365) and the Crohn's and Colitis Foundation (#477191).

REFERENCES

1. Sender R, Fuchs S, Milo R. (2016) Are we really vastly outnumbered? Revisiting the ratio of bacterial to host cells in humans. *Cell* **164**: 337–340, doi:10.1016/j.cell.2016.01.013.

2. Neu J, Rushing J. (2011) Cesarean versus vaginal delivery: Long-term infant outcomes and the hygiene hypothesis. *Clin Perinatol* **38**: 321–331, doi:10.1016/j.clp.2011.03.008.
3. Qin J, *et al.* (2010) A human gut microbial gene catalogue established by metagenomic sequencing. *Nature* **464**: 59–65, doi:10.1038/nature08821.
4. Hooper LV, Macpherson AJ. (2010) Immune adaptations that maintain homeostasis with the intestinal microbiota. *Nat Rev Immunol* **10**: 159–169.
5. Yang I, *et al.* (2016) The infant microbiome: Implications for infant health and neurocognitive development. *Nurs Res* **65**: 76–88, doi:10.1097/NNR.0000000000000133.
6. Lu J, Claud EC. (2019) Connection between gut microbiome and brain development in preterm infants. *Dev Psychobiol* **61**: 739–751, doi:10.1002/dev.21806.
7. Cowan CSM, Dinan TG, Cryan JF. (2020) Annual research review: Critical windows — the microbiota-gut-brain axis in neurocognitive development. *J Child Psychol Psychiatry* **61**: 353–371, doi:10.1111/jcpp.13156.
8. Cryan JF, *et al.* (2019) The microbiota-gut-brain axis. *Physiol Rev* **99**: 1877–2013, doi:10.1152/physrev.00018.2018.
9. Clarke G, *et al.* (2013) The microbiome-gut-brain axis during early life regulates the hippocampal serotonergic system in a sex-dependent manner. *Mol Psychiatry* **18**: 666–673, doi:10.1038/mp.2012.77.
10. Eck A, *et al.* (2020) Neonatal microbiota development and the effect of early life antibiotics are determined by two distinct settler types. *PLoS One* **15**: e0228133, doi:10.1371/journal.pone.0228133.
11. Aagaard K, *et al.* (2014) The placenta harbors a unique microbiome. *Sci Transl Med* **6**: 237rä65, doi:10.1126/scitranslmed.3008599.
12. Bäckhed F, *et al.* (2015) Dynamics and stabilization of the human gut microbiome during the first year of life. *Cell Host Microbe* **17**(5): 690–703, doi:10.1016/j.chom.2015.05.012.
13. Francino MP. (2014) Early development of the gut microbiota and immune health. *Pathogens* **3**: 769–790, doi:10.3390/pathogens3030769.
14. Kaplan JL, Walker WA. (2012) Early gut colonization and subsequent obesity risk. *Curr Opin Clin Nutr Metab Care* **15**: 278–284, doi:10.1097/MCO.0b013e32835133cb.

15. Centers for Disease Control and Prevention. (2017) *Method of Delivery*. https://www.cdc.gov/nchs/fastats/delivery.htm.
16. Makino H. (2018) Bifidobacterial strains in the intestines of newborns originate from their mothers. *Biosci Microbiota Food Health* **37**: 79–85, doi:10.12938/bmfh.18-011.
17. Shao Y, *et al.* (2019) Stunted microbiota and opportunistic pathogen colonization in caesarean-section birth. *Nature* **574**: 117–121, doi:10.1038/s41586-019-1560-1.
18. Bode L. (2012) Human milk oligosaccharides: Every baby needs a sugar mama. *Glycobiology* **22**: 1147–1162, doi:10.1093/glycob/cws074.
19. Chong CYL, Bloomfield FH, O'Sullivan JM. (2018) Factors affecting gastrointestinal microbiome development in neonates. *Nutrients* **10**: 274, doi:10.3390/nu10030274.
20. Smaill FM, Grivell RM. (2014) Antibiotic prophylaxis versus no prophylaxis for preventing infection after cesarean section. *Cochrane Database Syst Rev* **10**: CD007482, doi:10.1002/14651858.CD007482.pub3.
21. Dreyer JLL, Liebl AL. (2018) Early colonization of the gut microbiome and its relationship with obesity. *Human Microbiome J* **10**: 1–5, doi:10.1016/j.humic.2018.08.002.
22. Isolauri E. (2012) Development of healthy gut microbiota early in life. *J Paediatr Child Health* **48**(Suppl 3): 1–6, doi:10.1111/j.1440-1754.2012.02489.x.
23. Negi S, Das DK, Pahari S, Nadeem S, Agrewala JN. (2019) Potential role of gut microbiota in induction and regulation of innate immune memory. *Front Immunol* **10**: 2441, doi:10.3389/fimmu.2019.02441.
24. Erny D, *et al.* (2015) Host microbiota constantly control maturation and function of microglia in the CNS. *Nat Neurosci* **18**: 965–977, doi:10.1038/nn.4030.
25. Tanaka M, Nakayama J. (2017) Development of the gut microbiota in infancy and its impact on health in later life. *Allergol Int* **66**: 515–522, doi:10.1016/j.alit.2017.07.010.
26. Jeurink PV, *et al.* (2013) Human milk: A source of more life than we imagine. *Benef Microbes* **4**: 17–30, doi:10.3920/BM2012.0040.
27. Soto A, *et al.* (2014) Lactobacilli and bifidobacteria in human breast milk: Influence of antibiotherapy and other host and clinical factors. *J Pediatr Gastroenterol Nutr* **59**: 78–88, doi:10.1097/MPG.0000000000000347.

28. Brandtzaeg, P. (2003) Mucosal immunity: Integration between mother and the breast-fed infant. *Vaccine* **21**: 3382–3388, doi:10.1016/s0264-410x(03)00338-4.
29. Sjogren YM, *et al.* (2009) Influence of early gut microbiota on the maturation of childhood mucosal and systemic immune responses. *Clin Exp Allergy* **39**: 1842–1851, doi:10.1111/j.1365-2222.2009.03326.x.
30. Chichlowski M, De Lartigue G, German JB, Raybould HE, Mills DA. (2012) Bifidobacteria isolated from infants and cultured on human milk oligosaccharides affect intestinal epithelial function. *J Pediatr Gastroenterol Nutr* **55**: 321–327, doi:10.1097/MPG.0b013e31824fb899.
31. Sudo N. (2014) Microbiome, HPA axis and production of endocrine hormones in the gut. *Adv Exp Med Biol* **817**: 177–194, doi:10.1007/978-1-4939-0897-4_8.
32. Gareau MG, *et al.* (2011) Bacterial infection causes stress-induced memory dysfunction in mice. *Gut* **60**: 307–317, doi:10.1136/gut.2009.202515.
33. Diaz Heijtz R, *et al.* (2011) Normal gut microbiota modulates brain development and behavior. *Proc Natl Acad Sci USA* **108**: 3047–3052, doi:10.1073/pnas.1010529108.
34. Warner BB. (2019) The contribution of the gut microbiome to neurodevelopment and neuropsychiatric disorders. *Pediatr Res* **85**: 216–224, doi:10.1038/s41390-018-0191-9.
35. Ogbonnaya ES, *et al.* (2015) Adult hippocampal neurogenesis is regulated by the microbiome. *Biol Psychiatry* **78**: e7–e9, doi:10.1016/j.biopsych.2014.12.023.
36. Martin AM, Sun EW, Rogers GB, Keating DJ. (2019) The influence of the gut microbiome on host metabolism through the regulation of gut hormone release. *Front Physiol* **10**: 428, doi:10.3389/fphys.2019.00428.
37. Yoo BB, Mazmanian SK. (2017) The enteric network: Interactions between the immune and nervous systems of the gut. *Immunity* **46**: 910–926, doi:10.1016/j.immuni.2017.05.011.
38. Gareau MG, Jury J, Yang PC, MacQueen G, Perdue MH. (2006) Neonatal maternal separation causes colonic dysfunction in rat pups including impaired host resistance. *Pediatr Res* **59**: 83–88, doi:10.1203/01.pdr.0000190577.62426.45.

39. Henning SJ. (1986) Development of the gastrointestinal tract. *Proc Nutr Soc* **45**: 39–44.
40. Dalile B, Van Oudenhove L, Vervliet B, Verbeke K. (2019) The role of short-chain fatty acids in microbiota-gut-brain communication. *Nat Rev Gastroenterol Hepatol* **16**: 461–478, doi:10.1038/s41575-019-0157-3.
41. Silva YP, Bernardi A, Frozza RL. (2020) The role of short-chain fatty acids from gut microbiota in gut-brain communication. *Front Endocrinol (Lausanne)* **11**: 25, doi:10.3389/fendo.2020.00025.
42. Correa-Oliveira R, Fachi JL, Vieira A, Sato FT, Vinolo MA. (2016) Regulation of immune cell function by short-chain fatty acids. *Clin Transl Immunol* **5**: e73, doi:10.1038/cti.2016.17.
43. David LA, *et al.* (2014) Diet rapidly and reproducibly alters the human gut microbiome. *Nature* **505**: 559–563, doi:10.1038/nature12820.
44. Louis P, Flint HJ. (2009) Diversity, metabolism and microbial ecology of butyrate-producing bacteria from the human large intestine. *FEMS Microbiol Lett* **294**: 1–8, doi:10.1111/j.1574-6968.2009.01514.x.
45. Braniste V, *et al.* (2014) The gut microbiota influences blood-brain barrier permeability in mice. *Sci Transl Med* **6**: 263ra158, doi:10.1126/scitranslmed.3009759.
46. Fischer A, Sananbenesi F, Wang X, Dobbin M, Tsai LH. (2007) Recovery of learning and memory is associated with chromatin remodelling. *Nature* **447**: 178–182, doi:10.1038/nature05772.
47. Kratsman N, Getselter D, Elliott E. (2016) Sodium butyrate attenuates social behavior deficits and modifies the transcription of inhibitory/excitatory genes in the frontal cortex of an autism model. *Neuropharmacology* **102**: 136–145, doi:10.1016/j.neuropharm.2015.11.003.
48. Gagliano H, Delgado-Morales R, Sanz-Garcia A, Armario A. (2014) High doses of the histone deacetylase inhibitor sodium butyrate trigger a stress-like response. *Neuropharmacology* **79**: 75–82, doi:10.1016/j.neuropharm.2013.10.031.
49. Frost G, *et al.* (2014) The short-chain fatty acid acetate reduces appetite via a central homeostatic mechanism. *Nat Commun* **5**: 3611, doi:10.1038/ncomms4611.
50. Nankova BB, Agarwal R, MacFabe DF, La Gamma EF. (2014) Enteric bacterial metabolites propionic and butyric acid modulate gene expression,

including CREB-dependent catecholaminergic neurotransmission, in PC12 cells–possible relevance to autism spectrum disorders. *PLoS One* **9**: e103740, doi:10.1371/journal.pone.0103740.

51. Strandwitz P. (2018) Neurotransmitter modulation by the gut microbiota. *Brain Res* **1693**: 128–133, doi:10.1016/j.brainres.2018.03.015.
52. Borodovitsyna O, Flamini M, Chandler D. (2017) Noradrenergic modulation of cognition in health and disease. *Neural Plast* **2017**: 6031478, doi:10.1155/2017/6031478.
53. O'Donnell PM, Aviles H, Lyte M, Sonnenfeld G. (2006) Enhancement of in vitro growth of pathogenic bacteria by norepinephrine: Importance of inoculum density and role of transferrin. *Appl Environ Microbiol* **72**: 5097–5099, doi:10.1128/AEM.00075-06.
54. Freestone PP, *et al.* (2002) Growth stimulation of intestinal commensal Escherichia coli by catecholamines: A possible contributory factor in trauma-induced sepsis. *Shock* **18**: 465–470, doi:10.1097/00024382-200211000-00014.
55. Asano Y, *et al.* (2012) Critical role of gut microbiota in the production of biologically active, free catecholamines in the gut lumen of mice. *Am J Physiol Gastrointest Liver Physiol* **303**: G1288–G1295, doi:10.1152/ajpgi.00341.2012.
56. Kiraly DD, *et al.* (2016) Alterations of the host microbiome affect behavioral responses to cocaine. *Sci Rep* **6**: 35455, doi:10.1038/srep35455.
57. Foster JA, Rinaman L, Cryan JF. (2017) Stress & the gut-brain axis: Regulation by the microbiome. *Neurobiol Stress* **7**: 124–136, doi:10.1016/j.ynstr.2017.03.001.
58. Wikoff WR, *et al.* (2009) Metabolomics analysis reveals large effects of gut microflora on mammalian blood metabolites. *Proc Natl Acad Sci USA* **106**: 3698–3703.
59. Yano JM, *et al.* (2015) Indigenous bacteria from the gut microbiota regulate host serotonin biosynthesis. *Cell* **161**: 264–276, doi:10.1016/j.cell.2015.02.047.
60. Desbonnet L, Garrett L, Clarke G, Bienenstock J, Dinan TG. (2008) The probiotic bifidobacteria infantis: An assessment of potential antidepressant properties in the rat. *J Psychiatr Res* **43**: 164–174, doi:10.1016/j.jpsychires.2008.03.009.

61. Leinekugel X, *et al.* (1999) GABA is the principal fast-acting excitatory transmitter in the neonatal brain. *Adv Neurol* **79**: 189–201.
62. Bravo JA, *et al.* (2011) Ingestion of Lactobacillus strain regulates emotional behavior and central GABA receptor expression in a mouse via the vagus nerve. *Proc Natl Acad Sci USA* **108**: 16050–16055, doi:10.1073/pnas.1102999108.
63. Boonstra E, *et al.* (2015) Neurotransmitters as food supplements: The effects of GABA on brain and behavior. *Front Psychol* **6**: 1520, doi:10.3389/fpsyg.2015.01520.
64. Yoto A, *et al.* (2012) Oral intake of gamma-aminobutyric acid affects mood and activities of central nervous system during stressed condition induced by mental tasks. *Amino Acids* **43**: 1331–1337, doi:10.1007/s00726-011-1206-6.
65. Soderholm JD, *et al.* (2002) Neonatal maternal separation predisposes adult rats to colonic barrier dysfunction in response to mild stress. *Am J Physiol Gastrointest Liver Physiol* **283**: G1257–G1263, doi:10.1152/ajpgi.00314.2002.
66. Gareau MG, Jury J, MacQueen G, Sherman PM, Perdue MH. (2007) Probiotic treatment of rat pups normalises corticosterone release and ameliorates colonic dysfunction induced by maternal separation. *Gut* **56**: 1522–1528, doi:10.1136/gut.2006.117176.
67. Das UN. (2007) Vagus nerve stimulation, depression, and inflammation. *Neuropsychopharmacology* **32**: 2053–2054, doi:10.1038/sj.npp.1301286.
68. Perez-Burgos A, Mao YK, Bienenstock J, Kunze WA. (2014) The gut-brain axis rewired: Adding a functional vagal nicotinic "sensory synapse." *FASEB J* **28**: 3064–3074, doi:10.1096/fj.13-245282.
69. Bercik P, *et al.* (2011) The anxiolytic effect of bifidobacterium longum NCC3001 involves vagal pathways for gut-brain communication. *Neurogastroenterol Motil* **23**: 1132–1139, doi:10.1111/j.1365-2982.2011.01796.x.
70. Bercik P, *et al.* (2011) The intestinal microbiota affect central levels of brain-derived neurotropic factor and behavior in mice. *Gastroenterology* **141**: 599–609, doi:10.1053/j.gastro.2011.04.052.
71. Ntranos A, Casaccia P. (2018) The microbiome-gut-behavior axis: Crosstalk between the gut microbiome and oligodendrocytes modulates

behavioral responses. *Neurotherapeutics* **15**: 31–35, doi:10.1007/s13311-017-0597-9.

72. Williamson JM, Lyons DA. (2018) Myelin dynamics throughout life: An ever-changing landscape? *Front Cell Neurosci* **12**: 424, doi:10.3389/fncel.2018.00424.
73. Regenold WT, *et al.* (2006) Diffusion-weighted magnetic resonance imaging of white matter in bipolar disorder: A pilot study. *Bipolar Disord* **8**: 188–195, doi:10.1111/j.1399-5618.2006.00281.x.
74. Liu J, *et al.* (2012) Impaired adult myelination in the prefrontal cortex of socially isolated mice. *Nat Neurosci* **15**: 1621–1623, doi:10.1038/nn.3263.
75. Gacias M, *et al.* (2016) Microbiota-driven transcriptional changes in prefrontal cortex override genetic differences in social behavior. *Elife* **5**: e13442, doi:10.7554/eLife.13442.
76. Hoban AE, *et al.* (2016) Regulation of prefrontal cortex myelination by the microbiota. *Transl Psychiatry* **6**: e774, doi:10.1038/tp.2016.42.
77. Chen T, Noto D, Hoshino Y, Mizuno M, Miyake S. (2019) Butyrate suppresses demyelination and enhances remyelination. *J Neuroinflammation* **16**: 165, doi:10.1186/s12974-019-1552-y.
78. Huuskonen J, Suuronen T, Nuutinen T, Kyrylenko S, Salminen A. (2004) Regulation of microglial inflammatory response by sodium butyrate and short-chain fatty acids. *Br J Pharmacol* **141**: 874–880, doi:10.1038/sj.bjp.0705682.
79. Rao M, *et al.* (2015) Enteric glia express proteolipid protein 1 and are a transcriptionally unique population of glia in the mammalian nervous system. *Glia* **63**: 2040–2057, doi:10.1002/glia.22876.
80. Turnbaugh PJ, Gordon JI. (2009) The core gut microbiome, energy balance and obesity. *J Physiol* **587**: 4153–4158.
81. Turnbaugh PJ, *et al.* (2009) A core gut microbiome in obese and lean twins. *Nature* **457**: 480–484.
82. Halfvarson J, *et al.* (2017) Dynamics of the human gut microbiome in inflammatory bowel disease. *Nat Microbiol* **2**: 17004, doi:10.1038/nmicrobiol.2017.4.
83. Bernstein CN. (2014) Antibiotics, probiotics and prebiotics in IBD. *Nestle Nutr Inst Workshop Ser* **79**: 83–100, doi:10.1159/000360713.

84. Fedorak RN. (2008) Understanding why probiotic therapies can be effective in treating IBD. *J Clin Gastroenterol* **42**(Suppl 3, Pt 1): S111–S115.
85. Levy AN, Allegretti JR. (2019) Insights into the role of fecal microbiota transplantation for the treatment of inflammatory bowel disease. *Therap Adv Gastroenterol* **12**: 1756284819836893, doi:10.1177/1756284819836893.
86. Lobionda S, Sittipo P, Kwon HY, Lee YK. (2019) The role of gut microbiota in intestinal inflammation with respect to diet and extrinsic stressors. *Microorganisms* **7**: 271, doi:10.3390/microorganisms7080271.
87. Loh G, Blaut, M. (2012) Role of commensal gut bacteria in inflammatory bowel diseases. *Gut Microbes* **3**: 544–555, doi:10.4161/gmic.22156.
88. Ni J, Wu GD, Albenberg L, Tomov VT. (2017) Gut microbiota and IBD: Causation or correlation? *Nat Rev Gastroenterol Hepatol* **14**: 573–584, doi:10.1038/nrgastro.2017.88.
89. Bernstein CN. (2016) Psychological stress and depression: Risk factors for IBD? *Dig Dis* **34**: 58–63, doi:10.1159/000442929.
90. Ross SC, Strachan J, Russell RK, Wilson SL. (2011) Psychosocial functioning and health-related quality of life in paediatric inflammatory bowel disease. *J Pediatr Gastroenterol Nutr* **53**: 480–488, doi:10.1097/MPG.0b013e31822f2c32.
91. Palma JA, Kaufmann H. (2014) Autonomic disorders predicting Parkinson's disease. *Parkinsonism Relat Disord* **20**(Suppl 1): S94–S98, doi:10.1016/S1353-8020(13)70024-5.
92. Mulak A, Bonaz B. (2015) Brain-gut-microbiota axis in Parkinson's disease. *World J Gastroenterol* **21**: 10609–10620, doi:10.3748/wjg.v21.i37.10609.
93. Soykan I, Lin Z, Bennett JP, McCallum RW. (1999) Gastric myoelectrical activity in patients with Parkinson's disease: Evidence of a primary gastric abnormality. *Dig Dis Sci* **44**: 927–931, doi:10.1023/a:1026648311646.
94. Unger MM, *et al*. (2010) Real-time visualization of altered gastric motility by magnetic resonance imaging in patients with Parkinson's disease. *Mov Disord* **25**:623–628, doi:10.1002/mds.22841.
95. Goetze O, *et al*. (2006) Predictors of gastric emptying in Parkinson's disease. *Neurogastroenterol Motil* **18**: 369–375, doi:10.1111/j.1365-2982.2006.00780.x.

96. Lord C, Elsabbagh M, Baird G, Veenstra-Vanderweele J. (2018) Autism spectrum disorder. *Lancet (London, England)* **392**: 508–520, doi:10.1016/S0140-6736(18)31129-2.
97. Vuong HE, Hsiao EY. (2017) Emerging roles for the gut microbiome in autism spectrum disorder. *Biol Psychiatry* **81**: 411–423, doi:10.1016/j.biopsych.2016.08.024.
98. Adams JB, Johansen LJ, Powell LD, Quig D, Rubin RA. (2011) Gastrointestinal flora and gastrointestinal status in children with autism — comparisons to typical children and correlation with autism severity. *BMC Gastroenterol* **11**: 22, doi:10.1186/1471-230X-11-22.
99. de Magistris L, *et al.* (2010) Alterations of the intestinal barrier in patients with autism spectrum disorders and in their first-degree relatives. *J Pediatr Gastroenterol Nutri* **51**: 418–424, doi:10.1097/MPG.0b013e3181dcc4a5.
100. Sanctuary MR, *et al.* (2019) Pilot study of probiotic/colostrum supplementation on gut function in children with autism and gastrointestinal symptoms. *PLoS One* **14**: e0210064, doi:10.1371/journal.pone.0210064.
101. Hsiao EY, *et al.* (2013) Microbiota modulate behavioral and physiological abnormalities associated with neurodevelopmental disorders. *Cell* **155**: 1451–1463, doi:10.1016/j.cell.2013.11.024.

Chapter 3

Crosstalk Between Gut and Brain via Neurotransmitters

Andrea C. Guerrero[1,4,*], Viraj Shah[2,*], Dibyanshi Mishra[2], Jeenu Mittal[2] and Mehdi Mirsaeidi[1,3]

Abstract

Neurotransmitters such as serotonin, norepinephrine (NE), dopamine (DA), gamma-aminobutyric acid (GABA), and histamine play a crucial role in normal human physiology maintaining the sleep–wake cycle, fight or flight response, arousal states, moods, as well as many other neurological functions. However, multiple central nervous system (CNS) and enteric nervous system (ENS) disorders are associated with the dysregulation of these neurotransmitters. In this chapter, we discuss the role of neurotransmitters and how gut microbiota are implicated in modulating the endogenous levels of neurotransmitters through mechanisms such as direct

[1] Section of Pulmonary, Department of Medicine, Miami VA Health System, Miami, FL, USA.

[2] Department of Otolaryngology, University of Miami-Miller School of Medicine, Miami, FL, USA.

[3] Division of Pulmonary, Allergy, Critical Care and Sleep Medicine, University of Miami-Miller School of Medicine, Miami, FL, USA.

[4] Department of Psychiatry, Rutgers Medical School, Newark, NJ, USA.

* These authors contributed equally to this work.

Corresponding Author: Mehdi Mirsaeidi, MD, MPH, Division of Pulmonary, Critical Care, Sleep and Allergy, University of Miami-Miller School of Medicine, 1600 NW 10th Avenue, Miami, FL 33136, USA, Tel: (305) 243-9227, E-mail: msm249@miami.edu.

production, alteration of endogenous synthesis pathways, or alteration of host neurotransmitter receptors. This chapter explores the relevant literature on the effects of the neurotransmitters and their associations with nervous system disorders involving both the CNS and the ENS such as autism spectrum disorder (ASD), Parkinson's disease, Alzheimer's disease, schizophrenia (SCZ), irritable bowel syndrome (IBS) and Crohn's disease. Finally, the chapter explores future steps in ascertaining the specific molecular mechanisms by which the microbiome affects the CNS.

INTRODUCTION

Neurotransmitters are signaling messenger molecules that play a pivotal role in neuronal interaction and regulating the functions of CNS as well as human behavior.[1] The neuromodulatory function of these chemicals integrates the afferent input and coordinates the effector activity of the CNS to maintain homeostasis throughout the body.[2] Efficient communication between neurons involves the release of neurotransmitters and receptor binding on target cell to modulate its activity (Fig. 3.1). Neurotransmitters are com-

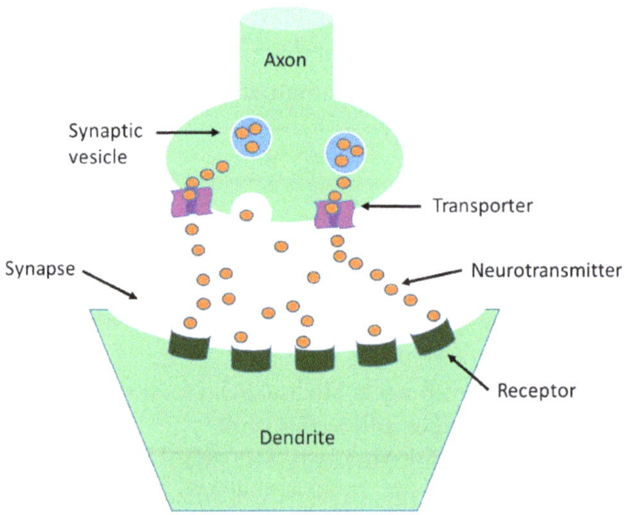

Fig. 3.1: Schematic representation of neurotransmitter release. Upon stimulation, neurotransmitters are released from synaptic cleft followed by binding to receptors on target cells or other neurons to exert their action.

posed of amines (serotonin, DA, NE, epinephrine, and histamine), amino acids (GABA, glutamate), acetylcholine, adenosine, among others.[2] One important target of these neurotransmitters is the gastrointestinal (GI) tract that despite having its own independent regulation by the ENS is also regulated by the CNS through the autonomic nervous system. In fact, interaction between the CNS and the GI tract is complex and has been described under the term "gut–brain axis" (GBA).[3] Distinct communication pathways have been described between the CNS and the GI tract including immune system, vagus nerve, and microbiome modulation.[4]

There is a growing evidence showing how GI and neuropsychiatric disorders are related, indicating how dysfunction of the GBA can cause disease.[5–7] IBS, depression, and Parkinson's disease are good examples of such disorders.[8] A number of bacteria have been reported to be able to produce and/or influence a range of neurotransmitters including DA, noradrenaline, serotonin, GABA, acetylcholine, histamine, and steroids (Table 3.1) (Fig. 3.2).[9] Neurotransmitters produced by bacteria are identical

Table 1: A summary of neurotransmitters produced by gut microbiota (adapted from Strandwitz[9])

Neurotransmitter	Bacteria	Source
Serotonin	Hafnia alvei	Ozogul[63]
	Klebsiella pneumoniae	Ozogul[63]
	Morganella morganii	Ozogul[63]
	Escherichia coli	Shishov *et al.*[65]
	Lactobacillus plantarum	Ozogul *et al.*[64]
	Lactococcus lactis subsp. cremoris	Ozogul *et al.*[64]
	Streptococcus thermophilus	Ozogul *et al.*[64]
	Monascus purpureus	Su *et al.*[73]
	Lactobacillus paracasei	Komatsuzaki *et al.*[74]
	Lactobacillus rhamnosus (YS9)	Siragusa *et al.*[75]
	Lactobacillus. Delbrueckii subsp. bulgaricus (PR1)	Siragusa *et al.*[75]

(*Continued*)

Table 1: (*Continued*)

Neurotransmitter	Bacteria	Source
GABA	Lactobacillus plantarum	Siragusa *et al.*[75]
	Lactobacillus buchneri (MS)	Cho, Chang, and Chang[71]
	Streptococcus salivarius subsp. thermophilus (Y2)	Yang *et al.*[70]
	Bifidobacterium adolescentis	Barrett *et al.*[69]
	Bifidobacterium dentium	Barrett *et al.*[69]
	Bifidobacterium infantis	Barrett *et al.*[69]
	Lactobacillus brevis	Barrett *et al.*[69]
	Lactobacillus reuteri (100–23)	Pokusaeva *et al.*[30]
	Bifidobacterium angulatum	Pokusaeva *et al.*[30]
Norepinephrine	Bacillus mycoides	Tsavkelova *et al.*[40]
	Bacillus subtilis	Tsavkelova *et al.*[40]
	Proteus vulgaris	Tsavkelova *et al.*[40]
	Serratia marcescens	Tsavkelova *et al.*[40]
	Escherichia coli (K-12)	Shishov *et al.*[65]
Dopamine	Bacillus cereus	Tsavkelova *et al.*[40]
	Bacillus mycoides	Tsavkelova *et al.*[40]
	Bacillus subtilis	Tsavkelova *et al.*[40]
	Escherichia coli	Tsavkelova *et al.*[40]
	Proteus vulgaris	Tsavkelova *et al.*[40]
	Serratia marcescens	Tsavkelova *et al.*[40]
	Staphylococcus aureus	Tsavkelova *et al.*[40]
	Hafnia alvei	Ozogul[63]
	Klebsiella pneumoniae	Ozogul[63]
	Morganella morganii	Ozogul[63]
	Escherichia coli (K-12)	Shishov *et al.*[65]
	Enterococcus faecium	Villageliú and Lyte[67]
	Citrobacter freundii	Kim *et al.*[66]
	Enterobacter spp.	Kim *et al.*[66]
	Oenococcus oeni	Landete, Ferrer, and Pardo[72]
	Lactobacillus hilgardii	Landete, Ferrer, and Pardo[72]

Table 1: (*Continued*)

Neurotransmitter	Bacteria	Source
Histamine	Lactobacillus mali	Landete, Ferrer, and Pardo[72]
	Pediococcus parvulus	Landete, Ferrer, and Pardo[72]
	Lactobacillus plantarum	Ozogul *et al.*[64]
	Lactococcus lactis subsp. cremoris	Ozogul *et al.*[64]
	Lactococcus lactis subsp. lactis	Ozogul *et al.*[64]
	Streptococcus thermophiles	Ozogul *et al.*[64]
	Lactobacillus Reuteri	Spinler *et al.*[68]

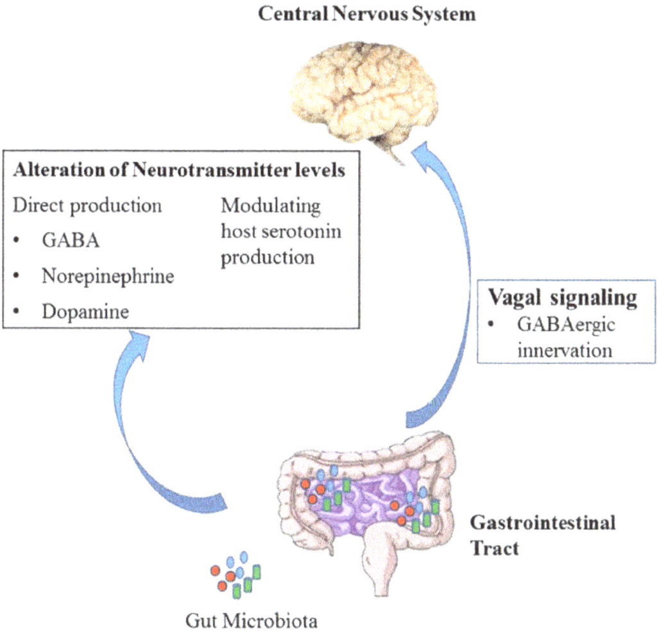

Fig. 3.2: The production of neurotransmitters by microbiota plays an important role in interaction between gut–brain axis (adapted from Strandwitz[9]).

to human neurotransmitters and can influence microbial growth and virulence. GBA has been demonstrated to be a bidirectional interaction[3] (Fig. 3.2). Neurotransmitters play a fundamental role in this bidirectional interaction (Table 3.2) and their modulation might

Table 2: A summary of neurotransmitters and their mode of action.

Neurotransmitter	Mode of action	Source
Serotonin	Regulates mood, stress response, appetite, CNS vascular tone, gastrointestinal secretion , motility Results in impairments in brain function involving mood, sleep, and behavior. Circulates tryptophan levels in germ free (GF) animals Has the ability to immune response regulation	O'Mahony *et al.*[10] Berger[11] O'Mahony *et al.*[10]; Strandwitz[9] Banskota, Ghia, and Khan[12] Luczynski *et al.*[13] Kennedy *et al.*[14] Reigstad *et al.*[15]; Yano *et al.*[16] Campbell *et al.*[17] Clarke *et al.*[18] Gao *et al.*[19] Tapia-Arancibia,[20] Wong, Bottiglieri, and Snead[21] Hyland[22] Wang and Kasper[23] Maestroni[24] Bjurstom[25] Mestecky[26] Mittal *et al.*[8] Lyte[27] Freestone *et al.*[28]
GABA	Regulates CNS disorders including anxiety, epilepsy and depression Can reduce visceral pain sensations Serves as a excitatory neurotransmitter Improves refractory epilepsy symptoms	Wong, Bottiglieri, and Snead[21] Strandwitz[9] Bravo *et al.*[29] Pokusaeva *et al.*[30] Liang *et al.*[31] Lv *et al.*[32] David *et al.*[33] Zheng *et al.*[34] Kootte *et al.*[35] Tillisch *et al.*[37] Hassan *et al.*[38]

Table 2: (*Continued*)

Neurotransmitter	Mode of action	Source
Norepinephrine	Augment motility, biofilm formation and virulence of bacterial species Inhibits local enteric inflammatory state and the effects of muscularis macrophages Inhibits human sequestration of iron, making iron more readily available for bacteria to sequester.	Strandwitz[9] Asano *et al.*[39] Freestone *et al.*[28] Mittal *et al.*[8] Tsavkelova *et al.*[40] Sperandio *et al.*[41] Moret and Briley[43] Petrosus *et al.*[42]
Dopamine	Reflects the caloric load of the infustates, indicating is an important caloric sensor Dictates the intake of calories in one's body Disrupts flavor-nutrient associations Diminishes one's ability to maintain constant calorie intake during intragastric feeding.	Mittal *et al.*[8] Ferreira[49] Koch[49]
Histamine	Influences the allergic response mainly through its actions on the H1R which leads to vasoconstriction, increased vascular permeability, and the H2R which leads to increased mucous secretion and can lead to a decrease in total peripheral resistance Can cause lung inflammation if mediated by bacterial release of histamine	Barcik *et al.*[54] Bieganski *et al.*[56]; Kroger *et al.*[57] White[58] Nuutinen and Panula[61]

(*Continued*)

Table 2: (*Continued*)

Neurotransmitter	Mode of action	Source
	Plays a role in brain functions such as arousal, control of pituitary hormone secretion Could be used to affect sleep-wake disorders	

provide a potential target for developing alternative therapies.[8] The aim of this chapter is to review our current understanding regarding the role of neurotransmitters in regulating GBA. A better knowledge about how neurotransmitters are regulated by the gut microbiome holds a great potential to enhance the future of medicine.

Serotonin

Serotonin (5-hydroxytryptamine or 5-HT) is a monoamine that regulates mood, stress response, appetite, CNS vascular tone, and GI secretion and motility (Table 3.2). Tryptophan is an essential amino acid and precursor of serotonin synthesis that takes place in both the CNS and the ENS. Almost 95% of this neurotransmitter is found mainly within the enterochromaffin cells of the gut and the other 5% in the brain. The dysfunction of serotonin in the GI system could result in impairments in brain function, such as those involved in mood, sleep, and behavior. The link between GI tract hyperalgesia in IBS and serotonin is another example of its role.[10,11] There has been an emerging evidence suggesting regulatory impact of gut microbiota on tryptophan metabolic pathway.[9,10] Bacterial direct utilization of tryptophan reduces its availability to eventually affect cognitive and local GI functions.[10] On the other hand, in the context of intestinal inflammation, mucosal 5-HT acting seems to be greatly implicated in immune cell activation and angiogenesis.[12]

Intestinal microbiota is becoming increasingly relevant in the CNS development and function since it influences neurotransmitter, synaptic, and neurotrophic signaling.[13] It was shown that the circulating tryptophan levels in germ-free (GF) animals were higher than animals with a normal microbiota, while the levels of 5-HT actually declined.[14] This decline agrees with studies that have shown the microbiotas' role in releasing small-chain fatty acids (SCFAs) that are critical in inducing GI 5-HT synthesis.[15,16] Furthermore, the kynurenine pathway that describes the metabolism of tryptophan into kynurenine, 5-HT, and other metabolites is regulated by inflammatory mediators in the immune system.[17] Microbiota-deficient animals have an immature immune system and as a result may also explain the low levels of 5-HT despite higher tryptophan levels in GF animals.[18] Large intestinal microbiota has been shown to affect hypothalamic neurotransmitter expressions in piglets, levels of 5-HT, and DA has been reported to decline in the hypothalamus after intestinal antibiotic infusion.[19] An increase in 5-HT and DA can promote the expression of brain-derived neurotrophic factor (BDNF), which is a neurotrophin that plays an important role in neuroplasticity, feeding behavior, among other functions. Decreased levels of BDNF have been associated with chronic depression.[20]

Literature also shows an association between the major inhibitory neurotransmitter of the CNS, GABA with CNS and ENS disorders, including behavioral, pain, and sleep disorders,[21] intestinal motility, gastric emptying, nociception, and acid secretion.[22] Additionally, there is also an association between the disruption of the serotonergic and kynurenine pathways and CNS pathologies such as Huntington's disease, Alzheimer's disease and dementia.[23] As mentioned before, neurotransmitters play a role in immune response regulation,[24] and GABA is not an exception since it has been found to mediate inhibition of T cell response.[25] On the other hand, catecholamines can alter the release of cytokines and immunoglobulins by influencing gut microbiota and B cells.[26] Furthermore, adrenergic receptors are expressed thorough in immune cells and can stimulate and inhibit inflammatory response.[8] Catecholamines are able to facilitate bacterial invasion by enhancing virulence factors and increase iron availability for bacterial survival.[27,28] All these studies suggest the crucial role of serotonin in regulating GBA.

Gamma-Aminobutyric Acid

Besides serotonin, GABA is the major inhibitory neurotransmitter in the CNS. Imbalances in GABA have been associated with a wide variety of CNS disorders including anxiety, epilepsy, and depression.[21] The microbiome has been shown to alter circulating GABA levels through consumption and production.[9] Multiple different bacteria including *Bifidobacterium* and *Lactobacillus* spp. have been reported to produce GABA. A study demonstrated reduced depressive symptoms and decreased anxiety-induced activity following administration of *Lactobacillus* spp. into mice that corroborated with increased levels of GABA.[29] However, the exact source of the increase in GABA was not determined by the study. Another study showed that bacteria engineered to produce GABA when administrated into rats could reduce visceral pain sensations.[30] While GABA production and consumption are one way the microbiome affects CNS activity, the microbiome can also induce alterations in GABA-A receptors in the hippocampus. It has been shown that gut microbiota disturbance in early life resulted in changes in emotional behavior and reduced expression of GABA-A receptor $a5$ and δ subunits in the hippocampus of adult rats.[31] These $a5$ and δ subunits are regulated by stress hormones and may be related to depression.[31] After administration of probiotics, significant improvement in the depressive behavior as well as receptor levels was observed.[31] GABA also plays a major role in the ENS where it acts as an excitatory neurotransmitter. Low levels of GABA due to imbalances in gut microbiota were associated with the pain sensation involved in Crohn's disease.[32] Diet has also been shown to affect the microbiome and GABA levels.[33] A ketogenic diet was shown to increase GABA levels in the CSF and lead to improvements in refractory epilepsy symptoms.[9]

A recent study showed that when gut microbiota was transferred from patients with SCZ to GF mice, the mice started exhibiting altered behavior as well.[34] They implicate the changes in behavior are associated with alterations in the glutamate-glutamine-GABA cycle whereby glutamate levels decreased, while glutamine and GABA levels increased in the SCZ microbiota recipient mice, as these changes have been implicated with SCZ in the past studies.[34] This revelation altering the microbiome in humans or animals leads to alterations in neurotransmitter levels like GABA, which is especially intriguing and becoming increasingly supported. Fecal microbiome transplantation from lean to obese patients showed a markedly elevated GABA levels in patient plasma.[35] Another study has pointed to GABA as being an important growth factor that is released by certain gut bacteria such as *Bacterioides*, *Parabacteroides* and *Escherichia* species.[36] When the fecal samples of 40 women were examined, the *Bacteroides* population was shown to be positively correlated with increased gray matter in the cerebellum, hippocampus, and frontal regions of the brain.[37] These women also exhibited less anxious behavior, distress and irritability when subjected to emotional testing.[37] Additionally, a study showed that mice with a high-fat diet showed lowered GABA levels, which was correlated to a decrease in *Bacteroides* numbers as well as increased depressive-like symptoms.[38] To further corroborate this, it was observed that *Bacteroides* numbers were negatively correlated with major depressive disorder (MDD)-associated brain signatures likely due to the alterations in GABA levels.[36]

The abovementioned studies show that circulating GABA levels are readily influenced by the gut microbiota. Furthermore, GABA levels are associated with a variety of disorders such as SCZ, MDD, anxiety and epilepsy. Hence, it is important to elucidate the relationship between the microbiota and GABA levels to find efficacious methods for combatting these neurological disorders. It is also interesting that transfer of microbiota from one individual to another can be a modality by which future therapies could be delivered.

Norepinephrine

NE is a principal neurotransmitter of the sympathetic nervous system and is instrumental in the arousal state as well as the fight or flight mechanism. While NE levels have been shown to be influenced by the gut microbiome, unlike GABA, it is not clear as to whether the gut microbiome directly synthesizes NE or influences intrinsic cellular process to increase or decrease NE production.[9] A recent study showed that GF mice have much lower levels of NE in the cecal lumen than normal mice, and that reintroduction of the microbiome in GF mice restored normal levels of NE.[39] NE also has a positive effect on bacterial growth *in vitro* such as in the case of *Escherichia coli*, *Klebsiella pneumoniae*, *Pseudomonas aeruginosa*, *Enterobacter cloacae*, *Shigella sonnei*, and *Staphylococcus aureus*.[9] Along with facilitating bacterial growth, catecholamines such as NE and DA appear to augment motility, biofilm formation, and virulence of these bacterial species.[28] The ability for NE to increase bacterial growth is thought to be associated with its inhibition of the local enteric inflammatory state and the effects of muscularis macrophages.[8] Another mechanism by which NE aids bacterial growth is its ability to inhibit human sequestration of iron, making iron more readily available for bacteria to sequester. This is helpful because survival of the bacteria depends on their ability to sequester iron.[8] There are also bacterial species that produce NE such as *E. coli*, *Proteus vulgaris*, *Serratia marcescens*, *Bacillus subtilis*, and *Bacillus mycoides*.[40] These bacteria secrete NE because it facilitates the quorum-sensing abilities of bacteria.[41]

It is hypothesized that most pathological bacteria are aerobic, and harmless bacteria are generally anaerobic. In a recent study, piglets treated with NE showed a shift in their gut microbiome to increased number of anaerobic bacteria though this is not certain as this could have been due to changes in the lactic acid bacteria population.[42] This could indicate that NE has the effect of increasing commensal bacteria and could be beneficial to the host. Microbiome influence on NE levels and vice versa is important because multiple pathologies arise from defects in the CNS that lead to issues with the ENS.[8] For example, in IBS, the alpha 2 adrenergic receptor is downregulated, which leads to increased noradrenergic activity.[8] Both elevated and depressed levels of NE are a risk factor for predisposition to developing neurological disorders. For example, lower levels of NE are associated with depression and increasing NE is one of the ways in which anti-depressive medications target the disease.[43] On the other hand, elevated NE levels are associated with anxiety.[8]

Dopamine

DA is a major neurotransmitter in reward-motivated behavior that prefaces other catecholamines, including epinephrine and NE, responsible for arousal and alertness as well as behavior and cognition.[44,45] Several bacteria, such as *E. coli*, *P. vulgaris*, *S. marcescens*, *B. subtilis*, and *B. mycoides*, have shown to produce DA and NE in their biomass.[8] It seems as if NE is produced as a quorum-sensing molecule in bacteria; however, the factors that lead to the production of DA still remains unclear.[8] Furthermore, accumulating evidence suggests that the synthesis of the neurotransmitters plays a role in host biosynthesis/catabolism. It was discovered through a recent study that without bacteria, substantially reduced levels of NE is produced in the cecal lumen of mice.[39] Furthermore, GF mice display increased turnover rate of DA and NE in the brain, reducing pools of systemic circulation independent of microbial production.[8]

DA has its own five receptors, D1–D5,[46] different from NE or epinephrine. The receptors are found in distinguishable locations; D1, D3 and D5 receptors can be found at nerve-ending layers of intestinal wall and gut mucosa, D4 receptors are found in mucosal layer, while D2 receptors are found only in the nerve-ending layer of intestinal wall.[8] DA regulates electrolyte absorption in the GI tract and increases Na^+ and Cl^- influx while decreasing Ca^{2+} influx.[47] In addition, changes in the flow of DA produced by the GI tract reflect the caloric load of the infusates, indicating that DA is a very important caloric sensor that dictates the intake of calories in one's body.[48] It disrupts flavor-nutrient associations and diminishes one's ability to maintain constant calorie intake during intragastric feeding. In a study, investigating the sensitivity of DA to intragastric transfusion of nutrients, mice had the ability to control their intake of calories, even in flavor-deficient nutrients.[49]

A disrupted gut-generated lipid signaling reduces dopaminergic function that occurs after chronic exposure to a high-fat diet. Restoring the signal pathway can increase the reward value of low-calorie foods.[50] A recent study suggested that, when obese rodents ingest food, they do not display DA signaling. The rodents on a high-fat diet have unusually low levels of the intestinal lipid messenger oleoylethanolamine. It was hypothesized that oleoylethanolamine be the missing link between excess dietary fat intake and DA deficiency.[50] When oleoylethanolamine was added to the diet of these rodents, mice began eating increasing amounts of low-fat foods. Nevertheless, after the GBA was surgically severed, the effects were abolished. It was concluded that communication between the GI tract and the brain's reward circuit constitutes one fundamental way by which animals attribute reward value to certain foods.[50]

Histamine

Histamine plays a well-known role in the allergic response as well as in inflammation.[51,52] It is secreted by a variety of cells including mast cells and basophils.[53] Histamine is synthesized from histidine by the enzyme histidine decarboxylase (HDC).[54] Along with bodily cells expressing this enzyme, multiple bacterial species also express HDC and release histamine.[55] There are four different histamine receptors that allow histamine to have its various systemic effects including immediate hypersensitivity response (H1R), modulation of H1R-mediated effects (H2R), controlling presynaptic neurotransmitter release (H3R), and cytokine production as well as chemoattractant secretion via H4R.[54] Histamine is then broken down by two enzymes: diamine oxidase (DAO), which is secreted extracellularly and plays a role in scavenging extracellular histamine, and N-methyl transferase (HNMT), which remains in the cytosol of the cell.[56,57] This is important because any excess bacterial histamine acts on these same four receptors and must be broken down by DAO and HNMT.[54]

Histamine influences the allergic response mainly through its actions on the H1R which, leads to vasoconstriction and increased vascular permeability, and the H2R, which leads to increased mucous secretion and a decrease in total peripheral resistance.[58]

Hence, in the case of asthma, it was shown that histamine-secreting bacteria populations, such as *E. coli*, *Morganella morganii*, *Lactobacillus vaginalis*, *Enterobacter aerogenes*, *Lactobacillus casei* and *Enterococcus durans*, are increased.[59] In a recent study, a recombinant strain of bacteria that secreted histamine was introduced into HDC knockout and H2R knockout mice, respectively, to observe whether endogenous histamine would influence the effect of bacterial-derived histamine, and whether H2R was the site of action for bacterially derived histamine.[54] It was found that while bacterial-derived histamine does not have to specifically lead to pathology, there is evidence suggesting that lung inflammation is mediated by bacterial release of histamine.[54]

While the above discussion suggests the histamine's role in allergic response, it is also a very important neurotransmitter and has widespread effects in the CNS.[60] Histamine plays a role in brain functions such as arousal and control of pituitary hormone secretion and is becoming increasingly noticed as a wake-promoting substance that could be used to affect sleep–wake disorders.[61] Histamine levels are shown to be low in Alzheimer's disease, while accentuated in Parkinson's disease and SCZ.[61] All these findings suggest that gut-induced modulation of histamine levels could have an influence on brain function and may be involved in neurological disorders. Further clinical studies using large cohort of patients will help in deciphering the role of histamine in regulating GBA.

CONCLUSIONS AND FUTURE DIRECTIONS

The gut microbiome's role in the pathophysiology of various neurological disorders, such as anxiety, Alzheimer's disease, Parkinson's disease, MDD and SCZ, as well as in ENS disorders, such as IBS and Crohn's disease, has been a long-standing area of research. In this chapter, we discussed neurotransmitters by which the gut bacteria can influence the GBA through multiple mechanisms. One mechanism is through the direct production of the neurotransmitters by the commensal bacteria. Another mechanism involves the induction of endogenous machinery to modulate neurotransmitter levels, as well as through the up or downregulation of neurotransmitter receptors such GABA. Many current and past studies have focused on deciphering the CNS marker changes in GF versus normal animals. While these studies have shown the association between the gut bacteria and the neurological disorders, they have not yet elucidated the precise underlying molecular mechanisms. While the fields of neuroscience, microbiology, and gastroenterology have worked to describe pathologies within their realm of study, a more interdisciplinary approach is necessary to observe and ascertain the specific mechanism by which all three fields interact.

Recent studies have shown promise in trying to bridge the gap between how specifically gut-produced factors such as neurotransmitters lead to their respective roles in the CNS. One hypothesis is that specialized CNS macrophages known as glial cells are the missing link between the gut microbiome and CNS diseases.[62] Not only have microglia been known to be important immune cells in the CNS and their dysfunction associated with various CNS disorders, but there is also evidence that gut bacteria control the maturation and function of microglial though the mechanism is currently unknown.[62] The importance of deciphering the mechanisms by which the gut and CNS interact is ultimately aimed at developing alternative courses of treatment. The emergence of new technologies will open up avenues to precisely understand the contribution of neurotransmitters in regulating GBA in pursuit of developing novel therapeutic strategies for GI and neurological disorders.

REFERENCES

1. D'Andrea G, Gucciardi A, Perini F, Leon A. (2019) Pathogenesis of Cluster Headache: From episodic to chronic form, the role of neurotransmitters and neuromodulators. *Headache*. **59**(9): 1665–1670.
2. Hyman SE. (2005) Neurotransmitters. *Curr Biol*. **15**(5): R154–R158.
3. Rhee SH, Pothoulakis C, Mayer EA. (2009) Principles and clinical implications of the brain-gut-enteric microbiota axis. *Nat Rev Gastroenterol Hepatol* **6**: 306–314.

4. Arneth BM. (2018) Gut-brain axis biochemical signalling from the gastrointestinal tract to the central nervous system: gut dysbiosis and altered brain function. *Postgrad Med J*. **94**(1114): 446–452.
5. Severance EG and Yolken RH (2020). Deciphering microbiome and neuroactive immune gene interactions in schizophrenia. *Neurobiol Dis*. **135**: 104331.
6. Ma Q, Xing C, Long W, Wang HY, Liu Q, Wang RF. (2019) Impact of microbiota on central nervous system and neurological diseases: The gut-brain axis. *J Neuroinflammation*. **16**(1): 53.
7. Breen DP, Halliday GM, Lang AE. (2019) Gut-brain axis and the spread of α-synuclein pathology: Vagal highway or dead end? *Mov Disord*. **34**(3): 307–316.
8. Mittal R, *et al*. (2017) Neurotransmitters: The critical modulators regulating gut-brain axis. *J Cell Physiol* **232**: 2359–2372.
9. Strandwitz P. (2018) Neurotransmitter modulation by the gut microbiota. *Brain Res* **1693**: 128–133.
10. O'Mahony SM, Clarke G, Borre YE, Dinan TG, Cryan JF. (2015) Serotonin, tryptophan metabolism and the brain-gut-microbiome axis. *Behav Brain Res* **277**: 32–48.
11. Berger M, Gray JA, Roth BL. (2009) The expanded biology of serotonin. *Annu Rev Med* **60**: 355–366.
12. Banskota S, Ghia JE, Khan WI. (2018) Serotonin in the gut: Blessing or a curse. *Biochimie* **161**: 56–64.
13. Luczynski P, McVey Neufeld KA, Oriach CS, *et al*. (2016) Growing up in a bubble: Using germ-free animals to assess the influence of the gut microbiota on brain and behavior. *Int J Neuropsychopharmacol* **19**(18): pyw020.
14. Kennedy PJ, Cryan JF, Dinan TG, Clarke G. (2017) Kynurenine pathway metabolism and the microbiota-gut-brain axis. *Neuropharmacology* **112**: 399–412.
15. Reigstad CS, *et al*. (2015) Gut microbes promote colonic serotonin production through an effect of short-chain fatty acids on enterochromaffin cells. *FASEB J* **29**: 1395–1403.
16. Yano JM, *et al*. (2015) Indigenous bacteria from the gut microbiota regulate host serotonin biosynthesis. *Cell* **161**: 264–276.
17. Campbell BM, Charych E, Lee AW, Moller T. (2014) Kynurenines in CNS disease: Regulation by inflammatory cytokines. *Front Neurosci* **8**: 12.

18. Clarke G, *et al.* (2013) The microbiome-gut-brain axis during early life regulates the hippocampal serotonergic system in a sex-dependent manner. *Mol Psychiatry* **18**: 666–673.
19. Gao K, Pi Y, Mu CL, Peng Y, Huang Z, Zhu WY. (2018) Antibiotics-induced modulation of large intestinal microbiota altered aromatic amino acid profile and expression of neurotransmitters in the hypothalamus of piglets. *J Neurochem* **146**: 219–234.
20. Tapia-Arancibia L, Rage F, Givalois L, Arancibia S. (2004) Physiology of BDNF: Focus on hypothalamic function. *Front Neuroendocrinol* **25**: 77–107.
21. Wong CG, Bottiglieri T, Snead OC. (2003) GABA, gamma-hydroxybutyric acid, and neurological disease. *Ann Neurol* **54**(Suppl 6): S3–S12.
22. Hyland NP, Cryan JF. (2010) A gut feeling about GABA: Focus on $GABA_B$ receptors. *Front Pharmacol* **1**: 124.
23. Wang Y, Kasper LH. (2014) The role of microbiome in central nervous system disorders. *Brain Behav Immun* **38**: 1–12.
24. Maestroni GJ. (2000) Dendritic cell migration controlled by alpha 1b-adrenergic receptors. *J Immunol* **165**: 6743–6747.
25. Bjurstom H. (2008) GABA, a natural immunomodulator of T lymphocytes. *J Neuroimmunol* **205**: 44–50.
26. Mestecky J, Russell MW. (2009) Specific antibody activity, glycan heterogeneity and polyreactivity contribute to the protective activity of S-IgA at mucosal surfaces. *Immunol Lett* **124**: 57–62.
27. Lyte M. (2004) Microbial endocrinology and infectious disease in the 21st century. *Trends Microbiol* **12**: 14–20.
28. Freestone PP, Williams PH, Haigh RD, Maggs AF, Neal CP, Lyte M. (2002) Growth stimulation of intestinal commensal Escherichia coli by catecholamines: A possible contributory factor in trauma-induced sepsis. *Shock* **18**: 465–470.
29. Bravo JA, *et al.* (2011) Ingestion of Lactobacillus strain regulates emotional behavior and central GABA receptor expression in a mouse via the vagus nerve. *Proc Natl Acad Sci USA* **108**: 16050–16055.
30. Pokusaeva K, *et al.* (2017) GABA-producing Bifidobacterium dentium modulates visceral sensitivity in the intestine. *Neurogastroenterol Motil* **29**(1): e12904.
31. Liang L, Zhou H, Zhang S, Yuan J, Wu H. (2017) Effects of gut microbiota disturbance induced in early life on the expression of extrasynaptic GABA-A receptor alpha5 and delta subunits in the hippocampus of adult rats. *Brain Res Bull* **135**: 113–119.

32. Lv K, *et al.* (2018) Neurotransmitter alterations in the anterior cingulate cortex in Crohn's disease patients with abdominal pain: A preliminary MR spectroscopy study. *Neuroimage Clin* **20**: 793–799.
33. David LA, *et al.* (2014) Diet rapidly and reproducibly alters the human gut microbiome. *Nature* **505**: 559–563.
34. Zheng P, *et al.* (2019) The gut microbiome from patients with schizophrenia modulates the glutamate-glutamine-GABA cycle and schizophrenia-relevant behaviors in mice. *Sci Adv* **5**: eaau8317.
35. Kootte RS, *et al.* (2017) Improvement of insulin sensitivity after lean donor feces in metabolic syndrome is driven by baseline intestinal microbiota composition. *Cell Metab* **26**: 611.e6–619.e6.
36. Strandwitz P, *et al.* (2019) GABA-modulating bacteria of the human gut microbiota. *Nat Microbiol* **4**: 396–403.
37. Tillisch K, *et al.* (2017) Brain structure and response to emotional stimuli as related to gut microbial profiles in healthy women. *Psychosom Med* **79**: 905–913.
38. Hassan AM, *et al.* (2018) High-fat diet induces depression-like behaviour in mice associated with changes in microbiome, neuropeptide Y, and brain metabolome. *Nutr Neurosci* **22**(12): 1–17.
39. Asano Y, *et al.* (2012) Critical role of gut microbiota in the production of biologically active, free catecholamines in the gut lumen of mice. *Am J Physiol Gastrointest Liver Physiol* **303**: G1288–G1295.
40. Tsavkelova EA, Botvinko IV, Kudrin VS, Oleskin AV. (2000) Detection of neurotransmitter amines in microorganisms with the use of high-performance liquid chromatography. *Dokl Biochem* **372**: 115–117.
41. Sperandio V, Torres AG, Jarvis B, Nataro JP, Kaper JB. (2003) Bacteria-host communication: The language of hormones. *Proc Natl Acad Sci USA* **100**: 8951–8956.
42. Petrosus E, Silva EB, Lay D, Jr., Eicher SD. (2018) Effects of orally administered cortisol and norepinephrine on weanling piglet gut microbial populations and Salmonella passage. *J Anim Sci* **96**: 4543–4551.
43. Moret C, Briley M. (2011) The importance of norepinephrine in depression. *Neuropsychiatr Dis Treat* **7**: 9–13.
44. Klein MO, Battagello DS, Cardoso AR, Hauser DN, Bittencourt JC, Correa RG. (2019) Dopamine: Functions, signaling, and association with neurological diseases. *Cell Mol Neurobiol* **39**: 31–59.
45. Liu C, Kaeser PS. (2019) Mechanisms and regulation of dopamine release. *Curr Opin Neurobiol* **57**: 46–53.

46. Bhatia A, Saadabadi A. (2019) Biochemistry, dopamine receptors, in *StatPearls*. Treasure Island, FL.
47. Eisenhofer G, Aneman A, Friberg P, Hooper D, Fåndriks L, Lonroth H, Hunyady B, Mezey E. (1997) Substantial production of dopamine in the human gastrointestinal tract. *J Clin Endocrinol Metab*.
48. de Araujo IE, Ferreira JG, Tellez LA, Ren X, Yeckel CW. (2012) The gut-brain dopamine axis: A regulatory system for caloric intake. *Physiol Behav* **106**: 394–399.
49. Ferreira JG, Tellez LA, Ren X, Yeckel CW, de Araujo IE. (2012) Regulation of fat intake in the absence of flavour signalling. *J Physiol*. **590**(4): 953–972.
50. Koch L. (2013) Gut-derived lipid signal links dopamine deficiency and high-fat feeding. *Nat Rev Endocrinol* **9**: 627.
51. Thurmond RL, Kazerouni K, Chaplan SR, *et al*. (2014) Peripheral Neuronal Mechanism of Itch: Histamine and Itch. In: Carstens E, Akiyama T, editors. Itch: Mechanisms and Treatment. Boca Raton (FL): CRC Press/Taylor & Francis; Chapter 10. Available from: https://www.ncbi.nlm.nih.gov/books/NBK200934/
52. Tiligada E, Ennis M. (2020) Histamine pharmacology: From Sir Henry Dale to the 21st century. *Br J Pharmacol*. **177**(3): 469–489.
53. Lieberman P. (2011) The basics of histamine biology. *Ann Allergy Asthma Immunol* **106**: S2–S5.
54. Barcik W, *et al*. (2018) Bacterial secretion of histamine within the gut influences immune responses within the lung. *Allergy* **74**: 899–909.
55. Barcik W, Untersmayr E, Pali-Scholl I, O'Mahony L, Frei R. (2015) Influence of microbiome and diet on immune responses in food allergy models. *Drug Discov Today Dis Models* **17–18**: 71–80.
56. Bieganski T, Kusche J, Lorenz W, Hesterberg R, Stahlknecht CD, Feussner K. D. (1983) Distribution and properties of human intestinal diamine oxidase and its relevance for the histamine catabolism. *Biochim Biophys Acta* **756**: 196–203.
57. Kroger S, *et al*. (2015) Effects of high levels of dietary zinc oxide on ex vivo epithelial histamine response and investigations on histamine receptor action in the proximal colon of weaned piglets. *J Anim Sci* **93**: 5265–5272.
58. White MV. (1990) The role of histamine in allergic diseases. *J Allergy Clin Immunol* **86**: 599–605.

59. Barcik W, *et al.* (2016) Histamine-secreting microbes are increased in the gut of adult asthma patients. *J Allergy Clin Immunol* **138**(5): 1491–1494.e7.
60. Alstadhaug KB. (2014) Histamine in migraine and brain. *Headache* **54**: 246–259.
61. Nuutinen S, Panula P. (2010) Histamine in neurotransmission and brain diseases. *Adv Exp Med Biol* **709**: 95–107.
62. Wang Y, *et al.* (2018) The gut-microglia connection: Implications for central nervous system diseases. *Front Immunol* **9**: 2325.
63. Ozogul, Fatih. (2004). Production of biogenic amines by Morganella morganii, Klebsiella pneumoniae and Hafnia alvei using a rapid HPLC method. European Food Research and Technology. 219. 465-469. 10.1007/s00217-004-0988-0.
64. Ozogul, Fatih & Kuley, Esmeray & Özogul, Yesim & Ozogul, Ilyas. (2012). The Function of Lactic Acid Bacteria on Biogenic Amines Production by Food-Borne Pathogens in Arginine Decarboxylase Broth. Food Science and Technology Research. 18. 795-804. 10.3136/fstr.18.795.
65. Shishov, V.A., Kirovskaya, T.A., Kudrin, V.S. et al. Amine neuromediators, their precursors, and oxidation products in the culture of Escherichia coli K-12. Appl Biochem Microbiol 45, 494–497 (2009). https://doi.org/10.1134/S0003683809050068
66. KIM, SHIN-HEE & Field, Katharine & Morrissey, Michael & PRICE, ROBERT & WEI, CHENG-I & AN, HAEJUNG. (2001). Source and Identification of Histamine-Producing Bacteria from Fresh and Temperature-Abused Albacore. Journal of Food Protection. 64. 1035-1044. 10.4315/0362-028X-64.7.1035.
67. Villageliú D, Lyte M. Dopamine production in Enterococcus faecium: A microbial endocrinology-based mechanism for the selection of probiotics based on neurochemical-producing potential. PLoS One. 2018 Nov 28;13(11):e0207038. doi: 10.1371/journal.pone.0207038.
68. Spinler JK, Sontakke A, Hollister EB, Venable SF, Oh PL, Balderas MA, Saulnier DM, Mistretta TA, Devaraj S, Walter J, Versalovic J, Highlander SK. From prediction to function using evolutionary genomics: human-specific ecotypes of Lactobacillus reuteri have diverse probiotic functions. Genome Biol Evol. 2014 Jun 19;6(7):1772-89. doi: 10.1093/gbe/evu137.
69. Barrett, E., Ross, R. P., O'Toole, P. W., Fitzgerald, G. F., & Stanton, C. (2012). γ-Aminobutyric acid production by culturable bacteria from the

human intestine. Journal of applied microbiology, 113(2), 411–417. https://doi.org/10.1111/j.1365-2672.2012.05344.x

70. Yang, S. Y., Lü, F. X., Lu, Z. X., Bie, X. M., Jiao, Y., Sun, L. J., & Yu, B. (2008). Production of gamma-aminobutyric acid by Streptococcus salivarius subsp. thermophilus Y2 under submerged fermentation. Amino acids, 34(3), 473–478. https://doi.org/10.1007/s00726-007-0544-x
71. Cho, Y. R., Chang, J. Y., & Chang, H. C. (2007). Production of gamma-aminobutyric acid (GABA) by Lactobacillus buchneri isolated from kimchi and its neuroprotective effect on neuronal cells. Journal of microbiology and biotechnology, 17(1), 104–109.
72. Landete, J. M., Pardo, I., & Ferrer, S. (2007). Tyramine and phenylethylamine production among lactic acid bacteria isolated from wine. International journal of food microbiology, 115(3), 364–368. https://doi.org/10.1016/j.ijfoodmicro.2006.10.051
73. Yuan-Chi Su, Jyh-Jye Wang, Tzu-Tsen Lin, Tzu-Ming Pan, Production of the secondary metabolites γ-aminobutyric acid and monacolin K by Monascus, Journal of Industrial Microbiology and Biotechnology, Volume 30, Issue 1, 1 January 2003, Pages 41–46.
74. Noriko Komatsuzaki, Jun Shima, Shinichi Kawamoto, Hiroh Momose, Toshinori Kimura, Production of γ-aminobutyric acid (GABA) by Lactobacillus paracasei isolated from traditional fermented foods, Food Microbiology, Volume 22, Issue 6, 2005, Pages 497-504, ISSN 0740-0020.
75. Siragusa S, De Angelis M, Di Cagno R, Rizzello CG, Coda R, Gobbetti M. Synthesis of γ-Aminobutyric Acid by Lactic Acid Bacteria Isolated from a Variety of Italian Cheeses. Appl Environ Microbiol. 15 nov 2007;73(22):7283-90.

Chapter 4

Dysbiosis, Inflammation, and the Blood–Brain Barrier

Jenna Bergman[1,*], Camron Davies* and Rebecca S. Eshraghi

Abstract

The brain is often thought of as a separate, immune-privileged organ, protected from the rest of the body due to the blood–brain barrier (BBB). On the contrary, current research disputes this concept and provides abundant evidence that suggests a bidirectional axis of communication between the central nervous system and periphery.

Indeed the gut–brain connection is one of the most exciting scientific discoveries in recent years. While there is a clear established relationship between the central nervous system and the gastrointestinal tract, the inner workings between them continue to be explored. One of the confirmed links is the connection between the neuroimmune system and gut microbiota. Our microbial diversity can either be neuroprotective or deleterious.

---

*These authors contributed equally to this work.

Corresponding Author: Rebecca S. Eshraghi, PhD, DNM, BBA, Hearing Research and Communication Disorders Laboratory, Department of Otolaryngology, Division of Neurotology, University of Miami-Miller School of Medicine, 1600 NW 10th Ave, Miami, FL 33136, USA; Quantum University for Integrative Medicine, Honolulu, HI 96813, USA, E-mail: rebeccaeshraghi@gmail.com; Telephone number: 1-305-720-9099

[1] Hearing Research and Communication Disorders Laboratory, Department of Otolaryngology, Division of Neurotology, University of Miami-Miller School of 30, Medicine, Miami, FL, USA

This chapter will discuss how microbial depletion and dysregulation can cause systemic inflammation leading to a breakdown of the BBB and subsequent neurologic dysfunction.

INTRODUCTION

The human microbiome consists of trillions of symbiotic bacterial cells with the majority residing in the gut.[1] The gut is densely populated with a diverse array of microorganisms such as bacteria, archaea, fungi, protozoans, and non-living viruses, all of which make up the human inner ecosystem.[2] This functioning ecosystem strongly depends on a state of homeostasis where dysregulation can cause systemic instability and have a profound impact on our health. Many chronic illnesses such as diabetes, cancer, depression, and autoimmune disease have an association with the physiologic status of the gut microbiota. Some chronic inflammation that leads to disease is believed to result from disturbances in the homeostasis of the gut microbiome. The full understanding of the gut–brain connection is still unknown, but recent research demonstrates a link between immune regulating gut bacteria and the functioning of the neuroimmune system. In this chapter, we will investigate the crosstalk between the immune system, central nervous system (CNS), the gut microbiota, and the gut's influence on the permeability of the blood–brain barrier (BBB).

The Human Microbiome, Dysbiosis, and the Immune System

The human gut microbiome is a key regulator of the immune system and plays a pivotal role in shaping many immune-mediated processes, including allergies and autoimmune disease.[3] Since 70%–80% of human immune cells are in gut-associated lymphoid tissue, lymphocyte accumulation, and differentiation in the gastrointestinal (GI) tract can be triggered in response to changes in microbiota composition.[4] A well-regulated immune system depends on optimal homeostasis between commensal bacteria and pathogenic bacteria that reside within the gut.[5]

Microglia are the primary cells of the CNS's immune system and are responsible for regulating neuroinflammatory reactions. Interestingly, recent studies have shown a correlation between microglia maturation and the microbiome.[6] One study found that germ-free (GF) mice raised in completely sterile environments displayed global deficiencies in microglia with changed cell proportions and an immature phenotype, consequently leading to compromised innate immune responses. The same study revealed that host microbiota constantly controls maturation and function of microglia in the CNS by demonstrating that temporal eradication of host microbiota severely changed microglia function, while recolonization with a complex microbiota partially restored their function.[7] Further mouse studies have shown that gut microbes produce compounds that prime cells to destroy harmful viruses in the brain and nervous system. Mice treated with antibiotics, which disturb the ecology of the gut microbiome before the onset of disease, were left defenseless and succumbed to disease because they had less microglia to aid in the removal of viruses by the immune system.[8]

Early microbial colonization during birth, infancy, and early childhood is of critical importance to establishing a healthy microbiome.[3] Many factors in our modern life, including mode of infant delivery, infant feeding, nutrition, genetics, and antibiotics, can significantly disrupt or shift this microbial homeostasis.[9–11]

Another way that the gut influences the CNS is through the production of neurotransmitters. Serotonin is a key neurotransmitter and 90% is synthesized by the metabolism of tryptophan in the GI tract.[12] Tryptophan is metabolized to serotonin by several bacteria species and impaired synthesis caused by dysbiosis in the GI tract has been implicated in several neurologic diseases. In a study by Yano *et al.*, GF mice produced significantly less serotonin than mice with intestinal microbes.[13] A deficiency in serotonin leads to many adverse symptoms such as depression, anxiety, and sleep disturbance. Notably, tryptophan metabolism has shown to be reduced in patients with autism who exhibit many of the adverse symptoms associated with low levels of serotonin.[14] In a study by Liang *et al.*, rats subjected to 21 days of restraint stress showed an improvement in anxiety, depression, and cognitive dysfunction when supplemented daily with *Lactobacillus helveticus* — a strain of bacteria associated with serotonin synthesis.[15] Furthermore, plasma levels of stress hormones such as corticosterone decreased over time rats who received the probiotic. However, it is important to note that the effects of serotonin are not yet clearly defined. Indeed, there is a difference between neuronal serotonin and mucosal serotonin. One study found that mucosal serotonin induces pro-inflammatory mediators from immune cells, while neuronal serotonin provides neuroprotection in the enteric nervous system.[16] Given the large quantity of serotonin produced in the gut mucosa, dysregulation in this area could serve as key mechanism inducing intestinal inflammation and permeability.

The gut microbiome also has a profound effect on antioxidants throughout the entire body. Vitamin B12 is an important cofactor in DNA methylation reactions and is produced by specific bacterial strains that compete with other microbes in the intestinal tract.[80] Furthermore, the oxidation of vitamin B12 serves as a signal to increase glutathione (GSH) synthesis — the primary intracellular antioxidant in the human body.[17] A lack of vitamin B12 increases oxidative stress, which has been implicated in systemic inflammation and neurologic dysfunction. Antioxidants and methylation-dependent epigenetic regulation are fundamental in maintaining healthy physiologic functioning throughout the body and especially the CNS.[18]

Many chronic diseases arise from disturbing intestinal microbial homeostasis, which is akin to disrupting a well-functioning ecosystem.[19] Systemic inflammation is a major result of gut dysbiosis, which results from the imbalance between beneficial gut microbes and pathogenic gut microbes.[20] Dysbiosis is a direct result of a dysregulated, pathogenic gut microbiome and may lead to the condition known as "leaky gut," which highlights the translocation of endotoxins and other pathogen-associated molecular pattern molecules (PAMPs) into the circulatory system further contributing to heightened immune activation and systemic inflammation.[21] Systemic inflammation can be triggered, sustained, and controlled by the lack and or imbalance of gut microbiome diversity.[21]

Role of SCFAs in Inflammation

Short-chain fatty acids (SCFAs), including acetate (C2), propionate (C3), and butyrate (C4), are some of the main biologically active by-products of the gut microbiome and critical regulators of inflammation. Generated as by-products of bacterial fermentation of prebiotics, that is, fibers (Fig. 4.1), SCFAs have numerous physiological effects as an energy source, epigenetic regulators, cell signaling molecules, and regulators of immune tolerance.[22,23] Of the three main SCFAs, butyrate is by far the best characterized. However, it is vital to consider them as a whole since they are not produced in isolation. The typical human gut produces acetate, propionate, and butyrate in approximately 60:20:20 ratios, respectively.[24] The precise amount and proportion of which depends on the species and abundance of microflora in the colon, the substrates in the colon, and transit time within the gut.[25] Moreover, nearly all the butyrate and a lesser amount of propionate are metabolized before they can reach the systemic circulation; consequently, many of their systemic effects come from their actions within the gut and portal circulation. Conversely, significant amounts of acetate do reach systemic circulation and are even found in the cerebrospinal fluid (CSF), giving acetate a more direct systemic profile. In the paragraphs below, we will provide an overview of the anti-inflammatory effects of the three predominant SCFAs with particular focus on butyrate.

A significant proportion of SCFAs systemic effects are the result of their actions in the gut. On a basic level, SCFA production, particularly acetate given its abundance, acidifies the colonic lumen, lowering the pH, and reducing the ability of dysbiotic species to replicate.[27] More important, however, are SCFAs' effects on the intestinal epithelial barrier itself. As by-products of prebiotic

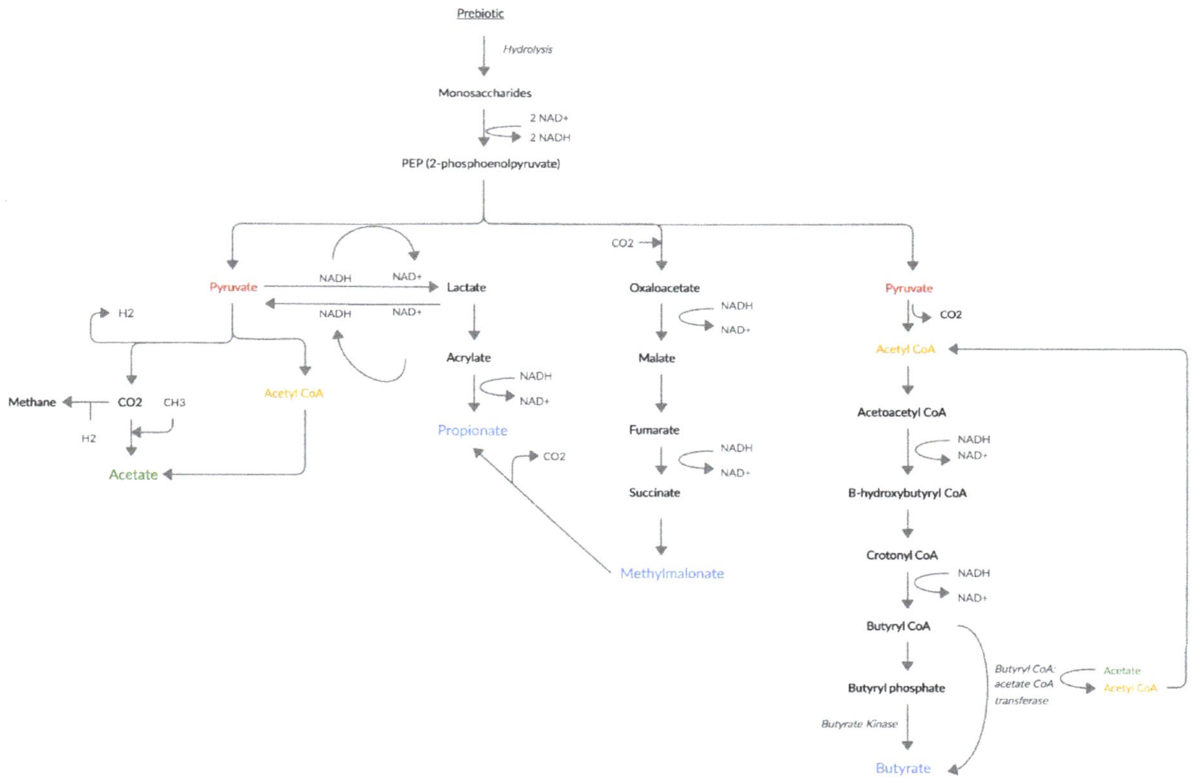

Fig. 4.1: Schematic representation of carbohydrates fermentation, which produces the short-chain fatty acids (SCFAs) acetate, propionate, and butyrate. The main products and intermediates of carbohydrate fermentation are highlighted in different colors and the main enzymes for butyrate production are italicized. Figure adapted from Venegas *et al*.[26]

fermentation, the presence of SCFAs shifts the microbial composition away from mucolytic bacteria while stimulating mucus production.[27–30] Maintenance of this mucus barrier is vital, as it is the first protective barrier separating the lumen from the epithelium.

Butyrate, in particular, has an even more significant role in maintaining the intestinal epithelium itself. Butyrate serves as the preferred energy source of colonocytes.[31,32] Additionally, butyrate metabolism increases colonocyte oxygen consumption, thereby creating physiologic hypoxia. Decreased oxygen concentration increases levels of hypoxia-inducible factor 1 (HIF-1), which is a key epigenetic regulator of intestinal barrier function.[33] Thus, butyrate synergistically functions to nourish colonocytes and produce the environment necessary for their function, making it fundamental in the maintenance of the intestinal epithelial barrier. Decrease in butyrate, as found in a low-fiber diet or after antibiotic administration, leads to increased translocation of bacteria and other PAMPs, such as lipopolysaccharide (LPS), across the gut barrier and, consequently, systemic inflammation. Indeed both butyrate and propionate have been demonstrated to increase trans-epithelial electrical resistance (TEER), a measure of intestinal permeability, via increasing zonula occludens-1, occludin, and E-cadherin expression, thereby decreases the translocation across colonic epithelium.[34–38]

SCFAs' regulation of inflammation does not stop in the gut. Butyrate, and to a lesser extent propionate, regulate inflammatory responses of both innate and adaptive immune system by functioning as histone deacetylases (HDAC) inhibitors.[26] By regulating histone acetylation, SCFAs can mediate the inflammatory response. Histone acetylation leads to the transcription of essential genes contributing to cell proliferation, differentiation, and the production of toll-like receptors (TLR) and interferons (IFN).[39,40]

Separate from their HDAC function, all three main SCFAs can regulate cellular immunity via the G-protein coupled receptors (GPCRs) free fatty acid receptors (FFARs) 2 and 3. FFARs are found throughout the body including on the colon, the heart, adipocytes, skeletal muscle, peripheral nervous system (PNS), the

BBB, and certain immune cells.[32,41–44] In passive immunity, acetate, propionate, and butyrate can regulate neutrophil chemotaxis and decrease the neutrophilic inflammatory response.[45–47] SCFAs are also powerful regulators of the adaptive immune response. Numerous studies have linked increased levels of butyrate to the inhibition of nuclear factor-kappa B (NF-κB) release from macrophages.[48] Combined butyrate and propionate have been shown to decrease LPS-induced pro-inflammatory markers (CXCL9–CXCL11) and cytokines ([interleukin] IL-6 and IL-12) from monocytes as well as increase Foxp3 and IL-10 release from naive CD4 T cells.[49,50] Finally, acetate, propionate, and butyrate combined have been found to increase prostaglandin E_2 (PGE2) and decrease LPS-induced tumor necrosis factor alpha (TNF-α) and IFN-γ release from peripheral blood mononuclear cells (PBMCs).[51] Key to these responses, especially butyrate's, were SCFAs HDAC inhibitors and GPCRs.[40,52] Given these effects, it is apparent that SCFAs have significant combined immunoprotective effects by reducing the transcription and release of pro-inflammatory molecules.

In addition to modulating the release of certain cytokines and chemokines for various immune cells, SCFAs are also fundamental in the differentiation of T cells into T helper 1 (Th-1), T helper 17 (Th-17), and T regulatory cells (Treg).[53–56] By controlling the direction of immune cell differentiation, SCFAs can alter the course of the inflammatory response. In this regard, butyrate's ability to induce Treg differentiation, by inhibiting the HDAC-suppressing FoxP3 is particularly important as it is an essential gene regulating the differentiation of naïve T cells into Tregs.[57,58] Well-known inducers of immunotolerance, Tregs are critical in inhibiting mucosal immune reactions and the development of food allergies, celiac disease, inflammatory bowel disease, and even inducing systemic tolerances.[59] Overall, these studies demonstrate the marked ability of acetate, propionate, and butyrate in maintaining intestinal homeostasis, decreasing intestinal permeability, and subsequently, systemic inflammation.

Despite these promising studies, it is important to note that SCFAs' systemic effects are still being elucidated, particularly regarding their bioavailability. Currently, it is known that after the majority of butyrate is metabolized by colonic epithelial cells, it passes through the portal circulation and is subsequently metabolized by hepatocytes. After metabolism in the liver, only 1–15 M of butyrate and propionate, and 100–200 M of acetate are in systemic circulation.[32] Moreover, only acetate has been found in the CSF at approximately 35 M.[32,60] Overall, this leaves inadequate amount of the most powerful anti-inflammatory SCFAs available in the systemic circulation. One hypothesis for their beneficial systemic effect is that butyrate and propionate serve to prime immune cells passing through portal circulation and, in this way, provide systemic immune tolerance without being present systemically.

T helper 17 cells

Neuroinflammation is inextricably linked to systemic and consequently intestinal inflammation. While many processes are implicated in neuroinflammation, the Th-17-mediated pathway stands out. Th-17 cells are characterized by production of cytokines, IL-17A, IL-17F and IL-22. Th-17 pathway and its eponymous cytokine IL-17 are highly pro-inflammatory and act synergistically with other PAMPs such as TNF-α.[61] This simultaneously makes it a powerful weapon against pathogens and a source of autoimmunity. Thus far, Th-17 cells have been implicated in autoimmune diseases including psoriasis, rheumatoid arthritis, and asthma. Importantly for this section, recent evidence has shown Th-17 cells play an essential role in many neurological disorders, including autism spectrum disorders (ASDs), Alzheimer's disease (AD), multiple sclerosis (MS), and ischemic strokes.[62] This has particular relevance to gut-induced systemic inflammation as there is evidence that these Th-17 cells originate systemically and extravasate into the CNS via C-C chemokine receptor 6 (CCR6) and lymphocyte function-associated antigen 1 (LFA-1) integrin.[63,64] Indeed, many studies have implicated gut-induced inflammation in Th-17 activation and Th-17 lymphocytes in increasing systemic inflammation, promoting BBB disruption and CNS inflammation.[65–68] Consequently, Th-17 cells may be primed for autoimmunity as the result of intestinal dysbiosis and a state of general inflammation for autoreactivity in the CNS. Here we will discuss Th-17 cell's putative role in a few neurological pathologies.

In models of MS, Th-17 cells induced a pro-inflammatory phenotype in astrocytes, critical regulators of neuronal homeostasis.[64] Indeed, ablation of the IL-17a pathway led to improvements in murine models of MS.[69] Similar results are seen in AD. Cytokine quantification in AD models indicates Th-17 polarization due to high levels of IL-2, TNF-α, IL-17, and GM-CSF.[70] It is hypothesized that amyloid-beta in AD acts as a chemoattractant for neutrophils by inducing the high-affinity conformation of LFA-1, which increases neutrophil and Th-17 extravasation into the CNS.[71,72] Once in the CNS, these cell types can act synergistically to increase IL-17 levels, which damage neurons and the BBB.[71] As in MS, the administration of IL-17 antibodies improves cognitive impairment and neuroinflammation in murine models.[62,73] Notably, these findings have found preliminary support in human studies as well, where increases in IL-17 and IL-23 have been observed in AD patients.[62]

Th-17 dysfunction, potentially exacerbated by dysbiosis in the gut, has also been implicated in the pathogenesis of ischemic strokes. Gut permeability and microbial composition are known to be altered in murine models of stroke.[74] In murine models of ischemic stroke, bacteria colonizing the lungs were found to be from translocation across the small intestine rather than due to aspiration from the oral cavity.[75] Moreover, fecal microbial transplant (FMT) from stroke models can induce pro-inflammatory Th-17 dysfunction in GF mice.[32,74] Finally, broad-spectrum antibiotic administration before ischemic events is associated with poorer outcomes.[32,76] These findings implicate the gut microbiome in the neuroinflammatory T cell response. Moreover, this T cell dysfunction, particularly of Th-17 cell, has significant implications in the post-ischemic environment. It is known that Th-17 cells and IL-17 levels are increased peripherally and in the brain parenchyma post-stroke while Tregs are decreased.[62,77–79] Supporting these findings, neutralization of the IL-17 axis is protective from ischemic stroke.[80] Moreover, abnormalities in Th-17 and IL-17 are also implicated in atherosclerotic plaque formation and instability.[62] These results of these heavily implicate Th-17 cells and the gut microbiome in post-ischemic neuroinflammation.

Finally, Th-17 dysfunction is also implicated in neurodevelopmental pathologies, including ASD. Murine models of maternal-immune-activation-induced ASD found that the maternal microbial composition was critical in inducing behavioral abnormalities in offspring. To this end, pre-conception microbiota transfer therapy altered the susceptibility to ASD. Th-17 cells were fundamental to maternal immune activation, and ablation of the IL-17a pathway was protective against ASD in offspring.[81,82]

Through these mechanisms, Th-17 cells can influence the inflammatory environment in the CNS.

Gut-Induced Inflammation and Disruption of the BBB

The long-established connection between the GI tract and the brain has largely focused on autonomic innervation of the digestive organs, otherwise known as the enteric nervous system. Recently, however, research has started to focus on the converse relationship; how the GI system influences the brain. The GI system is composed of trillions of microbes and is the first line of defense against any ingested pathogen. Commensal bacterial species within the gut inhibit the growth of pathogenic bacteria and produce abundant metabolites such as SCFAs and vitamin B12. These bacterial products maintain the integrity of the gut epithelial layer and exert anti-inflammatory effects. As a result, the gut microbiota serves as an essential part of the barrier that protects the body from toxins and infections.[18]

Inflammation is the attempt of the innate and adaptive immune systems to eliminate a source of harm. The immune response is a complex interaction between cytokines, chemokines, and leukocytes that can be both protective against infection and destructive to healthy tissues. The inflammatory response is robust in the GI tract since the majority of lymphocytes in the body are concentrated within gut-associated lymphoid tissues (GALT) located throughout the intestines. Dysbiosis has been shown to increase lymphocyte accumulation and differentiation in these tissues, consequently triggering a heightened inflammatory response.[83] Pro-inflammatory cytokines such as IFN-γ and TNF-α reorganize tight junction proteins within the GI mucosal epithelium leading to barrier dysfunction.[83] Increased permeability of the intestinal barrier allows for the escape of both pathogenic bacteria and inflammatory cytokines into the bloodstream and can cause a systemic immune response involving the CNS (Fig. 4.2).

When the defensive barrier of the GI tract is compromised from dysbiosis, the result is "leaky gut syndrome" — toxins, bacteria, and other inflammatory substances can enter the bloodstream due to increased intestinal permeability. Several studies have shown that invasion of inflammatory cytokines from the GI tract into systemic circulation are an important contributor to inflammatory conditions and have a significant effect on the brain and CNS.[18,85]

The function of the BBB is to support homeostasis in the CNS and protect the brain from invading pathogens. During homeostasis, the BBB is a highly selective, semipermeable border that separates cerebral capillary blood from the CSF — the interstitial fluid that envelops the brain. Despite the highly regulated mechanism and solid structure of the BBB, evidence suggests that peripheral leukocytes are capable of crossing and exerting their deleterious effects. A major consequence resulting from dysregulation of the BBB is the migration of peripheral immune cells such as lymphocytes, neutrophils, and monocytes into the brain parenchyma causing further tissue damage.[86] Once these inflammatory mediators release cytokines, they further increase the permeability of the BBB leading to the destruction of the cerebral endothelial vessels and CNS dysfunction.

The cerebral vasculature is especially susceptible to inflammatory damage during early development at the onset of myelin-

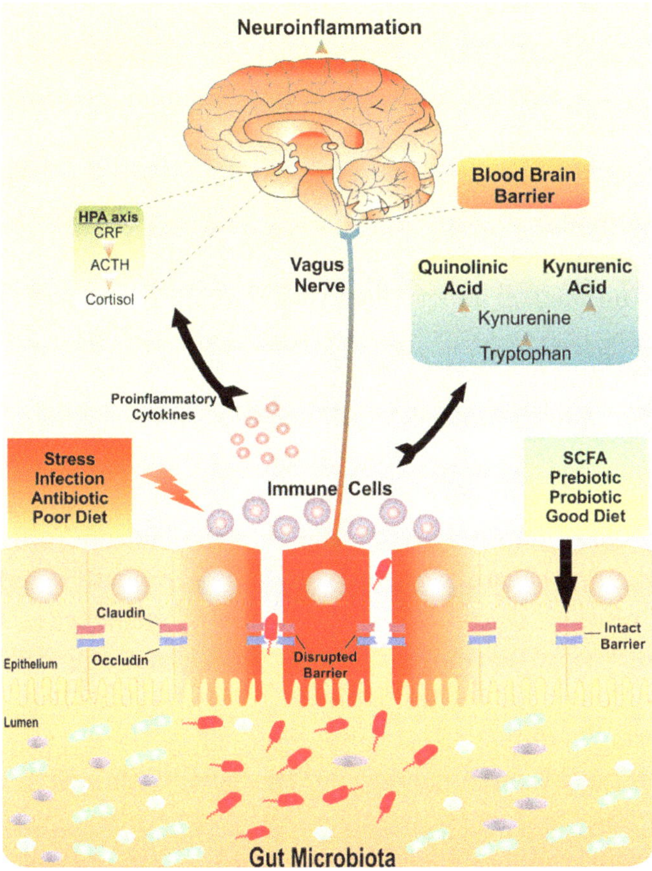

Fig. 4.2: Schematic representation of the gut–brain axis (GBA). This figure represents the hypothesized signaling pathways between the gut microflora, the gastrointestinal (GI) system, and the brain. Altered intestinal epithelial permeability, that is, a "leaky gut," is hypothesized to allow the translocation of bacteria and their virulence factors into systemic circulation, thereby creating a pro-inflammatory state, which induces neuroinflammation. Figure adapted from Kelly *et al.*[84]

ation.[87] Neuroinflammation that results from damage to the brain parenchyma and its vasculature activates resident immune cells such as microglia. Increased activation of microglia has been observed in patients with autism, and studies have revealed that their density is correlated with cellular damage.[87] ASD is a neurodevelopmental disorder associated with a heightened immune response and GI abnormalities. Patients with ASD have a distinct gut microbial composition with studies demonstrating significantly higher levels of *Clostridium* and *Bacteroides* with lower levels of *Coprococcus* and *Prevotella*.[88,89] The gut abnormalities seen in patients with ASD are accompanied by immune abnormalities. In post-mortem, brain tissue of individuals with ASD, there was a significantly higher incidence of pro-inflammatory and Th-1 cytokines compared to control patients. TNF-α promotes activation and proliferation of naïve and effector T cells, which in turn can activate meningeal macrophages and microglia. High levels of TNF-α are observed in patients with ASD and have an adverse effect on brain functioning.[90,91]

LPS, an immunogenic component of gram-negative bacteria, is frequently used to study the effect of systemic inflammation on the BBB due to its disruptive nature. When LPS is introduced into the bloodstream, it synthesizes prostanoids and nitric oxide, which in turn cause endothelial apoptosis, impaired tight junctions, glycocalyx degradation, and structural changes in astrocytic end-feet.[92,93] Furthermore, systemic LPS induces astrocyte gene transcription of pro-inflammatory astrocytes such as IL-6, TNF-α, and prostaglandins.[94]

The gut immune barrier and the BBB each have a complex set of cellular and molecular mechanisms to protect against infection. Recent emerging evidence reveals the important role gut microbiota play in the regulation of the immune system and brain physiology.[95] Dysregulation in the gut barrier can lead to systemic inflammation and a breakdown in the BBB, which in turn leads to neuropathologic states.

CONCLUSIONS

The brain is often thought of as a separate, immune-privileged organ, protected from the rest of the body due to the BBB. On the contrary, current research disputes this concept and provides abundant evidence that suggests a bidirectional axis of communication between the CNS and periphery. One of the most important interactions is that between the GI tract and brain. The microbiota in our gut, composed of a complex community of microorganisms, greatly influences the immune system through various mechanisms. Bacteria synthesize various substances such as vitamin B12 and butyrate, which are essential to all physiological functions in the body. A disturbance in homeostasis of the intestinal ecosystem can cause dysbiosis leading to depletion of vital nutrients and an overgrowth of pathogenic bacteria.

Intestinal inflammation that results from dysbiosis has been shown to cause damage to the intestinal barrier and leakage of pro-inflammatory cytokines and chemokines into the systemic circulation. This "leaky gut" can, in turn, cause a breakdown in the barrier on the other side of this axis. The mediators of systemic inflammation, including TNF-α, IFN-γ, and Th-17, eventually make their way to the BBB where they disrupt the tight junctions allowing the passage of blood-borne immune cells into the CNS. These leukocytes have demonstrated to be neurotoxic and cause activation of microglia and astrocytes leading to further neuroinflammation.

Many factors, including microbial constitution in early childhood, antibiotic use, diet, genetics, and acquired infections can disrupt the microbiome and induce systemic inflammation. Abundant research studies have implicated the inflammatory response to dysbiosis in the onset and progression of many chronic neurodegenerative diseases including AD, MS, Parkinson's disease, and autism. Further investigation is required to gain a deeper understanding of the gut–brain axis (GBA) and, in turn, develop interventions to aid in the management and prevention of neuroinflammatory conditions.

REFERENCES

1. Luke K, *et al.* (2012) Defining the human microbiome. *Nutr Rev* **70**: S38–S44.
2. Rogers, K. (2016, April 7). Human microbiome. Encyclopedia Britannica. https://www.britannica.com/science/human-microbiome.
3. Gensollen T, Iyer SS, Kasper DL, Blumberg RS. (2016) How colonization by microbiota in early life shapes the immune system. *Science* **352**: 539–544.
4. Alessandro Poggi RB, *et al.* (2019) Human gut-associated natural killer cells in health and disease. *Front Immunol* **10**: 961.
5. Shi N, Li N, Duan X, Niu H. (2017) Interaction between the gut microbiome and mucosal immune system. *Mil Med Res.* **4**: 14.
6. Wang Y, *et al.* (2018) The gut-microglia connection: Implications for central nervous system diseases. *Front Immunol* **9**: 2325.
7. Erny D, *et al.* (2015) Host microbiota constantly control maturation and function of microglia in the CNS. *Nat Neurosci* **18**(7): 965–977, doi:10.1038/nn.4030.
8. Brown DG, *et al.* (2019) The microbiota protects from viral-induced neurologic damage through microglia-intrinsic TLR signaling. *Elife* **8**: e47117.
9. Willing BP, Russell SL, Finlay BB. (2011) Shifting the balance: Antibiotic effects on host-microbiota mutualism. *Nat Rev Microbiol* **9**(4): 233–243, doi:10.1038/nrmicro2536.
10. Rothe M, Blaut M. (2013) Evolution of the gut microbiota and the influence of diet. *Benef Microbes* **4**(1): 31–37, doi:10.3920/BM2012.0029.
11. Akagawa S, *et al.* (2019) Effect of delivery mode and nutrition on gut microbiota in neonates. *Ann Nutr Metab* **74**(2): 132–139, doi:10.1159/000496427.
12. Whitaker-Azmitia PM. (2001) Serotonin and brain development: Role in human developmental diseases. *Brain Res Bull* **56**(5): 479–485, doi:10.1016/s0361-9230(01)00615-3.
13. Yano JM, *et al.* (2015) Indigenous bacteria from the gut microbiota regulate host serotonin biosynthesis. *Cell* **161**(2): 264–276, doi:10.1016/j.cell.2015.02.047.
14. Boccuto L, *et al.* (2013) Decreased tryptophan metabolism in patients with autism spectrum disorders. *Mol Autism* **4**(1): 16, doi:10.1186/2040-2392-4-16.

15. Liang S, *et al.* (2015) Administration of lactobacillus helveticus NS8 improves behavioral, cognitive, and biochemical aberrations caused by chronic restraint stress. *Neuroscience* **310**: 561–577, doi:10.1016/j.neuroscience.2015.09.033.
16. Shajib MS, Baranov A, Khan WI. (2017) Diverse effects of gut-derived serotonin in intestinal inflammation. *ACS Chem Neurosci* **8**(5): 920–931, doi:10.1021/acschemneuro.6b00414.
17. Herrmann W, Schorr H, Purschwitz K, Rassoul F, Richter V. (2001) Total homocysteine, vitamin B(12), and total antioxidant status in vegetarians. *Clin Chem* **47**(6): 1094–1101.
18. Michel L, Prat A. (2016) One more role for the gut: Microbiota and blood brain barrier. *Ann Transl Med* **4**(1): 15, doi:10.3978/j.issn.2305-5839.2015.10.16.
19. Durack J, Lynch SV. (2019) The gut microbiome: Relationships with disease and opportunities for therapy. *J Exp Med* **216**(1): 20–40, doi:10.1084/jem.20180448.
20. Saltzman ET, Palacios T, Thomsen M, Vitetta L. (2018) Intestinal microbiome shifts, dysbiosis, inflammation, and non-alcoholic fatty liver disease. *Front Microbiol* **9**: 61, doi:10.3389/fmicb.2018.00061.
21. Brandsma E, *et al.* (2019) A proinflammatory gut microbiota increases systemic inflammation and accelerates atherosclerosis. *Circ Res* **124**: 94–100.
22. Macfarlane S, Macfarlane GT. (2003) Regulation of short-chain fatty acid production. *Proc Nutr Soc* **62**(1): 67–72, doi:10.1079/PNS2002207.
23. Baxter NT, Schmidt AW, Venkataraman A, Kim KS, Waldron C, Schmidt TM. (2019) Dynamics of human gut microbiota and short-chain fatty acids in response to dietary interventions with three fermentable fibers. *mBio* **10**(1): e02566–e02518, doi:10.1128/mBio.02566-18.
24. den Besten G, van Eunen K, Groen AK, Venema K, Reijngoud DJ, Bakker BM. (2013) The role of short-chain fatty acids in the interplay between diet, gut microbiota, and host energy metabolism. *J Lipid Res* **54**(9): 2325–2340, doi:10.1194/jlr.R036012.
25. Wong JM, de Souza R, Kendall CW, Emam A, Jenkins DJ. (2006) Colonic health: Fermentation and short chain fatty acids. *J Clin Gastroenterol* **40**(3): 235–243, doi:10.1097/00004836-200603000-00015.

26. Venegas DP, *et al.* (2019) Short chain fatty acids (SCFAs)-mediated gut epithelial and immune regulation and its relevance for inflammatory bowel diseases. *Front Immunol* **10**: 277, doi:10.3389/fimmu.2019.00277.
27. Iacob S, Iacob DG. (2019) Infectious threats, the intestinal barrier, and its trojan horse: Dysbiosis. *Front Microbiol* **10**: 1676, doi:10.3389/fmicb.2019.01676.
28. Barcelo A, Claustre J, Moro F, Chayvialle JA, Cuber JC, Plaisancie P. (2000) Mucin secretion is modulated by luminal factors in the isolated vascularly perfused rat colon. *Gut* **46**(2): 218–224, doi:10.1136/gut.46.2.218.
29. Willemsen LE, Koetsier MA, van Deventer SJ, van Tol EA. (2003) Short chain fatty acids stimulate epithelial mucin 2 expression through differential effects on prostaglandin E(1) and E(2) production by intestinal myofibroblasts. *Gut* **52**(10): 1442–1447, doi:10.1136/gut.52.10.1442.
30. Jung TH, Park JH, Jeon WM, Han KS. (2015) Butyrate modulates bacterial adherence on LS174T human colorectal cells by stimulating mucin secretion and MAPK signaling pathway. *Nutr Res Pract* **9**(4): 343–349, doi:10.4162/nrp.2015.9.4.343.
31. Hamer HM, Jonkers D, Venema K, Vanhoutvin S, Troost FJ, Brummer RJ. (2008) Review article: The role of butyrate on colonic function. *Aliment Pharmacol Ther* **27**(2): 104–119, doi:10.1111/j.1365-2036.2007.03562.x.
32. Cryan JF, *et al.* (2019) The microbiota-gut-brain axis. *Physiol Rev* **99**(4): 1877–2013, doi:10.1152/physrev.00018.2018.
33. Kelly CJ, *et al.* (2015) Crosstalk between microbiota-derived short-chain fatty acids and intestinal epithelial HIF augments tissue barrier function. *Cell Host Microbe* **17**(5): 662–671, doi:10.1016/j.chom.2015.03.005.
34. Peng L, Li ZR, Green RS, Holzman IR, Lin J. (2009) Butyrate enhances the intestinal barrier by facilitating tight junction assembly via activation of AMP-activated protein kinase in Caco-2 cell monolayers. *J Nutr* **139**(9): 1619–1625, doi:10.3945/jn.109.104638.
35. Miao W, *et al.* (2016) Sodium butyrate promotes reassembly of tight junctions in Caco-2 monolayers involving inhibition of MLCK/MLC2 pathway and phosphorylation of PKCβ2. *Int J Mol Sci* **17**(10): 1696.
36. Tong LC, *et al.* (2016) Propionate ameliorates dextran sodium sulfate-induced colitis by improving intestinal barrier function and reducing inflammation and oxidative stress. *Front Pharmacol* **7**: 253, doi:10.3389/fphar.2016.00253.

37. Yan H, Ajuwon KM. (2017) Butyrate modifies intestinal barrier function in IPEC-J2 cells through a selective upregulation of tight junction proteins and activation of the Akt signaling pathway. *PLoS One* **12**: e0179586.
38. Zheng L, *et al.* (2017) Microbial-derived butyrate promotes epithelial barrier function through IL-10 receptor-dependent repression of claudin-2. *J Immunol* **199**(8): 2976–2984, doi:10.4049/jimmunol.1700105.
39. Shakespear MR, Halili MA, Irvine KM, Fairlie DP, Sweet MJ. (2011) Histone deacetylases as regulators of inflammation and immunity. *Trends Immunol* **32**(7): 335–343, doi:10.1016/j.it.2011.04.001.
40. Chang PV, Hao L, Offermanns S, Medzhitov R. (2014) The microbial metabolite butyrate regulates intestinal macrophage function via histone deacetylase inhibition. *Proc Natl Acad Sci USA* **111**(6): 2247–2252, doi:10.1073/pnas.1322269111.
41. Brown AJ, *et al.* (2003) The orphan G protein-coupled receptors GPR41 and GPR43 are activated by propionate and other short chain carboxylic acids. *J Biol Chem* **278**(13): 11312–11319, doi:10.1074/jbc.M211609200.
42. Le Poul E, *et al.* (2003) Functional characterization of human receptors for short chain fatty acids and their role in polymorphonuclear cell activation. *J Biol Chem* **278**(28): 25481–25489, doi:10.1074/jbc.M301403200.
43. Nilsson NE, Kotarsky K, Owman C, Olde B. (2003) Identification of a free fatty acid receptor, FFA2R, expressed on leukocytes and activated by short-chain fatty acids. *Biochem Biophys Res Commun* **303**(4): 1047–1052, doi:10.1016/s0006-291x(03)00488-1.
44. Nohr MK, *et al.* (2015) Expression of the short chain fatty acid receptor GPR41/FFAR3 in autonomic and somatic sensory ganglia. *Neuroscience* **290**: 126–137, doi:10.1016/j.neuroscience.2015.01.040.
45. Vinolo MA, *et al.* (2011) SCFAs induce mouse neutrophil chemotaxis through the GPR43 receptor. *PLoS One* **6**(6): e21205, doi:10.1371/journal.pone.0021205.
46. La Sala MS, *et al.* (2013) Modulation of taste responsiveness by the satiation hormone peptide YY. *FASEB J* **27**(12): 5022–5033, doi:10.1096/fj.13-228064.
47. Vieira AT, *et al.* (2017) Dietary fiber and the short-chain fatty acid acetate promote resolution of neutrophilic inflammation in a model of gout in mice. *J Leukoc Biol* **101**(1): 275–284, doi:10.1189/jlb.3A1015-453RRR.

48. Luhrs H, *et al.* (2002) Butyrate inhibits NF-kappaB activation in lamina propria macrophages of patients with ulcerative colitis. *Scand J Gastroenterol* **37**(4): 458–466, doi:10.1080/003655202317316105.
49. Smith PM, *et al.* (2013) The microbial metabolites, short-chain fatty acids, regulate colonic Treg cell homeostasis. *Science* **341**(6145): 569–573, doi:10.1126/science.1241165.
50. Nastasi C, *et al.* (2015) The effect of short-chain fatty acids on human monocyte-derived dendritic cells. *Sci Rep* **5**: 16148, doi:10.1038/srep16148.
51. Cox MA, *et al.* (2009) Short-chain fatty acids act as antiinflammatory mediators by regulating prostaglandin E(2) and cytokines. *World J Gastroenterol* **15**(44): 5549–5557, doi:10.3748/wjg.15.5549.
52. Yang G, *et al.* (2018) Implication of G protein-coupled receptor 43 in intestinal inflammation: A mini-review. *Front Immunol* **9**: 1434, doi:10.3389/fimmu.2018.01434.
53. Park J, *et al.* (2015) Short-chain fatty acids induce both effector and regulatory T cells by suppression of histone deacetylases and regulation of the mTOR-S6K pathway. *Mucosal Immunol* **8**(1): 80–93, doi:10.1038/mi.2014.44.
54. Asarat M, Apostolopoulos V, Vasiljevic T, Donkor O. (2016) Short-chain fatty acids regulate cytokines and Th17/Treg cells in human peripheral blood mononuclear cells in vitro. *Immunol Invest* **45**(3): 205–222, doi:10.3109/08820139.2015.1122613.
55. Tan J, *et al.* (2016) Dietary fiber and bacterial SCFA enhance oral tolerance and protect against food allergy through diverse cellular pathways. *Cell Rep* **15**(12): 2809–2824, doi:10.1016/j.celrep.2016.05.047.
56. Al Nabhani Z, *et al.* (2019) A weaning reaction to microbiota is required for resistance to immunopathologies in the adult. *Immunity* **50**(5): 1276–1288.e5, doi:10.1016/j.immuni.2019.02.014.
57. Arpaia N, *et al.* (2013) Metabolites produced by commensal bacteria promote peripheral regulatory T-cell generation. *Nature* **504**(7480): 451–455, doi:10.1038/nature12726.
58. Furusawa Y, *et al.* (2013) Commensal microbe-derived butyrate induces the differentiation of colonic regulatory T cells. *Nature* **504**(7480): 446–450, doi:10.1038/nature12721.

59. Luu M, Steinhoff U, Visekruna A. (2017) Functional heterogeneity of gut-resident regulatory T cells. *Clin Transl Immunol* **6**(9): e156, doi:10.1038/cti.2017.39.
60. Nagashima H, Morio Y, Meshitsuka S, Yamane K, Nanjo Y, Teshima R. (2010) High-resolution nuclear magnetic resonance spectroscopic study of metabolites in the cerebrospinal fluid of patients with cervical myelopathy and lumbar radiculopathy. *Eur Spine J* **19**(8): 1363–1368, doi:10.1007/s00586-010-1453-3.
61. Kirkham BW, Kavanaugh A, Reich K. (2014) Interleukin-17A: A unique pathway in immune-mediated diseases: Psoriasis, psoriatic arthritis and rheumatoid arthritis. *Immunology* **141**(2): 133–142, doi:10.1111/imm.12142.
62. Cipollini V, Anrather J, Orzi F, Iadecola C. (2019) Th17 and cognitive impairment: Possible mechanisms of action. *Front Neuroanat* **13**: 95, doi:10.3389/fnana.2019.00095.
63. Engelhardt B, Ransohoff RM. (2012) Capture, crawl, cross: The T cell code to breach the blood-brain barriers. *Trends Immunol* **33**(12): 579–589, doi:10.1016/j.it.2012.07.004.
64. Prajeeth CK, *et al.* (2017) Effectors of Th1 and Th17 cells act on astrocytes and augment their neuroinflammatory properties. *J Neuroinflammation* **14**(1): 204, doi:10.1186/s12974-017-0978-3.
65. Hania Kebir KK, *et al.* (2007) Human TH17 lymphocytes promote blood-brain barrier disruption and central nervous system inflammation. *Nat Med* **13**(10): 1173–1175.
66. Blaschitz C, Raffatellu M. (2010) Th17 cytokines and the gut mucosal barrier. *J Clin Immunol* **30**(2): 196–203, doi:10.1007/s10875-010-9368-7.
67. Shuttleworth S, *et al.* (2011) Progress in the development of small molecule therapeutics targeting Th17 cell function for the treatment of immune-inflammatory diseases. *Prog Med Chem* **50**: 109–133, doi:10.1016/B978-0-12-381290-2.00003-3.
68. Chewning JH, Weaver CT. (2014) Development and survival of Th17 cells within the intestines: The influence of microbiome- and diet-derived signals. *J Immunol* **193**(10): 4769–4777, doi:10.4049/jimmunol.1401835.
69. Kang Z, *et al.* (2010) Astrocyte-restricted ablation of interleukin-17-induced Act1-mediated signaling ameliorates autoimmune encephalomyelitis. *Immunity* **32**(3): 414–425, doi:10.1016/j.immuni.2010.03.004.

70. St-Amour I, *et al.* (2019) Peripheral adaptive immunity of the triple transgenic mouse model of Alzheimer's disease. *J Neuroinflammation* **16**(1): 3, doi:10.1186/s12974-018-1380-5.

71. Zenaro E, *et al.* (2015) Neutrophils promote Alzheimer's disease-like pathology and cognitive decline via LFA-1 integrin. *Nat Med* **21**(8): 880–886, doi:10.1038/nm.3913.

72. Verma NK, Kelleher D. (2017) Not just an adhesion molecule: LFA-1 contact tunes the T lymphocyte program. *J Immunol* **199**(4): 1213–1221, doi:10.4049/jimmunol.1700495.

73. Cristiano C, *et al.* (2019) Neutralization of IL-17 rescues amyloid-beta-induced neuroinflammation and memory impairment. *Br J Pharmacol* **176**(18): 3544–3557, doi:10.1111/bph.14586.

74. Singh V, *et al.* (2016) Microbiota dysbiosis controls the neuroinflammatory response after stroke. *J Neurosci* **36**(28): 7428–7440, doi:10.1523/JNEUROSCI.1114-16.2016.

75. Stanley D, *et al.* (2016) Translocation and dissemination of commensal bacteria in post-stroke infection. *Nat Med* **22**(11): 1277–1284, doi:10.1038/nm.4194.

76. Winek K, Dirnagl U, Meisel A. (2016) The gut microbiome as therapeutic target in central nervous system diseases: Implications for stroke. *Neurotherapeutics* **13**(4): 762–774, doi:10.1007/s13311-016-0475-x.

77. Li GZ, *et al.* (2005) Expression of interleukin-17 in ischemic brain tissue. *Scand J Immunol* **62**(5): 481–486, doi:10.1111/j.1365-3083.2005.01683.x.

78. Sutton CE, Lalor SJ, Sweeney CM, Brereton CF, Lavelle EC, Mills KH. (2009) Interleukin-1 and IL-23 induce innate IL-17 production from gammadelta T cells, amplifying Th17 responses and autoimmunity. *Immunity* **31**(2): 331–341, doi:10.1016/j.immuni.2009.08.001.

79. Dolati S, *et al.* (2018) Peripheral Th17/Treg imbalance in elderly patients with ischemic stroke. *Neurol Sci* **39**(4): 647–654, doi:10.1007/s10072-018-3250-4.

80. Gelderblom M, *et al.* (2012) Neutralization of the IL-17 axis diminishes neutrophil invasion and protects from ischemic stroke. *Blood* **120**(18): 3793–3802, doi:10.1182/blood-2012-02-412726.

81. Choi GB, *et al.* (2016) The maternal interleukin-17a pathway in mice promotes autism-like phenotypes in offspring. *Science* **351**(6276): 933–939, doi:10.1126/science.aad0314.

82. Lammert CR, *et al.* (2018) Cutting edge: Critical roles for microbiota-mediated regulation of the immune system in a prenatal immune activation model of Autism. *J Immunol* **201**(3): 845–850, doi:10.4049/jimmunol.1701755.
83. Bruewer M, *et al.* (2003) Proinflammatory cytokines disrupt epithelial barrier function by apoptosis-independent mechanisms. *J Immunol* **171**(11): 6164–6172, doi:10.4049/jimmunol.171.11.6164.
84. Kelly JR, Kennedy PJ, Cryan JF, Dinan TG, Clarke G, Hyland NP. (2015) Breaking down the barriers: The gut microbiome, intestinal permeability and stress-related psychiatric disorders. *Front Cell Neurosci* **9**: 392, doi:10.3389/fncel.2015.00392.
85. Eshraghi RS, *et al.* (2018) Early disruption of the microbiome leading to decreased antioxidant capacity and epigenetic changes: Implications for the rise in autism. *Front Cell Neurosci* **12**: 256, doi:10.3389/fncel.2018.00256.
86. Liebner S, Dijkhuizen RM, Reiss Y, Plate KH, Agalliu D, Constantin G. (2018) Functional morphology of the blood-brain barrier in health and disease. *Acta Neuropathol* **135**(3): 311–336, doi:10.1007/s00401-018-1815-1.
87. Stolp HB, *et al.* (2005) Breakdown of the blood-brain barrier to proteins in white matter of the developing brain following systemic inflammation. *Cell Tissue Res* **320**(3): 369–378, doi:10.1007/s00441-005-1088-6.
88. De Angelis M, *et al.* (2013) Fecal microbiota and metabolome of children with autism and pervasive developmental disorder not otherwise specified. *PLoS One* **8**(10): e76993, doi:10.1371/journal.pone.0076993.
89. Kang DW, *et al.* (2013) Reduced incidence of Prevotella and other fermenters in intestinal microflora of autistic children. *PLoS One* **8**(7): e68322, doi:10.1371/journal.pone.0068322.
90. Mehta AK, Gracias DT, Croft M. (2018) TNF activity and T cells. *Cytokine* **101**: 14–18, doi:10.1016/j.cyto.2016.08.003.
91. Eshraghi RS, *et al.* (2018) Early disruption of the microbiome leading to decreased antioxidant capacity and epigenetic changes: Implications for the rise in autism. *Front Cel Neurosci* **12**: 256.
92. Wiesinger A, *et al.* (2013) Nanomechanics of the endothelial glycocalyx in experimental sepsis. *PLoS One* **8**(11): e80905, doi:10.1371/journal.pone.0080905.

93. Fan L, Wang T, Chang L, Song Y, Wu Y, Ma D. (2014) Systemic inflammation induces a profound long term brain cell injury in rats. *Acta Neurobiol Exp (Wars)* **74**(3): 298–306.
94. Bush TG, *et al.* (1999) Leukocyte infiltration, neuronal degeneration, and neurite outgrowth after ablation of scar-forming, reactive astrocytes in adult transgenic mice. *Neuron* **23**(2): 297–308, doi:10.1016/s0896-6273(00)80781-3.
95. Daneman R, Rescigno M. (2009) The gut immune barrier and the blood-brain barrier: Are they so different? *Immunity* **31**(5): 722–735, doi:10.1016/j.immuni.2009.09.012.

Chapter 5

Bacterial Metabolites and Quorum Sensing Molecules Affect the Brain

Michael P. Moraskie, Gregory O'Connor, Sapna K. Deo and Sylvia Daunert*

Abstract

The symbiotic relationship between an organism and its microbiome involves a seemingly infinite complex web of interactions. The growing interest in elucidating just how influential our microbial counterparts are to our overall health has resulted in the development of numerous fields of investigation on the human microbiome. Studies looking into the existence of a microbiome–gut–brain axis have made significant progress, specifically those involving the interplay between bacterial-derived molecules and the brain. Here, we highlight research on gut bacteria production of short-chain fatty acids and their role in appetite suppression. We also discuss the regulation of serotonin production via the microbiome, host-mediated modulation of gut flora via manipulation of quorum sensing mechanisms, and permeability of the blood–brain barrier to gut bacterial metabolites.

*Department of Biochemistry and Molecular Biology, University of Miami-Miller School of Medicine, Miami, FL, USA.

Corresponding Author: Sylvia Daunert, PharmD, MS, PhD, Department of Biochemistry and Molecular Biology, University of Miami-Miller School of Medicine, Gautier Building, 1011 NW 15th Street, Miami, FL 33136-1019, USA, Telephone #: 305-243-6282, E-mail: sdaunert@med.miami.edu.

INTRODUCTION

On a brisk summer day of 1878 in Cassel, Germany, the prominent botanist Heinrich Anton de Bary stood before the Association of German Naturalists and Physicians to present what was presumed to be a lecture on contemporary methods used in botanical science and teaching. Prior to finding himself in front of the prestigious crowd, de Bary had spent the majority of his professional career dedicated to studying the life history of fungi and plant diseases. In fact, he had already published various field-defining books on the comparative anatomy of seed-producing plants and ferns, as well as on the morphology and physiology of fungi, lichens, and slime molds. Considered to be both, the father of plant pathology and the founder of modern mycology, de Bary was not lacking in life accomplishments by the time he walked onto the stage. Unbeknownst to him and those attending, his lecture would come to mark yet another moment in which de Bary etched his name into the history books of science.

He began by confiding to the spectators that, while pondering on topics to discuss, he had been studying the peculiar dual nature of lichens. These organisms displayed all of the typical morphological characteristics of fungi and had been accordingly classified as such, until it was discovered they contained cells with chlorophyll. These cells granted lichens the ability to assimilate carbon dioxide and live independently of organic compounds on desolate structures such as naked rocks, making them unique from typical fungi. de Bary uncovered that the photosynthetic cells were not part of the plant but were instead algae that live and grow in association with fungi that envelop them. A lichen was, therefore, not an individual organism, but rather, a composite one. He had thus established that lichen algae could live and thrive independently of the fungi, but the lichen fungi were entirely dependent on the algae for survival. The fungal spores quickly perished under favorable conditions if they did not engulf an algal partner, thus blurring the lines between host and tenant. This relationship puzzled de Bary, as it did not fit neatly into the established definitions of parasitism or commensalism. As such, de Bary proposed the creation of a broader category to describe the living together of differently named organisms: symbiosis.[1]

Similar to many profound scientific discoveries throughout history, de Bary's concept of symbiosis was quickly swept under the rug of mainstream biology and regarded as a mere biological curiosity. It would not be until the turn of the century and two world wars later that de Bary's concept of symbiosis would be re-visited and deservingly placed under the scientific spotlight by the evolutionary biologist Lynn Margulis (Fig. 5.1). A staunch proponent of symbiosis as an integral component of evolution, she would come to be known for her now widely accepted theory of endosymbiosis, in which she proposed the evolution of mitochondria and chloroplasts were the result of a symbiotic merging of primitive archaeal and prokaryotic organisms. This acceptance did not come without its sardonic fights though, as Margulis faced constant criticism from the popular Neo-Darwinists who posited evolution was solely driven by random genetic mutations.

In 1989, in an attempt to revitalize symbiosis more than a century after de Bary's speech, Margulis organized the first conference on Symbiosis as a Source of Evolutionary Innovation. It was a

Fig. 5.1: Heinrich Anton de Bary and Lynn Margulis.[2]

small gathering casting a wide international net, bringing together 20 evolutionary theorists and symbiosis biologists from nine countries to address the adequacy of the Neo-Darwinian concept of evolution and make an argument for the importance of symbiosis in the origin of evolutionary novelty. In her opening presentation, Margulis coined the term holobiont to refer to the assemblage of organisms that, together, make a whole.[3] The word has its roots in de Bary's description of symbiosis, with the prefix holo- derived from the Greek word *holos* meaning whole and -biont meaning unit of life, thus recasting the individual entity as only part of a larger picture.

Neither Margulis nor de Bary could have foreseen the explosion of technological and scientific advancements that emerged in the years following their respective presentations. Thanks in no small part to their early efforts, it is now well understood that humans are in fact, holobionts, sharing space and resources symbiotically with microbial counterparts collectively known as the microbiome. Early efforts to elucidate the microbiome focused on cataloging the bacterial species that could be cultured from different parts of the body and subsequently trying to understand their presence in the context of human physiology. Relying on culture-based techniques though, the only method available to researchers at the time was comparable to describing the contents of a room while looking through a keyhole; very few bacteria are, in fact, culturable. The arrival of non-culture-based molecular techniques to assess microbiota, such as next-generation sequencing and bioinformatic tools, bust the door wide open. They have enabled an ever-growing community of researchers to take a more complete look at this microbial world and begin to untangle the complex web of host–microbiome interactions.[4,5] With one discovery quickly following the next, new questions are continuously surfacing about basic human biological functions and revolutionizing our understanding of various bodily functions. Perhaps the most riveting questions are those pertaining to the possible link between the gut microbiome and the brain, a bidirectional relationship typically referred to as the microbiome–gut–brain (MGB) axis.

This chapter will focus on providing a review of investigations looking into the interplay of gut bacteria-derived molecules and the brain. The following summary of a comprehensive list of research papers will lend some context to the burning question at hand: Gut–brain connection, myth or reality?

BACTERIAL METABOLITES AND THE BRAIN: SHORT-CHAIN FATTY ACIDS

The human holobiont houses an astonishingly vast number of microbes, with the largest and most diverse communities residing in the gut.[6] There has been a growing appreciation for the various roles these microbiomes play in maintaining gut homeostasis, such as defending against pathogen colonization and maintenance of the immune system.[7] One particular role that has garnered wide-spread attention is their digestion of complex macronutrients in the host diet. Research has shown that the microbial composition of the gut fluctuates depending on the dietary intake of the host, indicating that resulting bacterial metabolites could play a role in the onset or prevention of several diseases.[8] The extent to which these metabolites affect gut homeostasis is still under intense investigation, but there has been some significant progress in identifying those possibly relevant to the MGB axis. One such family of metabolites garnering wide-spread attention within MGB research communities are short-chain fatty acids (SCFAs).

Every 5 years, the US Department of Health and Human Services (HHS) and the US Department of Agriculture (USDA) release the *Dietary Guidelines for Americans* based on the most current scientific evidence on nutrition.[9] The document emphasizes the importance of consuming a healthy diet, specifically one that provides a robust supply of essential nutrients. Among these nutrients deemed essential by government agencies is dietary fiber, a type of carbohydrate found exclusively in plant foods. While humans lack the innate capability to digest these important plant fibers, the human holobiont wields the bacterial cutlery necessary to extract the otherwise inaccessible energy from plant foods.

Within the umbrella of dietary fiber are soluble and insoluble fibers, each distinguished by their propensity to dissolve in water. Insoluble fibers pass through the gastrointestinal (GI) tract generally undigested, but soluble fibers serve as excellent sources of energy for enteric bacteria. When bacteria ferment soluble fibers the main by-products are SCFAs, which are saturated fatty acids ranging from one to six carbon atoms in length.[10] The three most common SCFAs produced via bacterial fermentation are acetate (C2), propionate (C3), and butyrate (C4) (Fig. 5.2). Work done by various researchers, summarized concisely by Koh *et al*., has found that bacteria can employ several biochemical pathways to extract energy from soluble fibers. So far, it is known that bacteria can synthesize acetate either via acetyl-CoA or via the Wood–Ljungdahl pathway. They can also produce propionate via either the succinate, acrylate, or propanediol pathways. Lastly, bacteria can make butyrate via the butryl-coA:acetate CoA-transferase route or the classical conversion of butyryl-CoA by phosphotransbutyrylase and butyrate kinase.[8]

While the existence of bacterial SCFAs and their importance for energy production has been known for quite sometime,[11] recent studies suggest they have direct interactions with the brain, influencing general psychological function as well as neurological and neuropsychiatric conditions.[12] One particular topic that has caught the imagination of many investigators is the possible connection between bacteria and obesity via SCFA modulation of

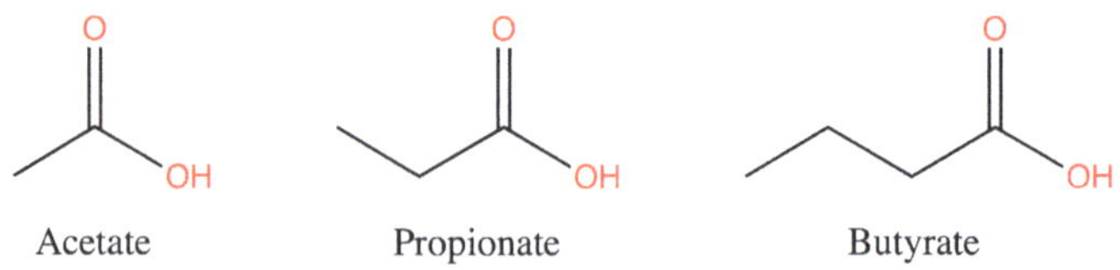

Fig. 5.2: Chemical structures of acetate, propionate, and butyrate.

the brain-derived sensation of appetite. Obesity, characterized by over-appetite, is an ongoing epidemic with incidence rates above 20% in Western countries, which presents a major burden on public health.[13] While genetic pre-dispositions certainly play a role in the onset of obesity, modern food processing has exacerbated this condition, as it has resulted in the mass production of foods high in refined sugars and low in essential nutrients such as dietary fiber. In fact, epidemiological studies have highlighted that diets rich in fiber inversely correlate with body weight,[14] and others have shown fiber-rich diets positively correlate with microbial diversity.[15] Precisely why this inverse relationship between consumption of fiber and body weight exists is a topic of intense debate, but notable progress addressing this matter is continuously being made.

Researchers working on rodent models of obesity fed a high-fat diet (HFD) have linked consumption of soluble fiber with lower dietary energy intake, body weight, and adiposity.[16–19] A separate, groundbreaking study by Frost *et al.*[20] hypothesized that SCFAs, specifically acetate due to its high colonic concentration and known proclivity to enter peripheral circulation, had a direct effect on appetite regulation. In it, mice fed an HFD supplemented with either cellulose (an insoluble fiber) or inulin (a soluble fiber) were observed for weight gain and food consumption. Similar to other studies, the mice supplemented with soluble fiber gained significantly less weight and consumed significantly less food than did the mice supplemented with insoluble fiber. Analysis of the colonic content of the mice further revealed a significant increase in the concentration of SCFAs, particularly acetate, in the mice supplemented with soluble fiber. Researchers then used positron emission topography (PET) to assess the biodistribution of acetate administered peripherally or colonically to mice in fed and fasted states. These experiments indicated that colonic acetate was up taken by the brain in similar quantities, albeit more gradually, compared to peripheral infusion of acetate. Furthermore, they found that acetate was taken up in higher quantities by the hypothalamus than other brain tissue. In a previous study, the investigators had observed that supplementation of soluble fiber resulted in hypothalamic neuronal activation, and, therefore, sought to determine whether intraperitoneal (IP) administration of acetate yielded similar results.[17] After observing an almost identical activation pattern, they proceeded to investigate specific changes in hypothalamic neuropeptide expression following IP administration of acetate. They found that levels of the anorectic (appetite suppressing) neuropeptide proopiomelanocortin (POMC) increased four-fold, while levels of the orexigenic (appetite inducing) neuropeptide agouti-related peptide (AgRP) were suppressed.

While these results revealed a promising explanation for the reduction in appetite as a result of consuming soluble fiber, the investigators wanted to go a step further to elucidate the mechanisms behind the diet-based change in neuropeptide expression. Other studies had previously shown that phosphorylation and inactivation of acetyl-CoA carboxylase (ACC) (an enzyme crucial for the biosynthesis of fatty acids) by an enzyme heavily involved in cellular energy homeostasis, adenosine monophosphate-activated protein kinase (AMPK), results in a decrease in levels of malonyl-CoA (a regulator of fatty acid synthesis). Furthermore, increased levels of malonyl-CoA in the hypothalamus had been associated with increased expression of the anorectic neuropeptide POMC and suppression of the orexigenic neuropeptide AgRP.[21,22] Given these findings, the investigators sought to determine whether appetite reduction by acetate was possibly mediated through these molecular channels. Analysis of mouse hypothalamic lysates following IP administration of acetate revealed a significant reduction in levels of phosphorylated AMPK and ACC. This finding suggests that acetate produced by bacterial fermentation of soluble fiber in the colon could mediate a reduction in appetite by making its way into the hypothalamus, inactivating hypothalamic AMPK, thereby increasing levels of ACC, which elevates levels of malonyl-CoA, and results in an increase of the anorectic neuropeptide POMC and a decrease of the orexigenic neuropeptide AgRP.

This novel insight into appetite suppression shows how acetate could serve as a potential anti-obesity therapeutic, given the implementation of eating regimens rich in dietary fiber to combat obesity has been difficult in affected communities.[20] Furthermore, it serves as an excellent example of how MGB research is revolutionizing how neuropsychiatric conditions such as over-appetite are studied. In the research discussed above, investigators show that consumption of soluble fibers by the host results in the production of SCFAs via bacterial fermentation, which in turn induce anorectic signals in the brain. In other words, bacterial metabolites derived from the gut might have *direct* interactions and downstream implications on the brain.

BACTERIAL METABOLITES AND THE BRAIN: SEROTONIN

Another subject that has also garnered much attention in the MGB community is the possible link between the gut microbiome and the production of 5-hydroxytryptamine (5-HT), commonly known as serotonin. Serotonin is perhaps most well known as an important neurotransmitter involved in the regulation of various brain functions such as mood, learning, and memory, among others.[23] People suffering from anxiety or major depressive disorder (MDD) are often prescribed selective serotonin reuptake inhibitors (SSRIs) to increase levels of serotonin in the brain as a form of treatment.[24] Despite its evident importance in mental health, the brain contains, relatively speaking, low concentrations of serotonin. It is estimated that more than 90% of the body's serotonin is actually synthesized in the gut, where it plays a pivotal role in transmitting vital GI information to the central nervous system (CNS).[25] Several studies have indicated that the gut microbiome plays an important role in post-natal development and maturation of the CNS,[26] thus leading to the hypothesis that gut flora metabolites play a pivotal neurodevelopmental role via direct regulation of serotonin production.

In one study on germ-free (GF) mice, Clarke *et al.*[27] found that the early-life microbiome regulates the hippocampal serotonergic system in a sex-dependent manner, which is of particular relevance to the onset of anxiety, stress, and depression.[28] A separate study on GF mice by Zheng *et al.*[29] highlighted another possible connection between the gut microbiome and onset of MDD. They found that fecal microbiota transplantation from MDD patients into GF mice resulted in an increase of depressive-like behaviors compared to GF mice harboring fecal microbiota from healthy individuals. While these groups were unable to unearth the underlying molecular mechanisms by which the gut microbiome was influencing the CNS, these studies suggest that it might have to do with interactions between the host and bacterial metabolism.

The majority of gut-derived serotonin is synthesized, stored, and subsequently released by a type of enteroendocrine cell in the intestinal mucosa called enterochromaffin (EC) cells. EC cells serve as sensory transducers to afferent nerves, relaying chemical and mechanical stimuli from within the GI tract. The biosynthesis of serotonin involves a two-step reaction (Fig. 5.3). First, the precursor amino acid tryptophan is catalyzed into 5-hydroxytryptophan via hydroxylation by the rate-limiting enzyme tryptophan hydroxylase (TPH). The resulting 5-hydroxytryptophan is then subsequently decarboxylated into serotonin by aromatic amino acid decarboxylase (AADC).[30]

Serotonin produced by EC cells has been found to make its way into the GI lumen and also the bloodstream, where it is taken up by blood platelets for distribution to other tissues in the body. Blood platelets lack the TPH enzyme and cannot produce serotonin. It is, therefore, believed that EC cell serotonin is the sole source of serotonin in the blood. This is of crucial importance, given that blood platelets carrying serotonin are vital for vasoconstriction and vasodilation throughout the body.[25] Additionally, EC cell serotonin mediates the perception of pain and nausea from

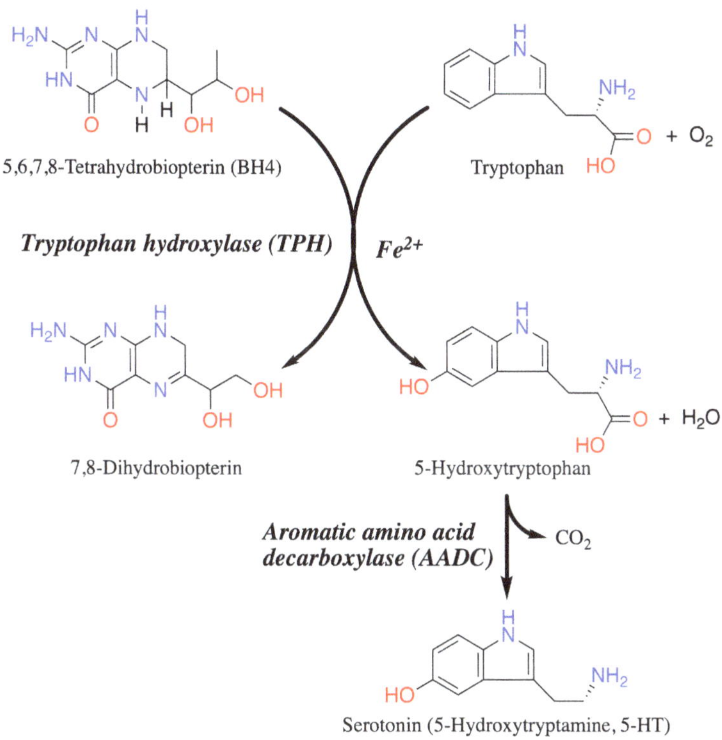

Fig. 5.3: Biosynthesis of serotonin.

the GI tract.[31] On account of the indispensability of EC cell serotonin, there has been a growing interest in looking into the interplay between EC cells and the gut microbiome.

In 2009, Wikoff *et al.*'s[32] paper galvanized the MGB research community. Inspired by other metabolomics-based investigations employing nuclear magnetic resonance (NMR) spectroscopy to study the microbiome, they set out to investigate the effect of the microbiome on plasma biochemistry. By comparing the metabolomic profiles of blood serum collected from GF mice and conventional mice fed a standardized autoclaved diet, the investigators discovered that numerous circulating molecules arose exclusively in the presence of a gut microflora. Of particular interest was the discovery that GF mice had significantly lower concentrations of serum serotonin and higher levels of serum tryptophan than did conventional mice. This discovery was a strong indication of the possible influence of the microbiome on the production of serotonin in the GI tract, paving the way for other studies to look closer at this potential connection.

One such study, published by Yano *et al.*,[33] took a closer look at how the gut microbiota affects pathways of serotonin metabolism in the GI tract. Similar to the investigation previously discussed, levels of serotonin in the blood of GF and conventional mice were compared, with the added analysis of fecal and GI tract samples. The investigators were not only able to corroborate the previous study's findings of lower levels of serotonin in the blood of GF mice, but also found lower levels of serotonin in their feces and colon relative to conventional mice. Interestingly, they did not observe significant differences in the small intestine, suggesting a specific role for gut flora in regulating colonic serotonin. To further support their findings, they found a decrease in serotonin localized to colonic EC cells in GF mice and not in small intestine EC cells. Additionally, quantification of EC cell number per area of colonic epithelial tissue did not reveal significant differences between GF and conventional mice, indicating that differences in colonic serotonin were due to a disruption in serotonin metabolism as opposed to a deficiency in EC cell development.

Having found that serotonin deficiencies in GF mice were localized to the colon, the researchers next investigated whether there were differences in the levels of key metabolites of the serotonin pathway between the colons of GF and conventional mice. They discovered that the colons of GF mice had significantly lower expression of *TPH*, the rate-limiting enzyme in the serotonin biosynthesis pathway, but no differences in the expression of enzymes involved in the packaging, release, and catabolism of serotonin. They also found there was an increase in the levels fecal and serum tryptophan, suggesting that a disruption in *TPH* expression could result in an accumulation of the substrate (offering a possible explanation for the findings of Wikoff *et al.*[32]). Oral supplementation of GF mice with the TPH product 5-hydroxytryptophan ameliorated levels of colon and fecal serotonin, while supplementation with tryptophan did not. In an attempt to further validate their findings that gut microbiota could possibly be promoting serotonin biosynthesis by increasing *TPH* expression in EC cells, Yano *et al.*[33] performed a set of experiments where GF mice were conventionalized with the microbiota of conventional mice at different stages of development. They found that at each age (post-natal day, weaning, and early adulthood) conventionalization of GF mice with the microbiota of conventional mice displayed restored serum and colon serotonin, as well as levels of *TPH* expression. They also found that eliminating the gut flora of conventional mice with antibiotics resulted in a decrease in levels of *TPH* expression and of serum and colon serotonin concentrations. All together, these findings provide a strong indication that the gut microbiome is heavily involved in the regulation of serotonin biosynthesis by EC cells.

The mechanism by which gut bacteria promote expression of *TPH* is still not well understood, but a study by Reigstad *et al.*[34] suggests that the same metabolites thought to regulate appetite might be involved. Similar to the Yano *et al.*[33] study, they observed levels of serotonin in GF mice and conventional mice, with the additional comparison to humanized (HM) mice. HM mice are simply GF mice colonized with human fecal microbiota. Not surprisingly, they also found decreased levels of serum and colon serotonin, as well as decreased expression of *TPH* in GF mice compared to conventional and HM mice. To further understand how bacteria might be inducing expression of *TPH*, they ran a set of experiments on a human-derived EC cell model. The investigators hypothesized that SCFAs and bacterial lipopolysaccharides (LPSs) might be involved, thus EC cells were treated with different concentrations of acetate, butyrate, and LPSs, and assessed for expression of *TPH* mRNA. They found that acetate and butyrate significantly increased levels *TPH* mRNA compared to untreated controls, whereas treatment with LPSs did not. This study further elevates the potential importance of bacterially derived SCFAs in

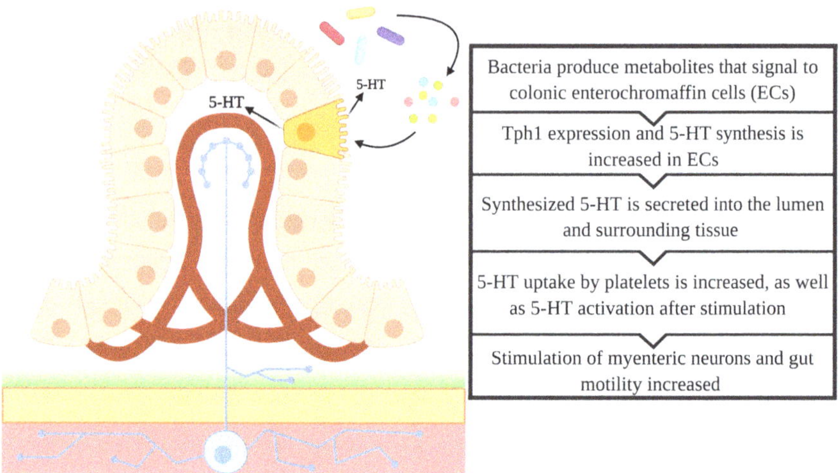

Fig. 5.4: Graphical summary of proposed serotonin regulation by gut bacteria.

the MGB axis by showing they can induce the expression of *TPH* in EC cells, thereby increasing levels of synthesized serotonin (Fig. 5.4).

BACTERIAL METABOLITES AND THE BRAIN: QUORUM SENSING MOLECULES

So far, the studies discussed in this chapter have focused on the influence of gut bacteria metabolites on their host. The MGB axis, though, is thought to function in a bidirectional manner, where host metabolites can influence the gut flora. One possible pathway through which this might occur is through the interaction between host metabolites and bacterial quorum sensing (QS) mechanisms. QS was first described in 1979 in a paper published by Nealson *et al.*[35] looking into the bioluminescence of two luminous marine bacterial species, *Vibrio fischeri* and *Vibrio harveyi*. These bacteria live symbiotically with a number of marine organisms, providing luminescent capabilities to the host in exchange for housing and nutrients in specialized light organs. What the investigators found was that light emission by the bacteria only occurred at high bacterial density in response to the accumulation of a secreted signaling molecule, suggesting the existence of cell–cell communication in bacteria. Similar to de Bary's proposition of symbiosis, the possible existence of cell–cell communication in bacteria was initially regarded as a biological curiosity. Back then, it was thought that only eukaryotes had the intricate ability to perform such a complex form of communication.

It is now understood that many, if not all, bacteria employ a kind of communication based on secreted chemical molecules to coordinate gene expression in a population-dependent manner.[36] The QS mechanism of *V. fischeri* has become the most extensively studied cell–cell communication in bacteria, regulating the expression of the luciferase structural operon *luxCDABE*. The QS circuitry of *V. fischeri* employs two regulatory proteins named LuxI and LuxR. LuxI produces and releases the chemical signal *N*-(3-oxohexanoyl)-homoserine lactone, known as an autoinducer, into the extracellular environment (Fig. 5.5). As the bacteria divide and grow in numbers so too does the concentration of autoinducer, which is freely diffusible across the cell membrane. Once the concentration of autoinducer reaches a certain threshold as a result of high cell density, it binds to LuxR. The *luxCDABE*

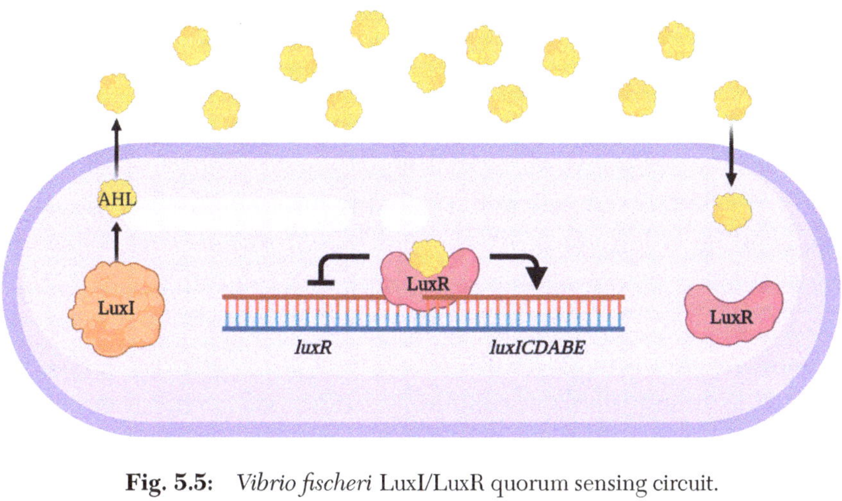

Fig. 5.5: *Vibrio fischeri* LuxI/LuxR quorum sensing circuit.

operon is under transcriptional control of LuxR, which is only able to bind the *luxCDABE* promoter when it is bound to the autoinducer. Therefore, at high cell density, LuxR binds the autoinducer and subsequently binds to the *luxCDABE* promoter to initiate transcription of the operon. Transcription of *luxCDABE* results in an exponential increase of autoinducer synthesis (via *luxI*) and light emission. Binding of LuxR to the autoinducer negatively regulates the expression of *luxR*, serving as a compensatory negative feedback loop.[37]

Research into the discovery of bacterial QS circuitry has opened the door to a new means of interpreting microbiome studies. QS mechanisms have been found to be involved in a vast array of relevant population-based bacterial phenomena, such as secretion of virulence factors and biofilm formation. Research has shown that bacterial species can employ a large variety of chemical signals for inter-species and even inter-kingdom communication, and that complex QS circuitries have evolved to cooperate and/or compete with each other. Queries into QS mechanisms employed by gram-negative bacteria have found that they typically employ proteins homologous to LuxI and LuxR, where a LuxI-like protein is responsible for biosynthesizing an autoinducer and a LuxR-like protein binds it. Furthermore, it is believed that gram-negative bacteria rely mostly on acylated homoserine lactones (AHLs) as autoinducers, while gram-positive bacteria employ secreted peptides as autoinducers in their QS circuitry.[38]

The recent explosion in investigations on bacterial QS resulted in some interesting studies relating to the MGB axis, such as the one published by Knecht *et al.*[39] looking into the possible interplay between host-derived serotonin and *Pseudomonas aeruginosa* QS mechanisms. *Pseudomonas aeruginosa* is a gram-negative pathogenic bacterium that causes infectious diseases in a vast range of organisms, including humans. *Pseudomonas aeruginosa* is characterized as being highly antibiotic-resistant and having nosocomial tendencies, often resulting in fatal outcomes upon infection in

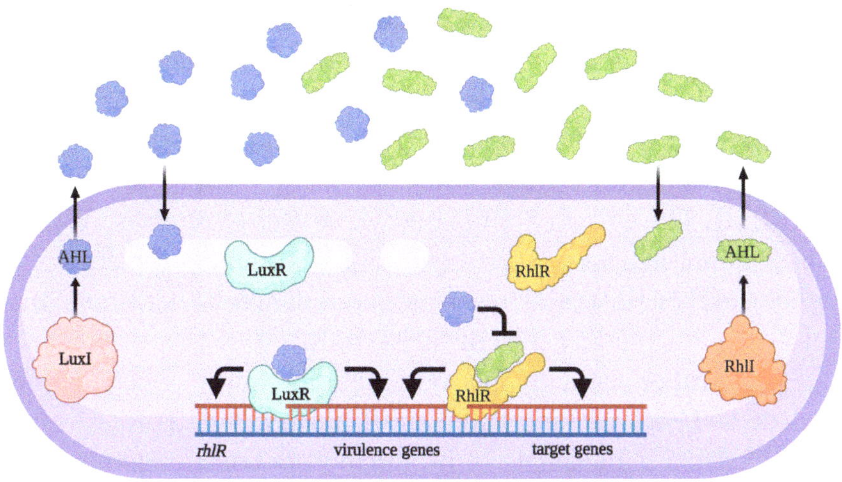

Fig. 5.6: *Pseudomonas aeruginosa* LasI/LasR-RhlI/RhlR quorum sensing circuit.

immunocompromised and/or young individuals.[40] The bacterium employs two pairs of LuxI/LuxR homologues, LasI/LasR and RhlI/RhlR, to control the expression of several virulence factors and proteins essential for biofilm formation (Fig. 5.6). LasI and RhlI catalyze the synthesis of the HSL autoinducers *N*-(3-oxododecanoyl)-homoserine lactone and *N*-(butryl)-homoserine lactone. At high cell density, LasR binds to its autoinducer and activates transcription of genes relevant to biofilm formation and several virulence factors, as well as *lasI*. The LasR–autoinducer complex also initiates the transcription of *rhlR*, thereby activating the second QS circuit. RhlR binds its cognate autoinducer and activates the expression of other virulence factors, in addition to *rhlI*. Interestingly, the autoinducer synthesized by LasI prevents RhlR from binding its cognate autoinducer. This is presumed to be a means by which the bacterium can initiate the QS cascades in the appropriate order.[37]

Using a *P. aeruginosa* intestinal infection mouse model, Knecht *et al.*[39] assessed whether IP administration of serotonin exacerbated the deleterious effects of infection. They found that infected mice supplemented with serotonin exhibited an increase in bacterial load, villi destruction, and epithelial injury than those without serotonin supplementation. Given that secretion of virulence factors and biofilm formation is mediated through QS in *P. aeruginosa*, the investigators performed *in vitro* experiments to determine whether serotonin was interacting with the LasI/LasR-RhlI/RhlR QS circuitry. Utilizing whole-cell-biosensors (WCBs) whose quantifiable bioluminescent signal was under the control of either LasR or RhlR, the investigators were able to determine that serotonin was binding selectively to LasR.

To further validate this observation, they inoculated mice with a *lasI/rhlI* mutant strain of *P. aeruginosa*. This avirulent mutant strain is incapable of producing the cognate autoinducer of LasR and RhlR, and, therefore, only re-establishes virulence in the presence of a molecule capable of binding these receptor proteins. Without supplementation of serotonin, the mutant strain caused no mortality in infected mice, while supplementation with serotonin resulted in half of the mice tested surviving. Analysis of the intestinal tissue of mice infected with the mutant strain was similar to healthy negative controls, while mice infected with the mutant strain and supplemented with serotonin displayed similar epithelial injury as those infected with the virulent strain. Their findings highlight the phenomena of social cheating in bacteria, where certain bacteria exploit sensing molecules produced by other organisms to activate their QS circuitry without expending their own energy.[38] Furthermore, this study shows that serotonin might play an important role in the onset or prevention of GI diseases by acting as an autoinducer, paving the way for future studies to look into other QS pathways activated by serotonin.

Similar studies have found that host-derived hormones epinephrine and norepinephrine can play a role in activating bacterial QS mechanisms. Enterohemorrhagic *Escherichia coli* (EHEC) is responsible for outbreaks of bloody diarrhea and hemolytic uremic syndrome, and it employs a QS mechanism to cause lesions in the intestinal tract and produce Shiga toxins.[41] The bacterium utilizes autoinducer-3 (AI-3) as a signaling molecule, which then binds to the sensor kinase QseC to initiate a complex signal cascade that activates the expression of virulence and motility genes.[5] Walters *et al.*[42] show that EHEC can sense luminal epinephrine and norepinephrine to initiate the QseC QS signal cascade and further exacerbate infection. They hypothesize that, as EHEC begins to disrupt the intestinal epithelium based on AI-3 signaling, the host is under stress and epinephrine and norepinephrine spills into the GI tract. This increase in epinephrine and norepinephrine can then be utilized by EHEC to increase the activation of virulence genes and cause further damage to the tissue, initiating a vicious cycle of infection. While further studies are necessary to test this hypothesis, these findings provide further support for the possible interplay between host-derived metabolites and bacterial QS mechanisms.

Other studies on the MGB axis with an emphasis on bacterial QS have made some intriguing insights into the possible direct interaction between QS metabolites and the brain. Studies predating Wynendaele *et al.'s*[43] work had shown that some neurological diseases are associated with an altered microbiota composition,[44] and that some peptides capable of crossing the blood–brain barrier (BBB) could potentially be used for therapeutic purposes.[45,46] Given this knowledge, Wynendaele *et al.*[43] set out to investigate whether QS peptides from gram-positive bacteria could play a role in the onset of neurological disorders. To do so, they tested the ability of three chemically diverse QS peptides to cross the BBB in an *in vivo* mouse model. The three chemically diverse peptides they tested were bacteriocin inducing peptide-2 (BIB-2), synthesized by *Streptococcus pneumonia*, PhrANTH2, synthesized by *Bacillus anthracis*, and PhrCACET1, synthesized by *Clostridium acetobutylicum*. Of the three peptides, PhrCACET1 displayed the most efficient crossing of the BBB followed by BIP-2. In contrast, PhrANTH2 did not show significant transport across the BBB. Based on these results, the investigators posit that some QS peptides, when present in the blood, can reach the brain tissue by crossing the BBB and effect the local CNS. Their study was a proof-of-concept demonstrating that it was possible for QS peptides to cross the BBB, allowing for potential to directly interact with the brain, offering a new avenue for studying the MGB axis.

In 2018, Janssens *et al.*[47] published a study partly inspired by the findings of Wynendaele *et al.*[43] in which they studied the biological effects of 85 chemically diverse QS peptides on different neuronal cell types. The cell lines they used were a combination of immortalized and cancer cell lines representative of each major brain cell type: general neuronal cells, hypothalamic neuronal cells, microglia, and astrocytes. The responses they screened for included toxicity, neurite differentiation, cytokine production, and morphological changes. In total, the robust screening found a total of 22 QS peptides capable of causing an *in vitro* effect on the different neuronal cell lines. These results are, of course, limited in that they utilized immortal cell lines and did not perform any dose-response assays. They do indicate, though, that QS peptides might directly affect brain tissue homeostasis, thus highlighting their potential role in the prevention or onset of neurodegenerative disorders.

CONCLUSIONS

The theory of symbiosis created by de Bary and elevated by Margulis was truly ahead of its time. Scoffed at by mainstream scientists of the past, symbiosis is now a given fact of life. No one questions the profound implications of the living together of differently named organisms, it's now a matter of investigating just how deep the connections truly are. The studies discussed in this chapter have provided a foundation on which to build knowledge of the MGB axis. They have shown that the SCFAs derived from bacterial digestion of dietary fiber could possibly regulate appetite by directly influencing the brain. They have also demonstrated that gut bacteria can possibly mediate the synthesis of serotonin in the gut, perhaps through SCFAs, thereby affecting a vast range of physio- and psychological functions. It has been demonstrated how the host might modulate the composition of gut flora via direct interactions with QS mechanisms through production of serotonin, and also how the psychological state of the host might exacerbate disease through the release of hormones and their subsequent initiation of pathogenic QS mechanisms. Finally, research has indicated that QS peptides can possibly cross the BBB and directly affect neuronal tissue.

Despite the significant discoveries made in studies such as the ones previously discussed, they are limited in that they most are based on *in vitro* and *in vivo* rodent model experiments. Rodent models and *in vitro* studies have proven invaluable for the advancement of scientific knowledge and elucidation of molecular mechanisms that would otherwise be impossible, if not immoral, to study in humans. The reality is, though, that when it comes to MGB axis research, there has been little convergence between animal and human studies.[12] Therein lies the issue at the core of MGB research. While extremely promising, more work needs to be done to develop investigative frameworks that allow for translational studies of the MGB axis in humans.

REFERENCES

1. Oulhen N, Schulz BJ, Carrier TJ. (2016). English translation of Heinrich Anton de Bary's 1878 speech, 'Die Erscheinung der Symbiose' ('De la symbiose'). *Symbiosis* **69**(3): 131–139.
2. Pedreira, J. Lynn Margulis (2005). https://commons.wikipdia.org/wiki/File:Lynn_Margulis_2005.jpg. https://creativecommons.org/licences/by/2.0/legalcode
3. Margulis L, Fester R. (1991). *Symbiosis as a Source of Evolutionary Innovation: Speciation and Morphogenesis*. Cambridge, MA, MIT Press, p. xii, 454 pp.
4. Turnbaugh PJ, Ley RE, Hamady M, Fraser-Liggett CM, Knight R, Gordon JI. (2007). The human microbiome project. *Nature* **449**(7164): 804–810.

5. Rhee SH, Pothoulakis C, Mayer EA. (2009). Principles and clinical implications of the brain-gut-enteric microbiota axis. *Nat Rev Gastroenterol Hepatol* **6**(5): 306–314.
6. Wang BH, Yao MF, Lv LX, Ling ZX, Li LJ. (2017). The human microbiota in health and disease. *Engineering-Prc* **3**(1): 71–82.
7. Pickard JM, Zeng MY, Caruso R, Nunez G. (2017). Gut microbiota: Role in pathogen colonization, immune responses, and inflammatory disease. *Immunol Rev* **279**(1): 70–89.
8. Koh A, De Vadder F, Kovatcheva-Datchary P, Backhed F. (2016). From dietary fiber to host physiology: Short-chain fatty acids as key bacterial metabolites. *Cell* **165**(6): 1332–1345.
9. DeSalvo KB, Olson R, Casavale KO. (2016). Dietary guidelines for Americans. *JAMA* **315**(5): 457–458.
10. Miller TL, Wolin MJ. (1996) Pathways of acetate, propionate, and butyrate formation by the human fecal microbial flora. *Appl Environ Microbiol* **62**(5): 1589–1592.
11. Cummings JH. (1981) Short chain fatty acids in the human colon. *Gut* **22**(9): 763–779.
12. Dalile B, Van Oudenhove L, Vervliet B, Verbeke, K. (2019) The role of short-chain fatty acids in microbiota-gut-brain communication. *Nat Rev Gastroenterol Hepatol* **16**(8): 461–478.
13. Swinburn BA, *et al.* (2011) The global obesity pandemic: Shaped by global drivers and local environments. *Lancet* **378**(9793): 804–814.
14. Maskarinec G, *et al.* (2006) Trends and dietary determinants of overweight and obesity in a multiethnic population. *Obesity (Silver Spring)* **14**(4): 717–726.
15. Smits SA, *et al.* (2017) Seasonal cycling in the gut microbiome of the Hadza hunter-gatherers of Tanzania. *Science* **357**(6353): 802–806.
16. Cani PD, Dewever C, Delzenne NM. (2004) Inulin-type fructans modulate gastrointestinal peptides involved in appetite regulation (glucagon-like peptide-1 and ghrelin) in rats. *Br J Nutr* **92**(3): 521–526.
17. Anastasovska J, *et al.* (2012) Fermentable carbohydrate alters hypothalamic neuronal activity and protects against the obesogenic environment. *Obesity (Silver Spring)* **20**(5): 1016–1023.
18. So PW, *et al.* (2007) Impact of resistant starch on body fat patterning and central appetite regulation. *PLoS One* **2**(12): e1309.

19. Pawlak DB, Kushner JA, Ludwig DS. (2004) Effects of dietary glycaemic index on adiposity, glucose homoeostasis, and plasma lipids in animals. *Lancet* **364**(9436): 778–785.
20. Frost G, *et al.* (2014) The short-chain fatty acid acetate reduces appetite via a central homeostatic mechanism. *Nat Commun* **5**: 3611.
21. Claret M, *et al.* (2007) AMPK is essential for energy homeostasis regulation and glucose sensing by POMC and AgRP neurons. *J Clin Invest* **117**(8): 2325–2336.
22. Hu ZY, Cha SH, Chohnan S, Lane MD. (2003) Hypothalamic malonyl-CoA as a mediator of feeding behavior. *Proc Natl Acad Sci USA* **100**(22): 12624–12629.
23. Lucki I. (1998) The spectrum of behaviors influenced by serotonin. *Biol Psychiat* **44**(3): 151–162.
24. Hyttel J. (1994) Pharmacological characterization of Selective Serotonin Reuptake Inhibitors (SSRIS). *Int Clin Psychopharm* **9**: 19–26.
25. Gershon MD, Tack J. (2007) The serotonin signaling system: From basic understanding to drug development-for functional GI disorders. *Gastroenterology* **132**(1): 397–414.
26. Cryan JF, O'Mahony SM. (2011) The microbiome-gut-brain axis: From bowel to behavior. *Neurogastroenterol Motil* **23**(3): 187–192.
27. Clarke G, *et al.* (2013) The microbiome-gut-brain axis during early life regulates the hippocampal serotonergic system in a sex-dependent manner. *Mol Psychiatry* **18**(6): 666–673.
28. Leonard BE. (2006) HPA and immune axes in stress: Involvement of the serotonergic system. *Neuroimmunomodulation* **13**(5–6): 268–276.
29. Zheng P, *et al.* (2016) Gut microbiome remodeling induces depressive-like behaviors through a pathway mediated by the host's metabolism. *Mol Psychiatry* **21**(6): 786–796.
30. Walther DJ, Bader M. (2003) A unique central tryptophan hydroxylase isoform. *Biochem Pharmacol* **66**(9): 1673–1680.
31. Mawe GM, Hoffman JM. (2013) Serotonin signalling in the gut-functions, dysfunctions and therapeutic targets. *Nat Rev Gastro Hepat* **10**(8): 473–486.
32. Wikoff WR, *et al.* (2009) Metabolomics analysis reveals large effects of gut microflora on mammalian blood metabolites. *Proc Natl Acad Sci USA* **106**(10): 3698–3703.

33. Yano JM, *et al.* (2015) Indigenous bacteria from the gut microbiota regulate host serotonin biosynthesis. *Cell* **161**(2): 264–276.
34. Reigstad CS, *et al.* (2015) Gut microbes promote colonic serotonin production through an effect of short-chain fatty acids on enterochromaffin cells. *FASEB J* **29**(4): 1395–1403.
35. Nealson KH, Hastings JW. (1979) Bacterial bioluminescence: Its control and ecological significance. *Microbiol Rev* **43**(4): 496–518.
36. Bassler BL, Losick R. (2006) Bacterially speaking. *Cell* **125**(2): 237–246.
37. Miller MB, Bassler BL. (2001) Quorum sensing in bacteria. *Annu Rev Microbiol* **55**: 165–199.
38. Whiteley M, Diggle SP, Greenberg EP. (2017) Progress in and promise of bacterial quorum sensing research. *Nature* **551**(7680): 313–320.
39. Knecht LD, *et al.* (2016) Serotonin activates bacterial quorum sensing and enhances the virulence of pseudomonas aeruginosa in the host. *EBioMedicine* **9**: 161–169.
40. Moura-Alves P, *et al.* (2019) Host monitoring of quorum sensing during Pseudomonas aeruginosa infection. *Science* **366**(6472): eaaw1629.
41. Reading NC, Torres AG, Kendall MM, Hughes DT, Yamamoto K, Sperandio V. (2007) A novel two-component signaling system that activates transcription of an enterohemorrhagic Escherichia coli effector involved in remodeling of host actin. *J Bacteriol* **189**(6): 2468–2476.
42. Walters M, Sperandio V. (2006) Autoinducer 3 and epinephrine signaling in the kinetics of locus of enterocyte effacement gene expression in enterohemorrhagic Escherichia coli. *Infect Immun* **74**(10): 5445–5455.
43. Wynendaele E, *et al.* (2015) Quorum sensing peptides selectively penetrate the blood-brain barrier. *PLoS One* **10**(11): e0142071.
44. Foster JA, McVey Neufeld KA. (2013) Gut-brain axis: How the microbiome influences anxiety and depression. *Trends Neurosci* **36**(5): 305–312.
45. Egleton RD, Davis TP. (2005) Development of neuropeptide drugs that cross the blood-brain barrier. *NeuroRx* **2**(1): 44–53.
46. Stalmans S, *et al.* (2014) Blood-brain barrier transport of short proline-rich antimicrobial peptides. *Protein Pept Lett* **21**(4): 399–406.
47. Janssens Y, *et al.* (2018) Screening of quorum sensing peptides for biological effects in neuronal cells. *Peptides* **101**: 150–156.

Chapter 6

Gut–Brain Axis: The Current State of Imaging Technologies, Their Clinical Implications, and Future Directions

David Tran[1,*], Drew Smith[1,*], Mario M. Perdomo[1], Viraj Shah[1], Dario Ebode[1], Rahul Mittal[1], Rita Ghose Bhatia[4] and Adrien A. Eshraghi[1,2,3]

Abstract

Recent research is increasingly suggesting bidirectional communication within the gut–brain axis (GBA). It is becoming apparent that the gut microbiome not only influences the predisposition to gastrointestinal (GI) disorders, but also affects the central nervous system (CNS) and many neurological conditions. While the association between the gut and brain is growing, we are limited in modalities that precisely map the gut and brain

[1] Department of Otolaryngology, University of Miami-Miller School of Medicine, Miami, FL, USA.

[2] Department of Neurological Surgery, University of Miami-Miller School of Medicine, Miami, FL, USA.

[3] Department of Biomedical Engineering, University of Miami, Coral Gables, FL, USA.

[4] Department of Radiology, Division of Neuro-radiology, University of Miami, Coral Gables, FL, USA.

* These authors contributed equally to this work.

Corresponding Author: Adrien A. Eshraghi, MD, MSc, FACS, Department of Otolaryngology, Neurological Surgery and Biomedical Engineering, University of Miami-Miller School of Medicine, 1600 NW 10th Avenue, Miami, FL 33136, USA, Tel: 305-243-8078, E-mail: aeshraghi@med.miami.edu.

together. For the brain, magnetic resonance imaging (MRI), computed tomography (CT), and positron emission tomography (PET) aid the visualization of internal anatomy and connectivity to some extent, but no clear modalities exist for the enteric nervous system and the gut. In this chapter, we discuss the structural and functional anatomy of the brain using MRI technique. We then go on to explore the current choice tools in mapping the brain, such as the functional and diffusion-weighted MRI. We also discuss the implications of MR spectroscopy using techniques such as point-resolved spectroscopy (PRESS) and stimulated echo acquisition mode (STEAM) for brain imaging. After discussing the limitations of current and newer technologies such as tractography, we highlight future implications and directions that need to be taken in brain modalities to provide accurate, non-invasive images. The latter section of this chapter focuses on the anatomy and physiology of the gut and the implications slow waves have on diagnosing various diseases that are difficult to otherwise identify. Current non-invasive gut imaging modalities include electrogastrography (EGG), which lacks accuracy in recording slow waves, while more trustable modalities such as high-resolution (HR) electrical mapping are invasive and require surgery. We summarize the current and the emerging gut imaging techniques, while indicating which improvements are necessary to better map the GBA.

INTRODUCTION

The gut–brain axis (GBA) refers to the interactions and connections between the central nervous system (CNS) and the enteric nervous system (ENS). Recently, the role of the microbiota's influence on the GBA through their effect on neurotransmitter production and alteration is coming to the forefront.[1] There are various CNS diseases that are now being linked to the microbiota–GBA such as Parkinson's disease, schizophrenia, Alzheimer's disease, and autism spectrum disorders.[2] The greatest issue with the GBA is that current imaging tools are either inaccurate or too invasive to provide accurate mapping capabilities to diagnose and identify the various GI and CNS disorders. However, the technologies used to image the brain and gut separately have the potential for advancement to image the GBA together (Fig. 6.1). For the brain, magnetic resonance imaging (MRI), computed tomography (CT), and positron emission tomography

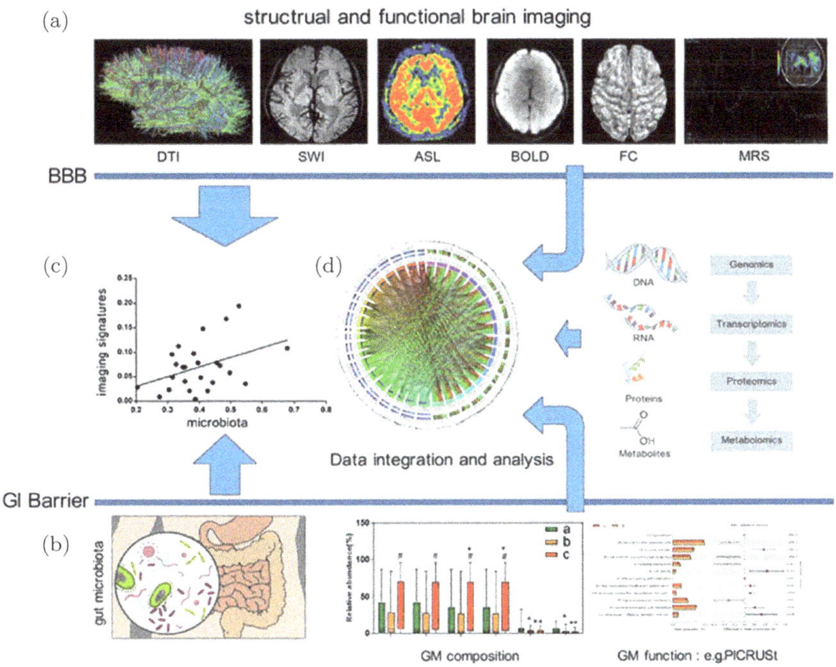

Fig. 6.1: Possible frameworks for exploring crosstalk between the GM and human brain. (a) Examples of distinct MRI maps: DTI, SWI, ASL, BOLD, FC, and MRS (from left to right). (b) Analysis of the GM, including composition and function. (c) Explorations of the direct association between the GM and human brain imaging. (d) Combining other "omics" techniques, such as metabolomics and proteomics, to generate a complete picture of host and microbial pathways. DTI, diffusion tensor imaging; SWI, susceptibility weighted imaging; ASL, arterial spin labeling; BOLD, blood-oxygenation level-dependent; FC, functional connection; MRS, magnetic resonance spectroscopy; BBB, blood–brain barrier; GI, gastrointestinal; GM, gut microbiota; PICRUSt, Phylogenetic Investigation of Communities by Reconstruction of Unobserved States. (Taken from Liu *et al.*[85] This is an open-access article distributed under the terms of the Creative Commons Attribution License (CC BY). The use, distribution, or reproduction in other forums is permitted, provided the original author(s) and the copyright owner(s) are credited and that the original publication in this journal is cited, in accordance with accepted academic practice.)

(PET) technologies can provide an overview of CNS structure (Table 6.1). In the gut, the interstitial cells of Cajal (ICC) initiate slow-wave activity, which are responsible for coordinating the gastrointestinal (GI) peristaltic movements.[3] These slow waves can be measured and disparities such as dysrhythmias can indicate gastric motility disorders such as gastroparesis.[4] The current imaging modalities of choice include electrogastrography (EGG) and high-resolution (HR) electrical mapping, which requires electrodes to be placed on the serosal surface of the gut requiring an invasive procedure that can be conducted for only a limited time period (Table 6.2).[5] New developments show promising wireless technologies that could function as pacemaker-like implants.[6] While there seem to be technologies available for both gut and the brain individually, there are no modalities that connect the two and provide a linkage between the GBA. This article aims to address these drawbacks in precise imaging techniques and proceed to review current technologies and the future direction of gut and brain imaging.

FUNCTIONAL ANATOMY OF THE BRAIN

To understand the functional processes of the human brain, it is imperative to understand its structural anatomy. A more comprehensive understanding of structure and how it relates to function would better illustrate pathological processes and potentially direct clinicians to intervene and manage certain diseases. It is well understood that the human cerebral cortex is essentially a mosaic of areas that serve different functions.[7] The process of mapping the cerebral cortex has been in existence for several years with a combination of many different techniques, including imaging, structural, and functional properties. This is referred to as cortical parcellation, a method in which the three-dimensional (3D) structure is analyzed in order to more accurately define borders between regions and correlate such with functional processes.

Table 6.2: Current Technologies to Image Gastrointestinal (GI) Tract

Current Gut-Imaging Modalities				
Technique	**Application**	**Limitations**	**Advantages**	**Citations**
Electro-gastrography (EGG)	Places abdominal electrodes to measure gastric slow waves	1. Provides a weighted summation of all gastric slow waves, preventing accurate detection of dysrhythmia, contractions or emptying 2. Patient movement during measurement induces artifacts 3. Cannot identify the specific regional activity of the stomach 4. Cannot conduct on obese patients due to increased abdominal wall thickness	1. Is non-invasive 2. Useful in measuring slow waves if accurate computerized spectral analysis is employed	1. Yin and Chen[104] 2. Farajivadar[6] 3. Simonian *et al.*[107]
High Resolution (HR) Electrical Mapping	Places dense array of electrodes on serosal surface to track bioelectrical sequences in fine spatiotemporal detail	1. Requires invasive surgery and 2. Placement of transabdominal wires prevents any long-term studies, increases chances of infection, discomfort and displacement of wires 3. Anesthesia use during surgery is shown to interfere with gastric slow-wave frequency	1. Allows for reading of normal and dysrhythmic gastric slow waves 2. Can record in spatiotemporal detail 3. More accurate slow-wave recordings than EGG	1. O'Grady *et al.*[3] 2. Farajivadar[6]

(Continued)

Table 6.2: (*Continued*)

Developing Future Imaging Modalities				
Technique	**Function**	**Improvement**	**Current Limitations and Future Indications**	
Mucosal HR Mapping	Electrodes are placed on the mucosal surface of the GI organs and the readings are comparable to slow-wave readings from serosal electrode placement	1. Less invasive technique without compromising the ability to accurately measure gastric slow waves	1. Development of an endoscopic route of implementation will make procedure even less invasive and safer to conduct on patients	1. Angeli *et al.*[106]
High Resolution Electro-gastrography (HR-EGG)	Utilization of cutaneous electrodes that can estimate slow-wave propagation direction and speed	1. Provides the spatiotemporal detail that HR mapping provides rather than the sum of all the gastric slow waves	1. Optimum number of electrodes for best readings still needs to be investigated 2. Results need to be compared against HR mapping from serosa of patients	1. Gharibans *et al.*[109] 2. Farajivadar[6]
Endoscopic High Resolution Electrical Mapping	Endoscopic technique to implant electrodes into the submucosal space	1. Less invasive solution for a previously invasive procedure	1. Flexible implants may be limiting in recording slow waves	1. Kim *et al.*[112] 2. Farajivadar[6]
Multi-channel Wireless Mapping	Uses a seven-multi-channel acquisition system to wirelessly acquire slow-wave measurements	1. Prevents the use bulky trans-abdominal cables 2. Allows for long-term studies to be conducted	1. Measurement of slow waves limited to five days 2. Improvements in channel counts, wireless technology, power efficiency can increase the time frame	1. Paskaranandavadivel *et al.*[4] 2. Wang *et al.*[105]

As such, MRI is currently the most widely used imaging modality for identifying important structures within the brain (Table 6.1). The MRI offers a non-invasive technique to display the structure of the cortex *in vivo*. Previously, the majority of understanding of the microstructure of the cortex was derived from post-mortem histological studies. Through the use of the MRI, researchers have been able to gain insight into the structure–function relationship of macroscopic brain anatomy and functional brain organization.[7] Microscopic structures, however, such as neurons and glial cells, cannot be directly imaged via *in vivo* MRI for humans. An understanding of these structures, however, have an implication in understanding the longitudinal progression of microstructure in combination with functional imaging and pathologic processes. The human neocortex is subdivided into macro-, meso-, and microstructural scale. Currently, the neocortex is systematized into general areas based on the work of early scientists like Brodmann, for example. The neocortex itself is organized into a six-layered framework, with cell body density and sizes varying over different regions. Parcellation of these areas have been possible with modern histological techniques. However, these parcellations have traditionally been performed via *ex vivo* histology. Recent advances have suggested a newer modality, termed *in vivo* histologic MRI (hMRI) as a methodology to map out features of cortical myelin, neuronal fibers, and iron deposition within the brain. These are all implicated in their own way for clinical translation.

One of these implications stems from the basic understanding of the mesoscopic structure and function of myelin. This allows definition of its function in brain development and pathology of myelin disease. Structural analysis of neuronal fibers is implicated in better understanding of behavioral function and dysfunction. Iron deposition is implicated in understanding the physiology of aging and potentially a better framework for age-related dementia. The use of *in vivo* mapping of the human cortex would allow for longitudinal research of living subjects that will aid in understanding how each of these topics contribute to physiology and pathology, and potentially advance treatment modalities in each respective arena. Previous studies have shown that hMRI is becoming more widely utilized.[8] One example where this technique was successfully demonstrated was via parcellation where *in vivo* mapping of myelin content was achieved using a myelin marker with functional MRI (fMRI) data.[7] While there has been substantial progress made in the *in vivo* imaging of the brain, specificity remains the most difficult issue to solve. Distinguishing between micro- and mesostructural origin of MRI signal intensity differences has been a challenge that is currently being investigated. Recent theories support using a multiple modality imaging approach with different types of MR combinations in order to achieve this understanding. In addition, adequate resolution has also been difficult to achieve.

An additional modality that has been used is diffusion tensor imaging (DTI). DTI has several clinical applications; however, one of the particular interest is demonstrating microstructural damage that occurs as a result of mild traumatic brain injury (mTBI). DTI and MRI are currently being investigated for use in detecting biomarkers that are specific for mTBI, which includes hemosiderin, white matter abnormalities, and the assessment of white matter integrity. Correlating these findings with clinical presentation would theoretically provide a quantitative measure that directly assesses neuroanatomy and improves prediction of neuropsychological outcomes.[9] *In vivo* biomarkers and imaging of these biomarkers are useful targets for finding causal relationships between pathology and structural abnormalities.[10]

BRAIN MAPPING

The choice tool for imaging and mapping the brain is the MRI scanner. The basic MRI consists of a magnetic coil and a radio frequency emitter and receiver, with the clinical MRIs having, commonly, a field strength of 3 T or below at this time. This combination of magnetic field and radio frequency allows for the detection of hydrogen within the body or tissues. For the purpose of the anatomical mapping of cortical structures, there are many techniques that can be applied using the MRI scanner; mainly diffusion-weighted MRI (dMRI) and fMRI are used. dMRI uses the diffusion of water molecules as an endogenous probe to map tissue microstructures. A key feature of dMRI is a high sensitivity to pore sizes with walls that are either semipermeable or non-permeable. dMRI may be ideal to map the distribution of axon diameter in each image voxel of the brain scan. The importance of knowing the axon diameter distribution is key, since diameter is directly related to transmission speed: an axon with a larger diameter conveys signals more rapidly than an axon with a smaller diameter.[11] A mathematical model of the nerve tissue signal composed of separate signals from axonal water, glial water, and extracellular water was designed to better understand this complexity. However, there are three main situations that cause problems when translating that mathematical model of an *in vitro* study to larger organs and neural networks: (1) variable orientations of a single fiber bundle; (2) multiple, separate fiber bundles crossing the same voxel; and (3) intra-voxel orientational scattering of a certain fiber bundle. Using high angular diffusion imaging (HARDI) with a basic model to minimize intricacy essentially solves the first situation. Revisions to HARDI address the second and third situations: within-voxel fiber crossings and within-voxel fiber-orientation scattering.[12,13] This newer modeling extends the segment of white matter over which one could acquire helpful axon diameter directories. More current refined models, which capture the time-dependence of extracellular diffusion[14–17] and non-standard pulse sequences, such as oscillating gradients,[18] double diffusion encoding (DDE),[19] and generalized waveforms.[20] Other tactics include q-space formulation,[21] convex optimization,[22] integration with fiber tracking,[23] and dictionary matching.[17]

The exchange of water among extracellular space and intra-axonal space owing to non-zero cell membrane permeability could theoretically result in overestimation. Other effects including intricacy of the extracellular diffusion process and alterations in the relaxation parameters of various water pools could also disturb the signal, but are frequently disregarded in axon diameter estimation models.[25] De Santis *et al*. showed that modeling time-dependence of the extra-axonal diffusivity can result in reduced estimation bias.[14] Horowitz *et al*. demonstrated an association between approximations of axon diameter distribution and inter-hemispheric apparent conduction velocity.[26] This was deduced from electrophysiological measurements, which suggests that these quantities could clarify the variances in individual brain function. Despite a lack of precision in characterizing the accurate axonal distribution, characterizing disease according to axon diameter maybe be conceivable through imaging biomarkers. If the overestimation of the axon diameter is nearly uniform, it is possible to conclude conduction velocity via estimates of axon diameter. However, whether this could accurately forecast actual conduction velocity from microstructural measurements *in vivo* is debatable.

To further estimate parameters for inclusion or exclusion, multi-compartment models have been formulated. Different conformations of these models have been tested in pre-clinical acquisitions and during *in vivo* human studies.[27–29] In large studies, these models show deviations in many neurological and psychiatric disorders.[30,31] There are challenges that emerge when synching these models with the dMRI data. It is necessary to locate the model parameters that produce the diffusion-weighted signals that are observed. Non-linear estimation procedures are often utilized for estimating parameters of multi-compartment biophysical models. Non-linear fitting systems attempt to locate an optimal value by progressively shifting through non-linear parameters, which can be laborious.

One "standard model," exists in which every section is defined by an anisotropic Gaussian function, known as a tensor.[32,33] This model was extended by Stanisz *et al.* and later advanced by Assaf *et al.*[11,34] Here, the measurement is presumed to surpass the resolution limit; to account for this, intra-neurite space is signified as sticks and tensors have zero perpendicular diffusivity. Extra-neurite space is defined by a prolate tensor with non-zero parallel and perpendicular diffusivities. To achieve accurate representation of this model, parameters in the model display the same value, by which it can be understood that parallel intra- and extra-neurite diffusivities are equal or are fixed to a given value.

Various clinical gradient systems exist for measuring microscopic diffusion anisotropy, including DDE and QTE.[35–37] Assuming that the material can be characterized by a wide range of diffusion tensors and utilizing an extension of the QTE signal, data on the covariance of the microenvironment tensors is available to be aquired.[38] Using this technique, one study discovered significant differences among such features in a small subset of healthy controls and individuals with schizophrenia. On the other hand, the DIVIDE framework utilizes a method of gamma distribution and relays such information to the power-averaged signal attained from various B-tensor shapes.[39]

Extra-strong gradients combined with enhanced waveforms is able to significantly decrease the TE across these experiments.[20,39–41] An explanation: the signal fraction pertaining to a still water compartment may be undervalued at the long TE attainable on clinical systems if the given T2 is shorter than that of further compartments.[42,43] Testing the signal at a shorter TE offers investigators a way to address a possible confounder. In addition, the shorter TE provided by the stronger gradient delivers an increase of signal-to-noise ratio (SNR) of approximately 50% when compared to that what is achievable by a system reaching 80 mT/m gradients.

A variety of MRI contrasts, relaxometry, susceptibility, and magnetization transfer are utilized to characterize tissue other than dMRI. However, white matter is so vastly complex that only one MRI contrast is not likely to appropriately provide a full characterization of its microstructure. It is possible that knowing the exact origin of the signal change is less important if tools that are sensitive to this change become available and are accessible. Still, unraveling significant additions of myelin, axons, and glial cells is necessary to comprehend the pathophysiology of neurological disorders subsequently result in the development of treatments. Besides pathology,[44] determining changes in white matter is also key for understanding mechanisms of the macroscopic developments taking place in the brain as an outcome of maturation,[45,46] training,[47] and lifestyle.[48] This reality can be counterbalanced through combination of additional methods, including multi-component relaxometry,[49] magnetization transfer,[50] and susceptibility imaging.[51–53]

The idyllic study would be one that reconstructs an N-dimensional spectrum, where the N dimensions represent various groupings of useful MR parameters. For the case of N = 2, (T1, T2), (T2, Diffusion), and (T1, Diffusion) are parameters of interest.[54] The complementary nature of such MRI modalities infers that division of dissimilar environments can be enhanced through examining each multi-dimensional correlation, associated by any of the one-dimensional (1D) spectra.[55] The earliest attempts of hybrid MRI tests in biological tissue attempted to establish the relationship between T2 species and diffusion behavior.[11] These trials were mostly performed *ex vivo*, utilizing a variant of the diffusion-weighted Carr–Purcell–Meiboom–Gill (CPMG) sequence, to examine whether compartments with shorter T2, hypothetically those surrounded by myelin, had diffusion characteristics consistent of this premise. This procurement involves the typical PGSE preparation trailed by a CPMG train of 180° pulses.[56] The T2 decay could, therefore, be quantified under capricious diffusion-weighting conditions, empowering the diffusion coefficient of the water compartments consistent with each T2 peak in the test. T1 compartmentalization was also evaluated when combined with diffusion by affixing inversion recovery to the DW-CPMG tests.[57] Recently, Benjamini *et al.* validated the recognition of seven distinct peaks by examining the T2-diffusion and T1-diffusion space from the spinal cord of a ferret, each linked with a different form of tissue including glial, axonal, neuronal, and myelin water.[58] An additional study combined inversion recovery and dMRI in order to map T1 along distinct white matter tracts voyaging inside a single voxel, presuming a mono-exponential T1 behavior per fiber populace.[15,16] The viability of this method *in vivo* was shown by changing the inversion time for a moderate b-value of 1000 s/mm^2.

It has never been possible to demonstrate microstructural anomalies of the crucial cortical deviations that are seen in the macroscopic view of various psychiatric and neural disorders, especially in exact laminae. Utilizing ultra-high resolution dMRI to do just this could result in novel insights into the pathophysiology of various disorders. Likewise, improved spatial resolution could not only benefit the quantification of microstructure diffusion properties, but could also meaningfully advance the precision of tractography reconstructions. Improving the spatial resolution enables dissimilar fiber pathways to be unraveled and allows other minor white matter connections to arise beside the major white matter fasciculi.[59] Intra-lobar connections as well as U-shaped fibers are indispensable short-range networks that address a vital role in spreading material throughout whole brain networks, yet are likely undervalued by many tractography and connectomic methods.[60]

It is imperative to determine neuronal activity in a non-invasive manner in order to better understand the lack of correct brain function during the disease process. It is possible to directly measure the neuronal current employing electroencephalography (EEG) and magnetoencephalography (MEG). Unfortunately, EEG and MEG are known to have limited spatial resolution, and the sensitivity of each is constrained to cortical activity. fMRI is an example of neurovascular coupling that is grounded on contrast of blood-oxygenation level-dependent (BOLD). Despite an ability to detect hemodynamic and metabolic changes within general areas of the brain, fMRI lacks spatial specificity due to the fact that BOLD contrast is detected in the capillaries, arteries, and veins in a relatively large area surrounding the site of neuronal activity.[61,62] Furthermore, the BOLD response is dependent on the density of vasculature, which is variable among sections of the brain and potentially compromised in some conditions, including stroke.[63] Neurovascular coupling can limit the ability to detect precise timing of neural activity. The hemodynamic response function (HRF) that defines BOLD signal change is characterized by a 1 to 2 seconds delay between the inception of neural activity and the period necessary for blood to reach the active region. This time that is variable and usually peaks between 6 and 12 seconds. Ideally, MRI spatial resolution would be combined to a technique that allows for high temporal resolution to detect timely and accurate information on brain activity. Currently, dMRI is able to signal neuronal activation by detecting cellular swelling. This technique has been proposed as a potential candidate to address the spatial and temporal weaknesses with fMRI.[64–66]

Researchers have not yet determined the type of contrast that would be used in diffusion fMRI. The intrinsically small SNR of the diffusion fMRI signal restricts the array of possible experimental exploitations. The advent of faster sequences like simultaneous multi-slice imaging can help improve signal acquisition, but only a significant boost in SNR will make these experiments practical.[67]

Magnetic resonance spectroscopy (MRS) can measure chemical contents in order to evaluate brain metabolism. It complements structural MR imaging and can evaluate levels of metabolites and neurobiological substrates in physiologic and pathologic states. MRS can assess biochemical changes in various neurological disorders including neoplasm, stroke, epilepsy, metabolic encephalopathy, hepatic encephalopathy, mitochondrial encephalopathy, as well as neurodegenerative diseases such as Alzheimer's disease and Parkinsonism, and psychiatric disorders like depression. The effectiveness of treatment can also be assessed by MRS.

MR spectra is obtainable from various body nuclei including 1-Hydrogen, 31-Phosphorus, and 19-Fluorine. Metabolic and physiologic data can be calculated from these nuclei. Typically, the 1H-MRS is the best option due to the high sensitivity of this nucleus, easy accessibility, and an abundance of hydrogen in nearly all metabolites. MRS are routinely performed on 1.5- and 3.0-T systems. Three-tesla field strength has significant clinical advantages including a SNR increase and an enhancement of the capability to deliver spectra results from minor voxels.[68,69]

The basic data obtained from MRS is vastly different when compared with MRI in that a spectrum is produced instead of an image. The MR spectrum entails of peaks, or spectra, that signify signal intensities as a function of frequency, which is indicated as parts per million (ppm). Spectra are attained from either one selected region of the brain, called single-voxel spectroscopy (SVS), or from multiple brain regions, which is known as multivoxel spectroscopy or chemical shift imaging (CSI).

The common SVS techniques, which are routinely applied for 1H-MRS, are identified as point-resolved spectroscopy (PRESS) and simulated echo acquisition mode (STEAM).[70] A STEAM pulse sequence utilizes a 90° refocusing pulse to gather signals in a manner similar to a gradient echo. STEAM is able to attain shorter echo times but sacrifices signal-to-noise (Fig. 6.2).

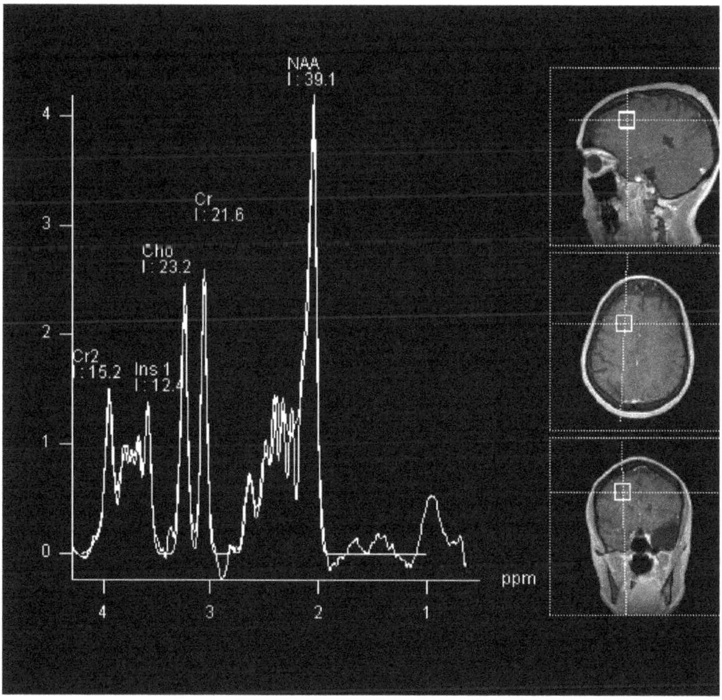

Fig. 6.2: Normal H-MR spectrum acquired at 3 T from a volume of interest in the frontal lobe of a healthy subject with the point-resolved spectroscopy (PRESS) sequence (TE 30 ms). Many metabolites can be quantified at short TEs; useful for myoinositol, glutamine/glutamate (Glx), amino acids, lactate, and lipids.

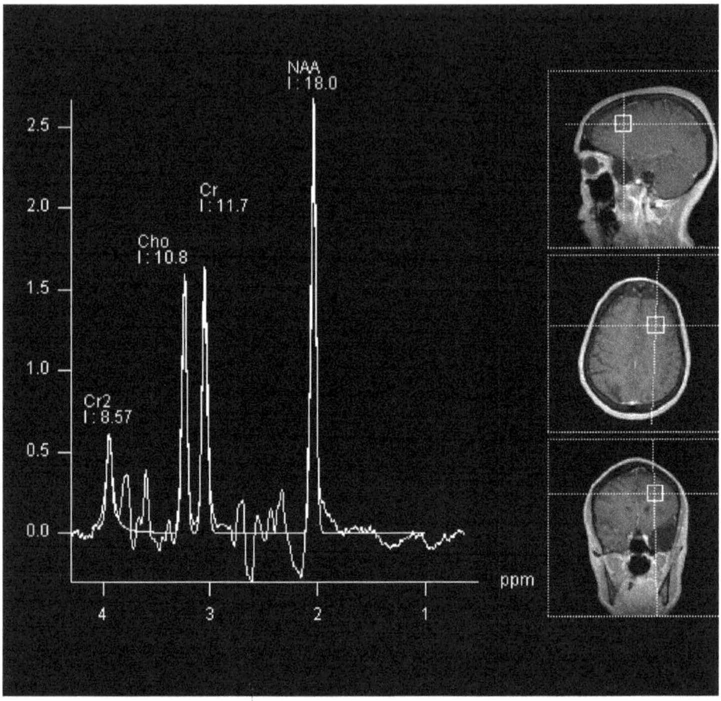

Fig. 6.3: Normal H-MR spectrum acquired at 3 T from a volume of interest in the frontal lobe of a healthy subject with the point-resolved spectroscopy (PRESS) sequence (TE 135 ms). Useful for differentiation of lactate/alanine peak from lipids at 1.3–1.5 ppm because inversion of lactate peak occurs at 1.3 ppm.

A PRESS sequence refocalizes each spin via a 180° rf pulse similar to a spin echo (Fig. 6.3).

To summarize, a single voxel, short TE technique is utilized to provide an initial diagnosis due to its high SNR as well as the representation of all metabolites. Multi-voxel, long TE techniques are utilized to portray other brain regions. In addition, multi-voxel, long TE techniques are employed to evaluate therapy response in several pathological conditions. Every metabolite emerges at a specific ppm, and each echoes precise biochemical and cellular processes. The peaks of each metabolite arise from several compositions as seen in Table 6.3. The amount of quantifiable metabolites is dependent upon a given pulse sequence and its parameters, as well as the SNR and spectral resolution.[71,72]

Table 6.3: Observable Proton Metabolites

ppm	Metabolite	Properties
0.9–1.4	Lipids	Brain damage
1.3	Lactate	Anaerobic glycolysis
2.0	NAA	Neuronal marker
2.2–2.4	Glutamine/GABA	Neurotransmitter
3.0	Creatine	Energy metabolism
3.2	Choline	Cell membrane marker
3.5	Myo-inositol	Glial cell marker, osmolyte hormone receptor mechanisms

Taken from http://spinwarp.ucsd.edu/neuroweb/Text/mrs-TXT.htm#:~:text=FUNDAMENTALS%20OF%20MR%20SPECTROSCOPY&text=MR%20spectroscopy%20provides%20a%20measure,than%20either%20sodium%20or%20phosphorus.

The most vital brain metabolites that yield an MRS signal include N-acetyl aspartate (NAA), creatine (Cr), and choline (Cho). Each is quantifiable at every clinical magnetic field strength and at nearly all practical TEs up to 280 ms.[73,74] Typically, short TEs of 25–35 ms for PRESS, 20 ms or less for STEAM, myoinsitol, amino acids glutamine or glutamate (Glx), lactate, and lipids are quantifiable.[75] Additionally, they can be used to monitor therapeutic responses in these neurological diseases. Intermediate TE of 135–144 ms offers data on less metabolites, but does hold advantages over short TE MRS including the ability to differentiate the lactate/alanine peak in lipids from 1.3 ppm to 1.5 ppm because intermediate TE can invert lactate at 1.3 ppm. It also better delineates the NAA peak at the 2.00–2.05 area and demonstrates a more precise lipid and choline peak. At field strengths equal to or greater than 3.0 T, other metabolites are distinguished at short TEs including gluthathione and g-aminobutyric acid, and the distinction of glutamate (Glu) and glutamine (Gln) is achievable.[76,77] At least 18 metabolites are quantifiable by short TEs and field strengths of 7.0–9.4 T.[77–79] NAA, available in neuronal cell types including oligodendrocytes, is a biomarker for neuronal activity that diminishes when pathology is unfavorably affected by the integrity of neurons. NAA levels can also help in determining mitochondrial (dys)function.[80]

Creatine is an organic compound found in the mitochondria, which functions as a biomarker for cell metabolism and provides a measure of energy stores. Choline is a precursor of phosphatidyl Cho (phospholipid of the cell membrane) and is a measure of cellular proliferation, greater membrane turnover, and has been shown to be increased in neoplasms and inflammation. Myoinositol, or Ins, is generally considered a glial marker and could have an impact on myelination. Elevated levels of Ins are seen during inflammation, astrocytosis, gliosis, and in other diseases like Alzheimer's.[81,82] Elevated lactate is suggestive of anaerobic glycolysis and could be heeded as an unspecific MRS biomarker for various pathologies.[83,84] To conclude, MRS has been proven to be a useful, non-invasive technique that is able to monitor biochemical variations due to changes in the gut microbiome.[85]

In summary, MRS has been used as an adjunct to routine MRI as both a research and clinical tool for evaluation of metabolic derangements that occur with common neurologic and neurosurgical conditions including tumors, demyelinating diseases, a variety of metabolic conditions, and neurodegenerative and psychiatric disorders. There is great hope in the future for its use as a potential tool in the understanding of the fascinating connection with the GBA through the combination of image-based biomarkers with radiomicrobiomics.

It has been asserted, however, that current modalities are not satisfactory. Tractography is a modality that is used to aid in visualization of the brain for surgical procedures, yet it is currently very unreliable.[86] Previous studies have been performed that demonstrate how standard MRI techniques are not helpful in identifying patients who are likely to have delayed recovery when suffering from mTBI.[87] This further emphasizes the need for developing new modalities that are capable of more precise imaging. In a study conducted by Farquharson, a comparison between tractography methodologies was utilized to visualize corticospinal tracts from 45 healthy controls and 10 pre-surgical patients. In this tensor-based study, constrained spherical deconvolution-based tractography methodologies were administered to all subjects. It was found that

DTI-based imaging methods were substantially underestimating the degree of tracts connected to the cortex of every subject within the control group. Interestingly, constrained spherical deconvolution produced a more accurate and biologically precise image of tract configuration. These results demonstrate the risk of neurological deficit following neurosurgical resection should these techniques be used independently, as safe margins of resection were vastly different in each scenario.[86] With this clinically misleading information, it is safe to assume that a more precise and biologically accurate form of imaging must be applied in order to reduce risk of harm to patients who have a need for precision imaging, such as in neurosurgical cases.

GI ANATOMY AND PHYSIOLOGY

After ingestion of any substance, digestion normally occurs. The substance bolus travels through the GI tract, where nutrients are absorbed and waste products are eventually eliminated from the body. The involuntary contraction and subsequent relaxation of smooth muscle, which drives the bolus through GI tract is termed peristalsis. To achieve peristalsis, GI slow waves are propagated from ICC, which are unique cells located along the GI tract. Networks of ICC exist both intramuscularly and in the myenteric plexus.[88] ICC, along with enteric neurons, are thought to be the main proponents in producing electrical signals that are ultimately responsible for the slow waves that lead to GI tract motility.[89] Our understanding of these waves is ultimately limited, however, there are several mechanisms that have been proposed.

As previously mentioned, there are two mechanisms proposed to explain how slow waves spread circumferentially and down the GI tract. The first is action potential (AP)-like propagation and the second is phase wave-based propagation. It has been shown that gut conditions determine the mechanism. To generate normal slow waves, the phase wave mechanism is dominant. The phase wave mechanism consists of the cyclical discharge of calcium as an oscillator, with connections of multiple oscillators. To visualize this, one can picture lined pendulums linked together by springs. When activated, these pendulums swing with a similar frequency but contain a short delay, resulting in a propagating wave. When the gut is unsettled, the AP-like propagation dominates. This AP-like mechanism consists of voltage-accelerated calcium waves that propagate sequentially. Single bundle gastric strips were used to identify these findings.[88] Additionally, a recent study examining the synchronization of oscillators in the phase wave mechanism measured the phase response curve via a new union of diameter mapping plus stimulation of a single pulse electrical field to show how isochrons cause delays to limit the wave cycle.[90]

Pendular contractions also exist throughout the GI tract. These consist of a back-and-forth movement of the intestine that lacks any propelling or peristaltic action, in order to thoroughly churn and mix the contents with gastric secretions. The mechanism of pendular contractions was shown in a 2005 study using recordings of high-resolution mechanical and electrical recordings. The recordings showed slow waves (not to be confused with ICC-generated slow waves) and spikes, which traveled in either a superior or inferior direction. A relaxation phase in one direction was accompanied with a contraction phase in the opposite direction, and vice versa. The slow-wave propagation pattern regulates the rhythm and the directionality of pendular contractions.[89]

Several types of circus movements, or re-entries, have also been discovered in the GI tract. While they have classically been seen in the stomach, recent studies have proved that they also exist in the small intestine.[91] Bioelectronic technology is now allowing scientists to further characterize these re-entries, as Angeli accomplished with his recent description of the circumferential and functional re-entries of the small intestine.[5] These re-entries, or arrhythmias, could lead to various pathologies and warrants further inspection. The increase in available gastrointestinal electrophysiology combined with bioelectronic technology will soon allow scientists to better comprehend the subcellular electrophysiology, and normal and abnormal electrophysiology of the GI tissue, which will result in an increased understanding of the patient pathology and the development of successful treatments.[92]

GI MAPPING

Early bioelectronics for mapping the GI tract activity have generally consisted of simplistic electrical recording devices, similar to an ECG. However, technology is rapidly improving. HR mapping of slow waves is now a core instrument to discover electrophysiological properties of the GI tract. Electrodes for high-resolution mapping are being fitted into various devices that can be used to chart areas of the tract. Such devices include endoscopes and implantables.[6]

Slow-wave mapping is implicated in pathology that is otherwise difficult to treat, including gastroparesis, slow transit constipation, and intestinal pseudo-obstruction. Most modalities that are used to map slow-wave activity have been either inefficient, are poor quality, or overly invasive in nature. The intended direction of future studies is to map the electrical activity of the gut with implantable devices that can be used to record activity over extended periods of times, similar to a Holter monitor, without inconvenience to the test subject. It is necessary to involve novel electrodes and acquisition systems in order to accomplish this.[6]

Current GI imaging modalities include HR electrical mapping and EGG.[5] EGG is a non-invasive procedure that utilizes abdominal skin electrodes to measure gastric slow waves.[93] GI HR mapping requires the usage of many dense electrode arrays to follow bioelectrical sequences in acceptable spatiotemporal detail.[94] However, with current technologies, to get accurate measurements, contact with the serosal surface is necessary. Hence, HR requires invasive surgical procedures such as laparotomy or laparoscopy using anesthesia.[95] This section will explore the different imaging modalities of the gut that are currently available, the limitations they face, and future gut imaging modalities that are developing such as endoscopic HR mapping, multi-channel wireless mapping, and HR-EGG.

EGG was first developed in the 1920s and has been used as a cutaneous non-invasive method of measuring gastric slow waves. Appropriate skin preparation and electrode placement is paramount to acquiring accurate slow-wave reading.[93] One of the limitations of EGG is that it provides a weighted summation of all gastric slow waves presented in the stomach and necessitates the use of computerized spectral analysis. This prevents it from accurately detecting dysrhythmia.[6] Other limitations include the presence of artifacts induced by patient movement during measurement, and the inability to detect the specific regional activity of the stomach. EGG also cannot be performed on patients with high BMIs as the increased abdominal wall thickness produces weak amplitude recordings.[96] In summary, EGG can be useful in measuring slow waves if the proper procedural measures are taken and accurate analyzation and interpretation occurs. It does not, however, detect gastric contractions or emptying, hence in these cases more invasive procedures provide better results.[93]

HR mapping is another currently used GI imaging modality that measures electrical activity in the gut in fine spatiotemporal detail.[94] With the development of a flexible printed circuit board, it became possible to place the electrodes in a way that conformed to the gut curvatures. This allowed for reading of normal and dysrhythmic gastric slow waves.[6] Dysrhythmias that are found in the gut are known to be associated with gastroparesis and dyspepsia, and the role of the HR mapping techniques is increasing in clinical settings to treat patients with these conditions.[6] The current issue is that to perform HR mapping, invasive surgical procedures performed under anesthesia are required. Not only does this limit studies to short-term projects, but also one complication that has been associated with anesthesia use is a possible decline in slow-wave frequencies.[94] Another limitation arises from the transabdominal wires that are necessary to record the serosal electrode readings.

This also prevents any long-term studies and increases chances for infection, discomfort, and displacement of the wires.[6] Hence, there have been developments to introduce less invasive, wireless technologies such as endoscopic HR mucosal mapping modality,[95] multi-channel wireless mapping,[97] and HR-EGG.[98]

Normally, HR mapping is performed by placing electrodes on the serosal surface and requires an invasive procedure. A new study by Angeli *et al.*[5] suggests that the electrodes can be placed on the mucosal surface of the GI organs and acquire comparable readings as with serosal electrode placement. A constellation catheter is a 64-channel electrophysiological device, which was originally used in cardiology, but provided a solution in GI mapping by allowing for uniform distribution of the electrodes.[99] This study placed the electrodes on the mucosal surface using a catheter through a slight incision via the gastric wall.[95] However, by further advancements, a catheter could be deployed using endoscopic techniques using visual guidance and endoscopic suction to increase mucosal surface contact.[95] There still needs to be improvements made in the analytical software to minimize the noise-to-signal ratio as well as modifications to the electrode design to fit mucosal imaging, and miniaturization of the electrode arrays to allow for passage through traditional endoscopic ports. If these improvements occur, mucosal HR mapping provides a minimally invasive method of acquiring spatiotemporal slow-wave information.

Another development has been the introduction of wireless mapping of the GI serosal slow waves.[97] In traditional HR mapping approach requires bulky cables and prevents long-term studies. These seven-multi-channel wireless acquisition system allows for comparable slow-wave measurements while decreasing discomfort, chances of infection, preventing displacement of the cables, and allowing for long-term studies.[97] Another 32-multi-channel acquisition system was also introduced.[100] Recently, slow waves were acquired wirelessly for an ambulatory system for 5 days.[94] Hence, the time frame for which the slow waves can be measured is increasing. Improvements that need to be made include increasing channel counts, wireless charging technology (to allow for longer recording times), and power efficiency.[97]

HR-EGG is another recent advancement that uses an array of cutaneous electrodes that can estimate the slow-wave propagation direction and speed.[98] This is in contrast to the traditional EGG, which provides a weighted summation of the gastric slow waves.[93] There is no need for an invasive procedure. There are still questions as to the optimum number of electrodes necessary to produce the best readings.[98] The results obtained in the aforementioned study, however, were not confirmed via HR mapping of the serosa within the patients.[6]

Going forward, the biggest challenge is posed by long-term chronic study of gastric slow waves. Multiple less invasive methods of slow-wave measurement were discussed such as the endoscopic mucosal HR approach and the HR-EGG, which is known to still be riddled by artifacts.[95,98] However, these do not answer how to develop a long-term chronic measurement solution. What is necessary is an implantable technology that can be inserted using an endoscopic procedure that can send wireless signals over a longer period, no matter if the patient is in a fasting or fed state.[6] Future advances must solve the size and power consumption limitations that are currently faced by HR mapping technologies. Application-specific integrated circuits (ASICs) and wireless power transfer (WPT) can be combined to provide that solution. There is an ASIC-linked system-on-chip (SoC) device that can communicate with a wireless wearable device passively, reducing battery usage considerably, and a new endoscopic technique that can implant electrodes into the submucosal space has been developed.[6] While some view flexible implants to be limiting in recording slow waves, the stretchable technology allows for longer longevity of the electrodes.[101] Overall, the various gut–brain imaging modalities are progressing toward a long-term implantable solution for HR mapping that requires less invasive procedures while still providing comparable readings to invasive serosal HR mapping technologies.

CONCLUSIONS AND FUTURE DIRECTIONS

GBA plays a crucial role in pre-disposition to neurological disorders such as autism spectrum disorders and other pathologies. Currently the technologies are limited to imaging of either the brain or GI tract alone. However, to fully understand these pathological links, we must be able to study the brain and the gut together. It will prove difficult to expand our knowledge of these connections if we are not able to view them as one entity versus separate powers. Fortunately, the vast technological advances that are currently occurring will likely make it possible to expand the scope of these techniques in mapping the GBA. While brain mapping research and subsequent technological advancements have been more prominent over the past decade, gut mapping is beginning to catch up. Researchers and biomedical engineers should place increased emphasis on the combination of these technologies to further grow our understanding of human physiology.

REFERENCES

1. Mittal R, *et al.* (2017) Neurotransmitters: The critical modulators regulating gut-brain axis. *J Cell Physiol* **232**(9): 2359–2372.
2. Zhu X, Han Y, Du J, Liu R, Jin K, Yi W. (2017) Microbiota-gut-brain axis and the central nervous system. *Oncotarget* **8**(32): 53829–53838.
3. O'Grady G, *et al.* (2010) Origin and propagation of human gastric slow-wave activity defined by high-resolution mapping. *Am J Physiol Gastrointest Liver Physiol* **299**(3): G585–G592.
4. Paskaranandavadivel N, Gao J, Du P, O'Grady G, Cheng LK. (2014) Automated classification and identification of slow wave propagation patterns in gastric dysrhythmia. *Ann Biomed Eng* **42**(1): 177–192.
5. Angeli TR, *et al.* (2013) Circumferential and functional re-entry of in vivo slow-wave activity in the porcine small intestine. *Neurogastroenterol Motil Official J Eur Gastrointestinal Motil Soc* **25**(5): e304–e314.
6. Farajidavar A. (2018) Bioelectronics for mapping gut activity. *Brain Res* **1693**(Pt B): 169–173.

7. Fischl B, Sereno MI. (2018) Microstructural parcellation of the human brain. *NeuroImage* **182**: 219–231.
8. Edwards LJ, Kirilina E, Mohammadi S, Weiskopf N. (2018) Microstructural imaging of human neocortex in vivo. *NeuroImage* **182**: 184–206.
9. Bigler ED. (2013) Neuroimaging biomarkers in mild traumatic brain injury (mTBI). *Neuropsychol Rev* **23**(3): 169–209.
10. Wallace EJ, Mathias JL, Ward L. (2018) The relationship between diffusion tensor imaging findings and cognitive outcomes following adult traumatic brain injury: A meta-analysis. *Neurosci Biobehav Rev* **92**: 93–103.
11. Stanisz GJ, Szafer A, Wright GA, Henkelman RM. (1997) An analytical model of restricted diffusion in bovine optic nerve. *Magn Reson Med* **37**(1): 103–111.
12. Zhang H, Dyrby TB, Alexander DC. (2011) Axon diameter mapping in crossing fibers with diffusion MRI. *MICCAI* **14**(Pt 2): 82–89.
13. Zhang H, Hubbard PL, Parker GJ, Alexander DC. (2011) Axon diameter mapping in the presence of orientation dispersion with diffusion MRI. *NeuroImage* **56**(3): 1301–1315.
14. De Santis S, Jones DK, Roebroeck A. (2016) Including diffusion time dependence in the extra-axonal space improves in vivo estimates of axonal diameter and density in human white matter. *NeuroImage* **130**: 91–103.
15. De Santis S, Barazany D, Jones DK, Assaf Y. (2016) Resolving relaxometry and diffusion properties within the same voxel in the presence of crossing fibres by combining inversion recovery and diffusion-weighted acquisitions. *Magn Reson Med* **75**(1): 372–380.
16. De Santis S, Assaf Y, Jeurissen B, Jones DK, Roebroeck A. (2016) T1 relaxometry of crossing fibres in the human brain. *NeuroImage* **141**: 133–142.
17. Sepehrband F, Alexander DC, Kurniawan ND, Reutens DC, Yang Z. (2016) Towards higher sensitivity and stability of axon diameter estimation with diffusion-weighted MRI. *NMR Biomed* **29**(3): 293–308.
18. Kakkar LS, *et al.* (2018) Low frequency oscillating gradient spin-echo sequences improve sensitivity to axon diameter: An experimental study in viable nerve tissue. *NeuroImage* **182**: 314–328.
19. Benjamini D, Basser PJ. (2016) Use of Marginal Distributions Constrained Optimization (MADCO) for accelerated 2D MRI relaxometry and diffusometry. *J Magn Reson* **271**: 40–45.

20. Drobnjak I, Zhang H, Hall MG, Alexander DC. (2011) The matrix formalism for generalised gradients with time-varying orientation in diffusion NMR. *J Magn Reson* **210**(1): 151–157.
21. Ong HH, Wehrli FW. (2010) Quantifying axon diameter and intra-cellular volume fraction in excised mouse spinal cord with q-space imaging. *NeuroImage* **51**(4): 1360–1366.
22. Daducci A, Canales-Rodriguez EJ, Zhang H, Dyrby TB, Alexander DC, Thiran JP. (2015) Accelerated Microstructure Imaging via Convex Optimization (AMICO) from diffusion MRI data. *NeuroImage* **105**: 32–44.
23. Girard G, *et al.* (2017) AxTract: Toward microstructure informed tractography. *Hum Brain Mapp* **38**(11): 5485–5500.
24. Nilsson M, Latt J, Stahlberg F, van Westen D, Hagslatt H. (2012) The importance of axonal undulation in diffusion MR measurements: A Monte Carlo simulation study. *NMR Biomed* **25**(5): 795–805.
25. Burcaw LM, Fieremans E, Novikov DS. (2015) Mesoscopic structure of neuronal tracts from time-dependent diffusion. *NeuroImage* **114**: 18–37.
26. Horowitz A, Barazany D, Tavor I, Bernstein M, Yovel G, Assaf Y. (2015) In vivo correlation between axon diameter and conduction velocity in the human brain. *Brain Struct Funct* **220**(3): 1777–1788.
27. Panagiotaki E, Schneider T, Siow B, Hall MG, Lythgoe MF, Alexander DC. (2012) Compartment models of the diffusion MR signal in brain white matter: A taxonomy and comparison. *NeuroImage* **59**(3): 2241–2254.
28. Bourne RM, Panagiotaki E, Bongers A, Sved P, Watson G, Alexander DC. (2014) Information theoretic ranking of four models of diffusion attenuation in fresh and fixed prostate tissue ex vivo. *Magn Reson Med* **72**(5): 1418–1426.
29. Ferizi U, *et al.* (2014) A ranking of diffusion MRI compartment models with in vivo human brain data. *Magn Reson Med* **72**(6): 1785–1792.
30. De Santis S, *et al.* (2017) Early axonal damage in normal appearing white matter in multiple sclerosis: Novel insights from multi-shell diffusion MRI. *Conference proceedings: Annual International Conference of the IEEE Engineering in Medicine and Biology Society IEEE Engineering in Medicine and Biology Society Annual Conference*, pp. 3024–3027.
31. Kamagata K, *et al.* (2017) Gray matter abnormalities in idiopathic Parkinson's disease: Evaluation by diffusional kurtosis imaging and neurite

orientation dispersion and density imaging. *Hum Brain Mapp* **38**(7): 3704–3722.

32. Kroenke CD, Ackerman JJ, Yablonskiy DA. (2004) On the nature of the NAA diffusion attenuated MR signal in the central nervous system. *Magn Reson Med* **52**(5): 1052–1059.
33. Jespersen SN, Lundell H, Sonderby CK, Dyrby TB. (2013) Orientationally invariant metrics of apparent compartment eccentricity from double pulsed field gradient diffusion experiments. *NMR Biomed* **26**(12): 1647–1662.
34. Assaf Y, Basser PJ. (2005) Composite Hindered and Restricted Model of Diffusion (CHARMED) MR imaging of the human brain. *NeuroImage* **27**(1): 48–58.
35. Hui ES, Jensen JH. (2015) Double-pulsed diffusional kurtosis imaging for the in vivo assessment of human brain microstructure. *NeuroImage* **120**: 371–381.
36. Lawrenz M, Brassen S, Finsterbusch J. (2016) Microscopic diffusion anisotropy in the human brain: Age-related changes. *NeuroImage* **141**: 313–325.
37. Lasič S, Szczepankiewicz F, Eriksson S, Nilsson M, Topgaard D. (2014) Microanisotropy imaging: Quantification of microscopic diffusion anisotropy and orientational order parameter by diffusion MRI with magic-angle spinning of the q-vector. *Front Phys* **2**: 11.
38. Westin CF, *et al.* (2016) Q-space trajectory imaging for multidimensional diffusion MRI of the human brain. *NeuroImage* **135**: 345–362.
39. Sjolund J, Szczepankiewicz F, Nilsson M, Topgaard D, Westin CF, Knutsson H. (2015) Constrained optimization of gradient waveforms for generalized diffusion encoding. *J Magn Reson* **261**: 157–168.
40. Drobnjak I, Alexander DC. (2011) Optimising time-varying gradient orientation for microstructure sensitivity in diffusion-weighted MR. *J Magn Reson* **212**(2): 344–354.
41. Drobnjak I, Siow B, Alexander DC. (2010) Optimizing gradient waveforms for microstructure sensitivity in diffusion-weighted MR. *J Magn Reson* **206**(1): 41–51.
42. Veraart J, Novikov DS, Fieremans E. (2018) TE dependent Diffusion Imaging (TEdDI) distinguishes between compartmental T2 relaxation times. *NeuroImage* **182**: 360–369.

43. Dhital B, Kellner E, Kiselev VG, Reisert M. (2018) The absence of restricted water pool in brain white matter. *NeuroImage* **182**: 398–406.
44. Bodini B, Ciccarelli O. (2014) Diffusion MRI in neurological disorders, in *Diffusion MRI.* San Diego, CA, Academic Press, pp. 241–255.
45. Sullivan EV, Pfefferbaum A. (2010) Diffusion tensor imaging in aging and age-related neurodegenerative disorders, in *Diffusion MRI.* Oxford, Oxford University Press, pp. 624–644.
46. Huppi PS, Dubois J. (2006) Diffusion tensor imaging of brain development. *Semin Fetal Neonat Med* **11**(6): 489–497.
47. Johansen Berg H. (2017) Imaging and stimulating adaptive brain plasticity. *European Neuropsychopharmacol* **27**: S1.
48. Ritchie SJ, *et al.* (2017) Risk and protective factors for structural brain ageing in the eighth decade of life. *Brain Struct Funct* **222**(8): 3477–3490.
49. MacKay A, Whittall K, Adler J, Li D, Paty D, Graeb D. (1994) In vivo visualization of myelin water in brain by magnetic resonance. *Magn Reson Med* **31**(6): 673–677.
50. Sled JG, *et al.* (2004) Regional variations in normal brain shown by quantitative magnetization transfer imaging. *Magn Reson Med* **51**(2): 299–303.
51. Duyn JH, Schenck J. (2017) Contributions to magnetic susceptibility of brain tissue. *NMR Biomed* **30**(4): e3546.
52. Duyn J. (2013) MR susceptibility imaging. *J Magn Reson* **229**: 198–207.
53. Shmueli K, de Zwart JA, van Gelderen P, Li TQ, Dodd SJ, Duyn JH. (2009) Magnetic susceptibility mapping of brain tissue in vivo using MRI phase data. *Magn Reson Med* **62**(6): 1510–1522.
54. Callaghan PT, Godefroy S, Ryland BN. (2003) Use of the second dimension in PGSE NMR studies of porous media. *Magn Reson Imaging* **21**(3–4): 243–248.
55. Celik H, Bouhrara M, Reiter DA, Fishbein KW, Spencer RG. (2013) Stabilization of the inverse Laplace transform of multiexponential decay through introduction of a second dimension. *J Magn Reson* **236**: 134–139.
56. van Dusschoten D, Moonen CT, de Jager PA, Van As H. (1996) Unraveling diffusion constants in biological tissue by combining Carr-Purcell-Meiboom-Gill imaging and pulsed field gradient NMR. *Magn Reson Med* **36**(6): 907–913.
57. Andrews TJ, Osborne MT, Does MD. (2006) Diffusion of myelin water. *Magn Reson Med* **56**(2): 381–385.

58. Benjamini D, Basser PJ. (2017) Magnetic resonance microdynamic imaging reveals distinct tissue microenvironments. *NeuroImage* **163**: 183–196.
59. Heidemann RM, Anwander A, Feiweier T, Knosche TR, Turner R. (2012) k-space and q-space: Combining ultra-high spatial and angular resolution in diffusion imaging using ZOOPPA at 7 T. *NeuroImage* **60**(2): 967–978.
60. Catani M, *et al.* (2017) Short parietal lobe connections of the human and monkey brain. *Cortex* **97**: 339–357.
61. Logothetis NK, Wandell BA. (2004) Interpreting the BOLD signal. *Annu Rev Physiol* **66**: 735–769.
62. Olman CA, Inati S, Heeger DJ. (2007) The effect of large veins on spatial localization with GE BOLD at 3 T: Displacement, not blurring. *Neuro Image* **34**(3): 1126–1135.
63. Vigneau-Roy N, Bernier M, Descoteaux M, Whittingstall K. (2014) Regional variations in vascular density correlate with resting-state and task-evoked blood oxygen level-dependent signal amplitude. *Hum Brain Mapp* **35**(5): 1906–1920.
64. Tasaki I, Iwasa K. (1982) Further studies of rapid mechanical changes in squid giant axon associated with action potential production. *Jpn J Physiol* **32**(4): 505–518.
65. Tasaki I, Iwasa K. (1982) Rapid pressure changes and surface displacements in the squid giant axon associated with production of action potentials. *Jpn J Physiol* **32**(1): 69–81.
66. Aso T, Urayama S, Poupon C, Sawamoto N, Fukuyama H, Le Bihan D. (2009) An intrinsic diffusion response function for analyzing diffusion functional MRI time series. *NeuroImage* **47**(4): 1487–1495.
67. Moeller S, *et al.* (2010) Multiband multislice GE-EPI at 7 tesla, with 16-fold acceleration using partial parallel imaging with application to high spatial and temporal whole-brain fMRI. *Magn Reson Med* **63**(5): 1144–1153.
68. Di Costanzo A, *et al.* (2003) High-field proton MRS of human brain. *Eur J Radiol* **48**(2): 146–153.
69. Kim J-H, *et al.* (2006) Comparison of 1.5T and 3T 1H MR spectroscopy for human brain tumors. *Korean J Radiol* **7**(3): 156.
70. Ross B, Coletti P, Lin A. (1840) Magnetic resonance spectroscopy of the brain: Neurospectroscopy, in *Clinical Magnetic Resonance Imaging*, Philadelphia, PA, Saunders Elsevier, p. 1907.

71. Gruetter R, *et al.* (1998) Resolution improvements in in vivo 1H NMR spectra with increased magnetic field strength. *J Magn Reson* **135**(1): 260–264.
72. Otazo R, Mueller B, Ugurbil K, Wald L, Posse S. (2006) Signal-to-noise ratio and spectral linewidth improvements between 1.5 and 7 Tesla in proton echo-planar spectroscopic imaging. *Magn Reson Med* **56**(6): 1200–1210.
73. Frahm J, Bruhn H, Gyngell ML, Merboldt KD, Hänicke W, Sauter R. (1989) Localized proton NMR spectroscopy in different regions of the human brain in vivo. Relaxation times and concentrations of cerebral metabolites. *Magn Reson Med* **11**(1): 47–63.
74. Mlynárik V, Gruber S, Moser E. (2001) Proton T_1 and T_2 relaxation times of human brain metabolites at 3 Tesla: Metabolite T_1 and T_2 in human brain at 3T. *NMR Biomed* **14**(5): 325–331.
75. Hofmann L, Slotboom J, Jung B, Maloca P, Boesch C, Kreis R. (2002) Quantitative 1H-magnetic resonance spectroscopy of human brain: Influence of composition and parameterization of the basis set in linear combination model-fitting. *Magn Reson Med* **48**(3): 440–453.
76. Mekle R, Mlynárik V, Gambarota G, Hergt M, Krueger G, Gruetter R. (2009) MR spectroscopy of the human brain with enhanced signal intensity at ultrashort echo times on a clinical platform at 3T and 7T. *Magn Reson Med* **61**(6): 1279–1285.
77. Tkáč I, Öz G, Adriany G, Uğurbil K, Gruetter R. (2009) In vivo ^{1}H NMR spectroscopy of the human brain at high magnetic fields: Metabolite quantification at 4T vs. 7T: Metabolite Quantification at 4T vs. 7T. *Magn Reson Med* **62**(4): 868–879.
78. Deelchand DK, *et al.* (2010) In vivo 1H NMR spectroscopy of the human brain at 9.4T: Initial results. *J Magn Reson* **206**(1): 74–80.
79. Marjańska M, Auerbach EJ, Valabrègue R, Van de Moortele P-F, Adriany G, Garwood M. (2012) Localized ^{1}H NMR spectroscopy in different regions of human brain *in vivo* at 7 T: T_2 relaxation times and concentrations of cerebral metabolites: Regional differences in T_2 and metabolite concentrations at 7T. *NMR Biomed* **25**(2): 332–339.
80. Bates TE, Strangward M, Keelan J, Davey GP, Munro PMG, Clark JB. (1996) Inhibition of N-acetylaspartate production: Implications for 1H MRS studies in vivo. *Neuroreport* **7**(8): 1397–1400.

81. Pouwels P, Kruse B, Korenke G, Mao X, Hanefeld F, Frahm J. (1998) Quantitative proton magnetic resonance spectroscopy of childhood adrenoleukodystrophy. *Neuropediatrics* **29**(5): 254–264.
82. Kantarci K, *et al.* (2008) Alzheimer disease: Postmortem neuropathologic correlates of antemortem ^{1}H MR spectroscopy metabolite measurements. *Radiology* **248**(1): 210–220.
83. Wardlaw JM, Marshall I, Wild J, Dennis MS, Cannon J, Lewis SC. (1998) Studies of acute ischemic stroke with proton magnetic resonance spectroscopy: Relation between time from onset, neurological deficit, metabolite abnormalities in the infarct, blood flow, and clinical outcome. *Stroke* **29**(8): 1618–1624.
84. Groenendaal F, Veenhoven RH, van der Grond J, Jansen GH, Witkamp TD, de Vries LS. (1994) Cerebral lactate and N-acetyl-aspartate/choline ratios in asphyxiated full-term neonates demonstrated in vivo using proton magnetic resonance spectroscopy. *Pediatr Res* **35**(2): 148–151.
85. Liu P, Peng G, Zhang N, Wang B, Luo B. (2019) Crosstalk between the gut microbiota and the brain: An update on neuroimaging findings. *Front Neurol* **10**: 883.
86. Farquharson S, *et al.* (2013) White matter fiber tractography: Why we need to move beyond DTI. *J Neurosurg* **118**(6): 1367–1377.
87. Hughes DG, Jackson A, Mason DL, Berry E, Hollis S, Yates DW. (2004) Abnormalities on magnetic resonance imaging seen acutely following mild traumatic brain injury: Correlation with neuropsychological tests and delayed recovery. *Neuroradiology* **46**(7): 550–558.
88. van Helden DF, Laver DR, Holdsworth J, Imtiaz MS. (2010) Generation and propagation of gastric slow waves. *Clin Exper Pharmacol Physiol* **37**(4): 516–524.
89. Lammers WJ. (2005) Spatial and temporal coupling between slow waves and pendular contractions. *Am J Physiol Gastrointest Liver Physiol* **289**(5): G898–G903.
90. Parsons SP, Huizinga JD. (2017) The phase response and state space of slow wave contractions in the small intestine. *Exp Physiol* **102**(9): 1118–1132.
91. Lammers WJ, Stephen B, Karam SM. (2012) Functional reentry and circus movement arrhythmias in the small intestine of normal and diabetic rats. *Am J Physiol Gastrointest Liver Physiol* **302**(7): G684–G689.

92. Lammers WJ. (2013) Arrhythmias in the gut. *Neurogastroenterol Motil* **25**(5): 353–357.
93. Yin J, Chen JD. (2013) Electrogastrography: Methodology, validation and applications. *J Neurogastroenterol Motil* **19**(1): 5–17.
94. Wang R, *et al*. (2018) A miniature configurable wireless system for recording gastric electrophysiological activity and delivering high-energy electrical stimulation. *IEEE J Emerg Sel Top Circuits Syst* **8**(2): 221–229.
95. Angeli TR, *et al*. (2016) Acute slow wave responses to high-frequency gastric electrical stimulation in patients with gastroparesis defined by high-resolution mapping. *Neuromodulation* **19**(8): 864–871.
96. Simonian HP, *et al*. (2004) (2004) Multichannel electrogastrography (EGG) in normal subjects: A multicenter study. *Dig Dis Sci* **49**(4): 594–601.
97. Paskaranandavadivel N, Wang R, Sathar S, O'Grady G, Cheng LK, Farajidavar A. (2015) Multi-channel wireless mapping of gastrointestinal serosal slow wave propagation. *Neurogastroenterol Motil* **27**(4): 580–585.
98. Gharibans AA, Kim S, Kunkel D, Coleman TP. (2017) High-resolution electrogastrogram: A novel, noninvasive method for determining gastric slow-wave direction and speed. *IEEE Trans Bio-Med Eng* **64**(4): 807–815.
99. Pappone C, Santinelli V. (2006) Multielectrode basket catheter: A new tool for curing atrial fibrillation? *Heart Rhythm* **3**(4): 385–386.
100. Springston CS, Rui B, Farajidavar A. (2016) A 32-channel wireless system for recording gastric electrical activity. *Conference Proceedings: Annual International Conference of the IEEE Engineering in Medicine and Biology Society IEEE Engineering in Medicine and Biology Society Annual Conference*, pp. 1971–1974.
101. Kim YS, *et al*. (2017) Scalable manufacturing of solderable and stretchable physiologic sensing systems. *Adv Mater* **29**(39): 1701312.

Chapter 7

The Gut–Brain Axis in the Context of Gut Microbiome and Its Metabolites

Santanu Banerjee

Abstract

Commensal microbiota and its homeostasis have been associated with general well-being of an individual for centuries. The alteration of microbial flora in gastrointestinal distress, infection and chronic metabolic conditions are associated with distinct comorbidities, which were known, and yet poorly understood till we started understanding the commensal microbial association with the diseases, their manifestation and pathogenesis. With extensive use of sequence-based molecular phylogeny, we now understand that the commensal microbiota is intimately integrated with the systemic life processes, continuously interacting with the host, both immunologically and biochemically in health and disease. With ample evidence that microbial metabolites are assimilated by the host and that they participate in host metabolic pathways, several phenomena hitherto ascribed to the "gut–brain axis" are becoming more apparent. In this chapter, we will describe major brain/central nervous system (CNS) pathways and how microbial changes (both in terms of abundance and "biochemical behavior") influence those pathways in health and disease.

Department of Surgery and Miami Integrative Metabolomics Research Center (MIMRC), University of Miami-Miller School of Medicine, FL, USA.
Corresponding Author: Santanu Banerjee, PhD, BRB 516, University of Miami, 1501 NW 10th Avenue, Miami, FL 33136, Tel: 305-243-2555, E-mail: santanu.banerjee@med.miami.edu.

INTRODUCTION

The human gastrointestinal (GI) tract harbors an upward of 100 trillion microbes (archaea, bacteria, fungi, protozoa, and viruses) and covers close to 95% of all commensal microbes hosted by an individual.[1] In terms of genes, microbial genes are 150 times larger (in Kilobases) than it's host's genetic compliment and contain close to 4 million distinct bacterial genes.[1,2] With the advent of sequence-based molecular phylogeny in the past 2 decades, it is increasingly becoming clear that commensal microbes are metabolically hyper-active, and microbial metabolites are routinely assimilated by the host, thereby influencing host metabolic pathways.[3] In fact, in the absence of a thriving microbiota, there would be a significant loss of circulating metabolites within the mammalian host.[1,4] It is now known that these metabolites circulate throughout the body, cross the blood–brain barrier (BBB), and constitute a complex bidirectional biochemical communication network, constantly shaping each other.[5–7] Hence, it is perceivable that a clear understanding of this network would provide a powerful and non-invasive tool to modulate the microbiome and achieving desired systemic effects in the host, including within distal organs/systems like brain and central nervous system (CNS).

Microbial metabolites from homeostatic and altered (disease-induced dysbiosis) microbiota are known to influence the host immune system and metabolism.[8,9] These are the main routes by which the microbiota is linked with brain and CNS disorders like the autism spectrum, Parkinson's disease (PD), Alzheimer's disease (AD), and psychiatric disorders, to name a few, leading to the coinage of the brain–gut–microbiome (BGM) axis.[7] This chapter will discuss the major metabolites produced by the gut microbiome, that directly participate, or indirectly contribute to the biochemical imbalances that affect major brain morbidities. While their microbial origin is the denominator, less emphasis will be placed on the identity of microbes that produce these metabolites to maintain the focus in our narrative.

Commensal Microbes and Their Metabolites

The gut microbiome is a vibrant ecosystem with high metabolic activity. Almost all of the metabolic reactants available to the microbes are diet-derived, and the rest are host-derived. The gut microbes either transform these metabolites into a different metabolic entity (e.g., dietary fibers to short-chain fatty acids; SCFA),[10] or in the process of energy metabolism, to a slightly modified product (e.g., primary to secondary bile formation).[11] In many cases, including the two examples above, mammalian biochemistry is incapable of synthesizing these metabolites by themselves. The products are readily assimilated by the host and proceed to participate in a plethora of host systemic signaling pathways. The mammalian host has several checks and balances in place that restrict the entry and exit of cells and molecules for privileged compartments and organ systems in the body (brain being one of them).[12] The one separating systemic circulation and brain/CNS is the BBB. Hence, to understand the role of microbial metabolites in the manipulation of brain metabolic pathways, a clear understanding of BBB and its interaction with the metabolites is required.

Microbial Metabolites and BBB

The BBB is not a singular entity. Rather, it is a functionality established by a diverse group of cells and macromolecules, including brain microvascular endothelial cells (BMECs), the extracellular matrix (ECM), and pericytes within the vascular system, which interact continuously with the astrocytes, neurons and microglial cells within the brain/CNS for oversight, surveillance and maintenance of BBB.[13] One of the primary functions of the BBB is to preserve the ionic homeostasis of the brain, which requires the restriction of paracellular movement across the BMEC membrane. This is achieved by enforcing a highly regulated tight-junction between the BMECs, which are wrapped around by the pericytes (brain-side of the barrier). This unit is then connected to the neurons by the astrocytes, thus forming a tight BBB.[14] A range of microbial metabolites exhibit metabolic and signaling properties, which can modulate brain function by compromising or modifying BBB integrity and acting on cells within the CNS and brain.

Short-Chain Fatty Acids

One of the most well-studied groups of microbial metabolites are the SCFAs. These are small monocarboxylic acids and microbial fermentation products of dietary fibers that are almost completely assimilated by the host.[15] The major SCFAs are acetate, propionate, and butyrate, collectively present at 100 mM concentration in the colon at an approximate ratio of 60:25:15, respectively.[16] Once assimilated, SCFAs maintain a healthy blood concentration of 22–42 μM (acetate), 0.9–1.2 μM (propionate), and 0.3–1.5 μM (butyrate) that have ready access to the BBB for modulation and signaling.[17] All studies so far, point toward a favorable effect of SCFAs on mammalian cells. For example, butyrate and propionate are known nuclear factor kappa B (NF-κB) inhibitors, thereby ameliorating neuroinflammation in aging mice.[18] The SCFAs achieve this by inhibiting the enzyme histone deacetylase (HDAC). Within the nucleus of a cell, the DNA coils around the histone proteins and acetylation and deacetylation of histones are one major step in regulation of gene expression. In PD, schizophrenia and depression, excessive levels and activation of HDAC is seen,[19,20] and its inhibition has shown positive prognosis in several CNS diseases including dementia, autoimmune encephalitis, and general improvement in cognitive functions.[21,22] However, there are other reports showing exacerbation of autism symptoms due to high brain propionate levels and this can be reversed by enhancing butyrate levels in the brain.[23] This shows that relative proportions of individual components of microbial SCFAs are important while evaluating their positive effects in brain function and BBB integrity. Apart from the activities mentioned above, SCFAs are also known to signal through the CNS to cause systemic effects. SCFAs activate specific G protein-coupled receptors (GPCRs), GPR41 and GPR43 (also known as free fatty acid receptors; FFAR 2 and 3, respectively) on the sympathetic ganglia, increasing the sympathetic outflow and resulting in optimal energy utilization in the body.[24] Interestingly, GPR41 is abundantly expressed in CNS and this is the only function it is known to perform. This study links high fiber intake and by association, enhanced SCFA flux in the host to favorable energy metabolism in the context of improved basal metabolic rate and control of obesity.

Bacterial Trimethylamines

Several bacterial species, predominantly from Firmicutes and Proteobacterial phyla can convert dietary choline and phosphatidylcholine into trimethylamine (TMA). Upon assimilation, host liver enzymes (flavin-containing monooxygenases; FMOs) convert TMA to trimethylamine-N-oxide (TMAO),[25] which is capable of crossing the BBB.[26] TMA has beneficial as well as detrimental effects depending on the general physiology of an individual. One of the beneficial effects is a partial or significant reduction of diet-induced impaired glucose tolerance. This feeds into AD and dementia, where obesity, impaired glucose tolerance, and diabetes are the major pre-disposing factors. Increased TMAO in circulation sets up a cascade, where it reduces endoplasmic reticulum (ER) stress in adipocytes, thereby reducing lipogenesis. This leads to increased insulin secretion in the pancreas and optimal energy utilization in the body.[5,27] Additionally, in AD, increased TMAO results in proper microtubule assembly (inability of mutant tau proteins to do this, in addition to neuronal death are the major pathological conditions in AD).[28]

There are several reports of adverse effects of TMAO including the risk of developing cardiovascular diseases and atherosclerosis since TMAO is known to modulate cholesterol metabolism.[29] Since cardiovascular disease is a strong risk factor for dementia and AD, further studies are required to understand the dynamics of TMAO for each underlying condition to maximize the benefits and mitigate the adverse effects on brain/CNS morbidities.

Lactic Acid

Increased levels of D-lactate producing microbes are seen in patients with intestinal hyperpermeability and bacterial translocation across the gastric mucosa. These are primarily patients that have undergone abdominal surgery or have underlying conditions that induce a "leaky gut." Enhanced serum level of D-lactate is a hallmark of patients with chronic fatigue syndrome (CFS), small intestine bacterial overload (SIBO) and neurocognitive disorder.[30] Reports suggest that in almost all cases, increased serum levels of D-lactate is associated with myalgic encephalomyelitis/CFS.[31] The remedial measures are often therapies that reduce gut barrier compromise and carefully designed prebiotic or probiotic cocktails that preferentially suppress D-lactate producing microbial flora.[1]

Amino Acids and Derivatives

The gut microbiota has a significant metabolic contribution to both excitatory and inhibitory arms of neurostimulation in the host. Microbial metabolites either feed into pathways that produce these neurotransmitters, or they produce the neurotransmitters directly for assimilation by the host. One of the prime examples for direct neurotransmitter production is the gut microbial synthesis of gamma-aminobutyric acid (GABA), a major inhibitory neurotransmitter in mammalian CNS, from diet or host-derived amino acid glutamate.[32] Several health benefits of GABA include optimization of growth hormone secretion, prevention of acute renal failure, and controlled hypertension, to name a few.

Another major amino acid, tryptophan is metabolized at tandem by microbes and the host to synthesize a plethora of neurotransmitters that signal within the brain and CNS. Members of the phylum Actinobacteria, especially genus *Bifidobacterium*, increase tryptophan levels within host serum, potentially mitigating depression in a rat model.[33] Other microbes, like *Clostridia* can convert dietary tryptophan to tryptamine, a neurotransmitter that signals the release of serotonin in the enteric nervous system. Interestingly, this conversion happens outside the bacterial cell due to secreted decarboxylase enzymes acting on diet-derived tryptophan.[34] Tryptophan entering the bacterial cells, on the other hand, get metabolized into a plethora of compounds with known beneficial and adverse effects on BBB and brain. Prime examples are quinolinic acid, a neurotoxic metabolite, and kynurenic acid, which is neuroprotective and anti-inflammatory metabolite with systemic beneficial effects.[5]

Ammonia

Bacterial ureases utilize urea in the gut and convert it into ammonia, which is readily absorbed across the gastric mucosa. In healthy individuals, the liver processes this acquired ammonia and clears it through various mechanisms.[35] In individuals with cirrhotic liver, ammonia escapes hepatic processing, increasing its circulatory levels, resulting in hepatic encephalopathy (HE) and also exposing BBB to toxic levels of ammonia.[36] This leads to a "decoupling" of BMEC tight junctions, compromising the integrity of BBB. A "leaky" BMEC layer allows aberrant intake of macromolecules, including aromatic amino acids (AAA), reducing the synthesis of cerebral serotonin and dopamine (and instead producing octopamine and phenylethylamine), resulting in neurotoxicity and cognitive dysfunction.[37]

Vitamins

Most vitamins, an essential micronutrient, cannot be produced by humans and have to be obtained from the diet and commensal microbial metabolism. The levels of vitamins are tightly regulated in our body and both excess and deficiency result in severe pathological conditions. Compared to other nutritional components (carbohydrates, proteins, fat, etc.), the discovery of vitamins and the understanding of their importance, is relatively recent (late 19th to early 20th century).[38] Imbalance in certain key vitamin levels within the CNS is associated with mild to severe neurocognitive disorders. For example, vitamin K deficiency is a known risk factor in both AD and PD, while excess vitamin K is correlated with mild memory lapses in the elderly.[39,40] Vitamin K is produced by several bacteria of Phyla Firmicutes, Bacteriodetes and Proteobacteria. In addition to vitamin K, these bacteria also produce vitamin B2, vitamin B9, and vitamin B12 and their assimilation primarily occurs in the colon, in contrast with dietary vitamins that are predominantly absorbed in small intestine.[41] Like vitamin K, vitamins B6/B9/B12 also influence the synthesis of homocysteine within the CNS and perturbations in their levels within brain significantly influence neurotoxicity and cognitive disorders.[42]

Microbial Metabolites, Immunity, and Brain/CNS Function

Before reaching the BBB, bacterial metabolites influence the proximal mucosa and the resident immune cells in the GI surface. Several mechanisms have been described how these metabolites may alter the activation states of these immune cells, which can then migrate to the brain/CNS and contribute to neurotoxicity. A few scenarios are presented here.

(A) Any disease or insult that compromises the integrity of the gut mucosal barrier, results in passive diffusion of gut microbes and microbial products, triggering the innate immune system. For example, opioid use or HIV infection, result in gut barrier compromise and cause bacterial translocation and gastric inflammation by triggering toll-like receptor (TLR) activation.[43,44] Activated TLR signaling produces a plethora of inflammatory cytokines and chemokines that enter circulation and result in systemic inflammation. Opioids, through a micro-RNA-mediated mechanism, do not allow a timely resolution of this inflammation (although the gut barrier repairs itself after a while).[45] As a result of this persistent inflammation, there is a compromise in BBB integrity leading to neurotoxicity and cognitive deficits.[46]

(B) Another scenario is where microbial metabolites alter the enteric nervous system by either activating it or by altering its response to natural ligands, thereby disrupting its communication with CNS. For example, as mentioned earlier, tryptamine (produced by microbial decarboxylation of tryptophan) induces the secretion of serotonin, stimulating GI motility and neurogenesis.[47] At the same time, TLR activation and pro-inflammatory cytokine release acts on the enteric neurons and modulate the vagus nerve signaling to brain.[5] Innervation or loss of enteric neuron functionality eventually leads to disruption or complete loss of gut–brain neuronal communications. This, in addition to direct effects mentioned above (e.g., GABA production and action of other metabolites) make the gut microbiome and its metabolites an integral component of gut and brain/CNS health.

(C) The third scenario is where activated immune cells, with a backdrop of systemic inflammation mentioned above, cross BBB and foster neuroinflammation. Circulating pro-inflammatory cytokines compromise BBB integrity by down-regulating tight-junction proteins at the BMEC layer. At the same time, adhesion molecules and matrix metalloproteases are up-regulated, facilitating the extravasation of activated immune cells from the systemic circulation into the privileged brain/CNS compartment.[48]

As evident from the scenarios presented above, microbial metabolites (and metabolites modified by microbial metabolism) interact with host cellular components to foster a variety of responses that can be beneficial or adverse for the host (Table 7.1). It is also clear that the beneficial metabolites mostly

Table 7.1: Major bacterial metabolites affecting host metabolic pathways in neurocognitive disorders

Metabolite	Description	Reference
Short chain fatty acids (acetate, Butyrate and propionate)	Bacterial fermentation product of dietary fibers. NF-κB and HDAC inhibitor. Improves neurocognition.	16–20
TMA	Bacterial conversion product of dietary choline/phosphatidylcholine. Converts to TMAO in host and crosses BBB. Both beneficial and detrimental effects on CNS.	25, 26, 28, 29
Lactate	Largely associated with compromised mucosal barrier. Associated with increased SIBO, CFS and neurocognitive disorders.	30
GABA	Inhibitory neurotransmitter produced directly by gut bacteria from dietary or host-derived glutamate.	32
Tryptophan	Gut bacteria convert tryptophan to various neurotranmitters like tryptamine, quinolinic acid and kyenurenic acid which have neurocognitive effects.	5, 34
Ammonia	Bacterial conversion product of gut urea.	37
Vitamins	Gut bacteria-derived vitamins (K, B6, B9, B12) get assimilated in the colon and modulate homocysteine synthesis in the CNS.	41, 42
Lipopolysaccharides and Lipoteichoic acid	Bacterial cell membrane/wall products that trigger innate immune response through TLRs. Always associated with gut barrier compromise.	44
Serotonin	Induced by increased production of Tryptamine by gut bacteria. Induces gut motility and neurogenesis.	47

mitigate the adverse effects of the others. Since manipulation of the microbiome is minimally invasive, this provides a powerful tool to develop strategies to mitigate brain/CNS disorders or to aid therapies for the diseases.

CONCLUSIONS

In the past 15–20 years, intensive research has gone into understanding the molecular basis of host-microbe association with particular emphasis on the gut microbiome and its metabolic activity. The influence of gut microbial metabolites on brain function in health and disease is a relatively new area of research (Fig. 7.1). This is despite the fact that this association was known and postulated for the better part of the last century. The availability and implementation of sensitive and high-throughput state-of-the-art techniques and validation in human subjects and small animal models have conclusively established a direct line between a healthy homeostatic microbiome and brain/CNS health at the molecular level. As we study diseases using genetic associations, transcriptomic changes, proteomics, and enter the realm of metabolites, there is exponential increase in complexity with each stage. The behavior of cells (constituents of BBB, immune cells, neurons and gastric epithelial, glial and enteric cells) are well characterized in many neurologic diseases, the big picture is slowly becoming clear as we learn more about the microbiome and its contribution to these phenomena. The microbiota has a dual character of being a tightly regulated ecosystem within an organism, and for most practical purposes, it behaves like an organ. Once we have a better understanding of its internal molecular dynamics, modulation of the microbiome will be a promising tool to manage morbidities and aiding therapy in a variety of human diseases, including those of the brain and CNS.

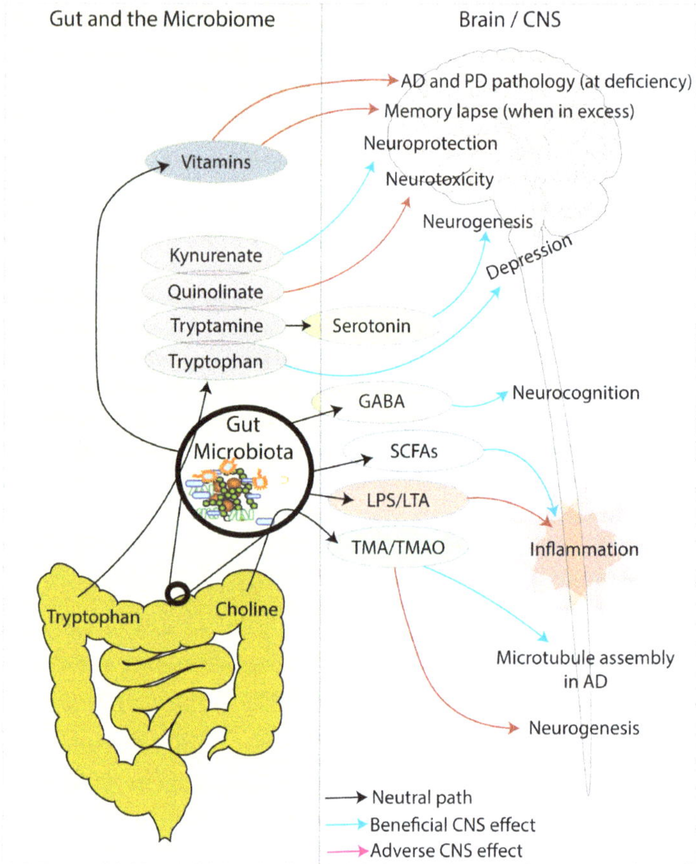

Fig. 7.1: Microbial metabolites and its influence on the brain/central nervous system (CNS) pathology. Gut microbiome produce and modify diet- or host-derived metabolites that are known to influence the Brain/CNS pathways in health and disease (also see Table 7.1). In most cases, these metabolites can have either adversarial or beneficial effects, depending on their absolute concentration within the CNS, as well as their relative concentration compared to other metabolites and neurotransmitters.

REFERENCES

1. Galland L. (2014) The gut microbiome and the brain. *J Med Food* **17**(12): 1261–1272.
2. Qin J, *et al.* (2010) A human gut microbial gene catalogue established by metagenomic sequencing. *Nature* **464**(7285): 59–65.
3. Jacobs JP, Braun J. (2014) Immune and genetic gardening of the intestinal microbiome. *FEBS Lett* **588**(22): 4102–4111.

4. Wikoff WR, *et al.* (2009) Metabolomics analysis reveals large effects of gut microflora on mammalian blood metabolites. *Proc Natl Acad Sci USA* **106**(10): 3698–3703.
5. Parker A, Fonseca S, Carding SR. (2020) Gut microbes and metabolites as modulators of blood-brain barrier integrity and brain health. *Gut Microbes* **11**(2): 135–157.
6. Martinez KB, Leone V, Chang EB. (2017) Microbial metabolites in health and disease: Navigating the unknown in search of function. *J Biol Chem* **292**(21): 8553–8559.
7. Martin CR, Osadchiy V, Kalani A, Mayer EA. (2018) The Brain-Gut-Microbiome axis. *Cell Mol Gastroentero Hepatol* **6**(2): 133–148.
8. Hooper LV. (2004) Bacterial contributions to mammalian gut development. *Trends Microbiol* **12**(3): 129–134.
9. Jacobsen UP, *et al.* (2013) The chemical interactome space between the human host and the genetically defined gut metabotypes. *ISME J* **7**(4): 730–742.
10. Morrison DJ, Preston T. (2016) Formation of short chain fatty acids by the gut microbiota and their impact on human metabolism. *Gut Microbes* **7**(3): 189–200.
11. Dawson PA, Karpen SJ. (2015) Intestinal transport and metabolism of bile acids. *J Lipid Res* **56**(6): 1085–1099.
12. Engelhardt B, Vajkoczy P, Weller RO. (2017) The movers and shapers in immune privilege of the CNS. *Nat Immunol* **18**(2): 123–131.
13. Hawkins BT, Davis TP. (2005) The blood-brain barrier/neurovascular unit in health and disease. *Pharmacol Rev* **57**(2): 173–185.
14. Abbott NJ, Rönnbäck L, Hansson E. (2006) Astrocyte-endothelial interactions at the blood-brain barrier. *Nat Rev Neurosci* **7**(1): 41–53.
15. Salminen S, *et al.* (1998) Functional food science and gastrointestinal physiology and function. *Br J Nutr* **80**(Supp l1): S147–S166.
16. Ganapathy V, Thangaraju M, Prasad PD, Martin PM, Singh N. (2013) Transporters and receptors for short-chain fatty acids as the molecular link between colonic bacteria and the host. *Curr Opin Pharmacol* **13**(6): 869–874.
17. Hoyles L, *et al.* (2018) Microbiome–host systems interactions: Protective effects of propionate upon the blood–brain barrier. *Microbiome* **6**(55): 1–13.

18. Matt SM, Allen JM, Lawson MA, Mailing LJ, Johnson RW. (2018,) Butyrate and dietary soluble fiber improve neuroinflammation associated with aging in mice. *Front Immounol* **9**: 1832.
19. Mahgoub M, Monteggia LM. (2013) Epigenetics and psychiatry. *Neurotherapeutics* **10**: 734–741.
20. Harrison IF, Dexter DT. (2013) Epigenetic targeting of histone deacetylase: Therapeutic potential in Parkinson's disease? *Pharmacol Ther* **140**(1): 34–52.
21. Konsoula Z, Barile FA. (2012) Epigenetic histone acetylation and deacetylation mechanisms in experimental models of neurodegenerative disorders. *J Pharmacol Toxicol Methods* **66**(3): 215–220.
22. Graff J, Tsai L. (2013) The potential of HDAC inhibitors as cognitive enhancers. *Annu Rev Pharmacol Toxicol* **53**: 311–330.
23. Macfabe DF, *et al*. (2007) Neurobiological effects of intraventricular propionic acid in rats: Possible role of short chain fatty acids on the pathogenesis and characteristics of autism spectrum disorders. *Behav Brain Res* **176**: 149–169.
24. Kimura I, Inoue D, Maeda T, Hara T, Ichimura A, Miyauchi S. (2011) Short-chain fatty acids and ketones directly regulate sympathetic nervous system via G protein-coupled 41 (GPR41). *Proc Natl Acad Sci USA* **108**(19): 8030–8035.
25. Bennett B.J, *et al*. (2013) Trimethylamine-N-Oxide, a metabolite associated with atherosclerosis, exhibits complex genetic and dietary regulation. *Cell Metab* **17**(1): 49–60.
26. Del D, *et al*. (2017) The Gut Microbial metabolite trimethylamine-N-oxide is present in human cerebrospinal fluid. *Nutrients* **9**: 1053.
27. Lupachyk S, Watcho P, Stavniichuk R, Shevalye H, Obrosova IG. (2013) Endoplasmic reticulum stress plays a key role in the pathogenesis of diabetic peripheral neuropathy. *Diabetes* **62**(3): 944–952.
28. Tseng H, Graves DJ. (1998) Natural methylamine osmolytes, trimethylamine N-oxide and betaine, increase tau-induced polymerization of microtubules. *Biochem Biophys Res Commun* **250**: 726–730.
29. Koeth RA, *et al*. (2013) Intestinal microbiota metabolism of L-carnitine, a nutrient in red meat, promotes atherosclerosis. *Nat Med* **19**(5): 576–585.

30. Sheedy JR, *et al.* (2009) Increased D-Lactic acid intestinal bacteria in patients with chronic fatigue syndrome. *In Vivo (Brooklyn)* **23**: 621–628.
31. Wallis A, Ball M, Mckechnie S, Butt H, Lewis DP, Bruck D. (2017) Examining clinical similarities between myalgic encephalomyelitis/chronic fatigue syndrome and D-lactic acidosis: A systematic review. *J Transl Med:* **15**(1): 129.
32. Barrett E, Ross RP, Toole PWO, Fitzgerald GF, Stanton C. (2012) γ-Aminobutyric acid production by culturable bacteria from the human intestine. *J Appl Microbiol* **113**(2): 411–417.
33. Desbonnet L, Garrett L, Clarke G, Bienenstock J, Dinan TG. (2009) The probiotic Bifidobacteria infantis: An assessment of potential antidepressant properties in the rat. *J Psychiatr Res* **43**(2): 164–174.
34. Williams BB, *et al.* (2014) Article discovery and characterization of gut microbiota decarboxylases that can produce the neurotransmitter tryptamine. *Cell Host Microbe* **16**(4): 495–503.
35. Olde Damink SWM, Deutz NEP, Dejong CHC, Soeters PB, Jalan R. (2002) Interorgan ammonia metabolism in liver failure. *Nerochem Int* **41**(2–3): 177–188.
36. Qureshi MO, Khokhar N, Shafqat F. (2014) Ammonia levels and the severity of hepatic encephalopathy. *J Coll Physicians Surg Pak* **24**(3): 160–163.
37. Skowronska M, Albrecht J. (2012) Alterations of blood brain barrier function in hyperammonemia: An overview. *Neurotoxin Res* **21**: 236–244.
38. Semba RD. (2012) The discovery of the vitamins. *Int J Vitam Nutr Res* **82**(5): 310–315.
39. Luna F, *et al.* (2013) Vitamins K interact with N-terminus α-synuclein and modulate the protein fibrillization in vitro. Exploring the interaction between quinones and α-synuclein. *Neurochem Int* **62**(1): 103–112.
40. Soutif-veillon A, *et al.* (2016) Increased dietary vitamin K intake is associated with less severe subjective memory complaint among older adults. *Maturitas* **93**: 131–136.
41. Leblanc JG, Milani C, De Giori GS, Sesma F, Van Sinderen D, Ventura M. (2013) Bacteria as vitamin suppliers to their host: A gut microbiota perspective. *Curr Opin Biotechnol* **24**(2): 160–168.
42. Kennedy DO. (2016) B Vitamins and the brain: Mechanisms, dose and efficacy — A review. *Nutrients* **8**: 1–29.

43. Meng J, *et al.* (2013) Morphine induces bacterial translocation in mice by compromising intestinal barrier function in a TLR-dependent manner. *PLoS One* **8**(1): e54040.
44. Meng J, Banerjee S, Zhang L, Sindberg G, Ramakrishnan S, Roy S. (2020) Opioids impair intestinal epithelial repair in HIV-infected humanized mice. *Front Immunol* **10**: 1–14.
45. Banerjee S, *et al.* (2013) Morphine induced exacerbation of sepsis is mediated by tempering endotoxin tolerance through modulation of miR-146a. *Sci Rep* **3**: 1977.
46. Dutta R, *et al.* (2012) Morphine modulation of toll-like receptors in microglial cells potentiates neuropathogenesis in a HIV-1 model of coinfection with pneumococcal pneumoniae. *J Neurosci* **32**(29): 9917–9930.
47. De Vadder F, Grasset E, Mannerås L, Karsenty G, Macpherson AJ. (2018) Gut microbiota regulates maturation of the adult enteric nervous system via enteric serotonin networks. *Proc Natl Acad Sci USA* **115**: 6458–6463.
48. Herisson F, *et al.* (2018) Direct vascular channels connect skull bone marrow and the brain surface enabling myeloid cell migration. *Nat Neurosci* **21**(9): 1209–1217.

Chapter 8

Modulation of Gut–Brain Axis Through Prebiotics and Psychobiotics

Hannah Marwede[1,*], Timothy Yassa[1,*], Elijah Horesh[1], Jeenu Mittal[1], Rahul Sinha[1], Aditi Kumar[1] and Sudhir Aggarwal[2,3]

Abstract

Prebiotics are naturally occurring, non-digestible food components that have been associated with the growth of beneficial microbes in the gut. Psychobiotics are live microbes that confer neurological and gastrointestinal health benefits when ingested in adequate amounts. Prebiotics and psychobiotics have wide ranging spread effects varying from emotional to cognitive to systemic effects, and have complex mechanisms of action. The existing body of evidence suggests that both prebiotics and psychobiotics modulate the gut–brain axis and have implications in health and disease. It is imperative to understand the molecular mechanisms

[1]Hearing Research and Communication Disorders Laboratory, Department of Otolaryngology, University of Miami-Miller School of Medicine, Miami, FL, USA.

[2]Medical Director Lower, Merion Neurology, 250 W Lancaster Ave, Suite 220, Paoli, PA 19301, Paoli, PA, USA.

[3]Assistant Professor Department of Neurology, Lewis Katz School of Medicine, Temple University, Philadelphia, PA, USA.

*These authors contributed equally to this work.

Corresponding Author: Sudhir Aggarwal, MD, PhD, Medical Director, Lower Merion Neurology, Paoli, PA, USA; Assistant Professor Department of Neurology, Lewis Katz School of Medicine, Temple University, Philadelphia, PA, USA, Telephone: 4844132572. E-mail: saggarwal@lowermerionneurology.com.

by which they modulate the gut–brain axis, and this will, in turn, help us utilize their beneficial effects by developing therapeutic approaches. This chapter discusses the potential of prebiotics and psychobiotics to enhance functionality of the gut–brain axis.

INTRODUCTION

The gut–brain axis is a bidirectional communication between the central and enteric nervous systems. The means of communication between the central and gastrointestinal systems involve links between neural, endocrine, immune, and humoral subsystems.[1,2] A bidirectional communication implies that changes in the gut microenvironment may result in changes in the central nervous system, and vice versa. The imbalances along the gut–brain axis can have significant physiological and pathological ramifications. It has become evident that the microenvironment of the gastrointestinal system is central to an individual's overall health, well-being, and immune function. Therefore, the effect of the gut microenvironment on a person's immune, mental, and physical health is an important subject in current research.[3–8]

There exist colonies of microbes in the gastrointestinal system, predominantly in the large intestine. The variety and quantity of these bacteria (gut microbiota) are numerous. However, the distribution of microbiota in the gastrointestinal tract is unique for each individual.[9] Studies have shown that healthy individuals have microbial phylotype overlaps, indicating that specific types of microbes in the gut may be critical to a person's health.[10–12] Firmicutes and Bacterioidetes, for example, make up 75% of the most prominent phyla in these healthy individuals. These bacteria, among others, are crucial to the digestion of foods we consume and, therefore, are crucial to the functioning of our numerous physiological systems. Previous research has shown that a person's lifestyle — from the type of food they eat, to their mode of delivery (vaginal versus cesarean section) — all have an effect on the quality of their gut microbiota. Current research is now suggesting that the microbiota in the gut play a crucial role in health and disease, rather than merely facilitating in the digestion of food. Gut microbiota play a major role in mental and emotional health as well as host immunity.[13,14] These findings have generated a great deal of interest in research on modulating the microbiome–gut–brain axis through supplementation with prebiotics and psychobiotics that provide health benefits. Prebiotics are naturally occurring, non-digestible food components, whereas psychobiotics include live bacteria that are capable of affecting the gut–brain axis. This chapter will aim to address the influence of prebiotics and psychobiotics on the development of the brain and behavior as well as highlight their potential to be one of the major frontiers for modulating the gut–brain axis.

Psychobiotics, Gut Health, and Mood

As a popular self-enacted medical intervention, people attempt to improve their digestive health by ingesting probiotics through means such as pills or cultured yogurt. These attempts are not in vain. However, there is more to know and explore about these "probiotics." In fact, probiotics belong to a larger class of microbes called psychobiotics, which are currently a subject of profound interest and investigation for their effects on the gut–brain axis. The psychobiotics have been defined as a broader class of ingestible microbes, including probiotics.[15–17] Prebiotics are defined as selectively fermented ingredients that result in specific changes, in the composition and/or activity in the gastrointestinal microbiota, thus conferring benefit(s) upon host health.[18]

Psychobiotics are capable of impacting the gut microbiota and spurring health benefits to patients that suffer from psychiatric illnesses, by modulating the gut–brain axis.[19] Thus, psychobiotics affect higher order brain activity patterns, and have the potential to treat various psychiatric disorders such as anxiety and depression. The mechanisms by which bacteria in the gut exert these beneficial effects include affecting tryptophan metabolism as well as producing neurotransmitters such as 5-hydroxytryptamine (5-HT), dopamine, serotonin, acetylcholine, gamma-aminobutyric acid (GABA), adrenaline, and kynurenine.[20–22]

Certain psychobiotics possess anti-depressant and anxiolytic properties[17] allowing them to suppress detrimental cortical and hormonal reactions associated with homeostatic imbalances. Psychobiotics are capable of delivering neurotransmitters such as GABA and serotonin to the central nervous system synapses.[23,24] These molecules are essential for mental and emotional health, and are linked with the gut–brain axis. Psychobiotics have been shown to affect the psychophysiological markers of anxiety and depression in rodent and human models.[25,26] In a study, healthy adult male BALB/c mice that were anxious and sensitive to stress, but otherwise normal, were fed probiotic *Lactobacillus rhamnosus* JB-1.[23] It was observed that the mice fed with the probiotic showed a reduced expression of GABA mRNA in the amygdala and hippocampus, along with an increase in GABA mRNA expression in the prelimbic and cingulate regions as compared with the control group. These findings demonstrate the effect of the probiotics in modulating the regional neuroexcitation in the brain, thus regulating and reducing anxiety and depressive behaviors. In the same study, mice fed with another probiotic *Mycobacterium vaccae* showed reduced anxiety levels in a maze-learning task compared to the controls. In another study, *Bifidobacterium infantis* was fed to Sprague-Dawley rats and their stress responses, inflammation levels, and monoaminergic activity in a forced swim test were assessed.[20] When compared with controls, the probiotic did not result in any behavioral changes in the performance of the swim test among those rats. However, there was a significant increase in the serotonin precursor, tryptophan, in the plasma of rats fed with the probiotic. Serotonin is known to be a mood-modulating neurotransmitter, and an increase in its levels is associated with a decrease in depression and anxiety. The probiotic-fed rats also showed reduced concentrations of pro-inflammatory cytokines, another indicator of decreased physiological stress due to the consumption of a psychobiotic.

The psychophysiological effects of prebiotics have also been an area of interest for some investigators. Galacto-oligosaccharides (GOS) and fructo-oligosaccharides (FOS) are prebiotics that are a source of food for *Bifidobacteria* and *Lactobacteria* as well as are capable of spurring the spread of probiotics in the gut. In a study, GOS and FOS were fed to adult male Sprague-Dawley rats.[27] Compared with the controls, the rats that consumed these prebiotics demonstrated increased expression of brain-derived neurotrophic factor (BDNF) and N-methyl-D-aspartate receptor (NMDAR) subunits in the hippocampus. These findings suggest an increased hippocampal memory formation induced by prebiotics.

Besides prebiotics, traditional fermented milk beverages, such as kefir, have been shown to provide health benefits by modulating the microbiome–gut–brain axis.[28–31] Kefir is prepared by adding to milk a kefir grain that consists of exopolysaccharide matrices harboring symbiotic microbial communities, including bacteria and yeasts, leading to the fermentation of the milk. Kefir supplementation in experimental animal models has been associated with anti-inflammatory effects,[32,33] reduction of high-fat diet-induced obesity in mice,[34–36] and decreased hypertension in spontaneously hypertensive rats.[37] A recent study demonstrated that supplementation of mice with two distinct kefirs (Fr1 and UK4) led to improvements in repetitive and reward-associated behaviors.[38] UK4 promoted the predominance of Treg cells and an increase in interleukin-10 (IL-10) levels, whereas Fr1 prevented stress-induced enhancement in neutrophil as well as Chemokine (C-X-C motif) ligand 1 (CXCL1) levels. Both kefirs significantly changed the composition and functional capacity of the host microbiota in terms of production of neurotransmitters such as GABA, therefore, modulating the gut–brain axis. Similar beneficial effects of kefir and other probiotics have been observed on mood in human subjects. Kefir supplementation has also been associated with a reduction in bloating and an improvement in mood in patients having gastrointestinal disorders such as inflammatory bowel disease (IBD).[39] In another study, human participants who consumed a mixture of probiotics showed a significant decline in self-reported negative mood and distress as compared to the control group.[40] The participants who consumed the probiotics also showed a decreased level of urinary free cortisol indicating reduced stress, thus suggesting the importance of the gut–brain axis. These findings suggest that psychobiotic treatments are capable of providing significant benefit from several disorders, both mental and physical.

In a study on gut health-promoting bacteria, *Bifidobacterium infantis*, a probiotic with therapeutic applications, was investigated for its effect on the gut–brain axis. These bacteria are known to have the ability to ameliorate symptoms of depression and chronic fatigue syndrome. In this double-blind study involving healthy, middle-aged woman volunteers, the effect of bidaily ingestion of a fermented milk product (FMPP) was compared to a non-fermented milk product and a non-dairy product over a course of 4 weeks. FMPP contained *Bifidobacterium animalis* ssp. *lactis*, *Streptococcus thermophiles*, *Lactobacillus bulgaricus*, and *Lactococcus lactis* ssp. *lactis*, and these bacteria are considered to be gut health-promoting bacteria. Before and after the 4-week FMPP (or control) ingestion, each subject underwent a functional magnetic resonance imaging (fMRI) to measure the brain's response to emotional stimuli. It was found that FMPP intake was associated with reduced reactivity to an emotional attention task in the peri-aqueductal gray, prefrontal cortex, precuneus, basal ganglia, parahippocampal gyrus, insular cortex and the primary somatosensory cortex. These are the regions of the brain that control the awareness of internal body sensations (interoceptive) and external body sensations (somatosensory). In summary, the findings of these studies demonstrate that psychobiotics, by their action on the gut microbiota, affect not only the gut but also other physiological processes in our body, including senses, immunity and emotions.

Depression is associated with the presence of biomarkers of inflammation including increased IL-6, tumor necrosis factor alpha (TNF-α), and C-reactive protein (the acute phase reactant). Stress and anxiety result in elevated levels of these pro-inflammatory cytokines.[41] The reduction in the levels of these biomarkers and the resulting reduction in inflammation with the ingestion of psychobiotics shows promise for the use of psychobiotics as an effective treatment for an array of mood disorders. While IL-6 is pro-inflammatory, and is associated with depression, IL-10 has an anti-inflammatory effect. Previous studies have suggested that plasma synthesis of IL-10 is induced by the ingestion of commensal bacteria in specific pathogen-free (SPF) rats.[24] These observations suggest that gut flora is capable of directly combating the physical symptoms of depression by virtue of the anti-inflammatory and anti-nociceptive properties of IL-10. However, the effect of gut flora on the behavioral symptoms of depression remains unknown. Although our understanding of these processes is evolving, it can be reasonably expected that these findings will lay the foundation for developing effective interventions for behavior and mood disorders in the future.

Serotonin is another neurotransmitter that modulates well-being and happiness in the brain. It has been found in significantly higher amounts in conventional mice compared to germ-free (GF) mice (mice without microorganisms especially those found in the gut). Ninety percent of the body's serotonin is produced in the gut. Studies have suggested that some of the gut microbes including *Escherichia*, *Bacillus* and *Saccharomyces spp*. help in the production of serotonin.[42,43] While the role of serotonin in the pathophysiology of depression is widely known, recent research has led to a clearer explanation of the correlation between low serotonin levels and depression.[44] It has been observed that patients with severe forms of depression often have low plasma levels of the amino acid tryptophan, a precursor of serotonin. Accelerated breakdown of tryptophan is also known to be associated with inflammation. Therefore, in vulnerable individuals, inflammation could result in lower plasma tryptophan levels, resulting in diminished serotonin availability and activity in the brain, consequently causing depression.[44] Therefore, the role of some of psychobiotics in increasing both tryptophan and serotonin levels can be explored to develop effective treatment modalities for mood disorders.

Microbiota and Hypothalamic–Pituitary–Adrenal Axis

The hypothalamic–pituitary–adrenal (HPA) axis is a neuronal network that regulates the interplay between the central nervous system (hypothalamus) and the endocrine system (pituitary and adrenal glands). Both environmental stress and elevated levels of inflammatory cytokines activate the HPA axis, resulting in the release of corticotropin-releasing factor (CRF) from the hypothalamus and adrenocorticotropic hormone (ACTH) from the pituitary (Fig. 8.1). These cascades result in the release of the body's main stress hormone, cortisol, which is found in elevated levels in many patients with depression.[45] The HPA axis can be considered

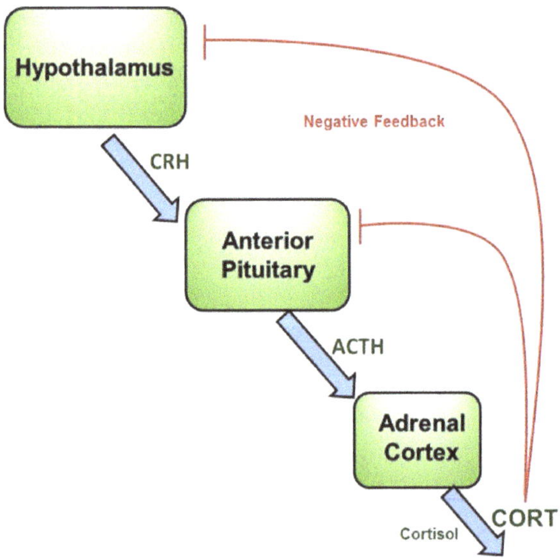

Fig. 8.1: Schematic representation of hypothalamic–pituitary–adrenal (HPA) axis.

the main efferent stress axis of the body, coordinating the internal responses of the body to stressors including regulation of temperature, digestion, immune function, mood and sexuality.[45–47] This system is closely associated with physical conditions such as chronic fatigue syndrome and irritable bowel syndrome (IBS), and mental disorders including attention deficit disorder (ADD), post-traumatic stress disorder (PTSD), anxiety, and major depressive disorder. Therefore, it may be possible to target the HPA axis through the modulation of the gut microbiome in order to discover novel treatments for mood disorders.[48]

Multiple studies have shown that patients with anxiety and depressive disorders have altered (increased or decreased) responsiveness of the HPA axis.[49–53] Studies have shown that depressed patients have a higher HPA axis activation compared to normal individuals.[54] Elevated cortisol and corticotropin-releasing hormone (CRH) levels, a blunted ACTH response to CRH, and a reduced hippocampal volume are all frequently found in depressed patients. Thus, HPA axis activation is associated with increased relapse rates of depression, and sometimes even suicide, and could have a prognostic value in these conditions.[55]

Recent studies have shown that psychobiotics have the ability to combat a hyperactive stress response through the suppression of the HPA axis, and thus have the potential to ameliorate the detrimental effects of anxiety and depression. Known activators of the HPA axis include psychological stress, hyperactive amygdala, noradrenergic neurons in the brainstem (nucleus of tractus solitarius), pro-inflammatory cytokines, prostaglandins, glutamate, norepinephrine, 5-HT, and glucagon-like peptide 1.[56,57] It has been observed that patients suffering from major depression have increased activity of the glutamatergic system, as well as increased levels of pro-inflammatory cytokines including IL-2.[57] This provides an obvious target for the treatment of depression. As the HPA axis is activated by secondary by-products of itself, if there is a way to decrease the levels of these activators of the HPA axis, this should cause the entire cycle to stop, thus helping to treat the mood disorders. To explore this hypothesis, a probiotic *Bifidobacterium pseudocatenulatum* was administered to mice starting 2 days after birth (post-natal day 2) and they were subjected to the stress of maternal separation.[58] It was observed that on post-natal day 41, the probiotic had completely eliminated any stress-induced increase in the levels of corticosterone in the stool as well as the catecholamine levels in the hypothalamus in these mice. Anxiety-like behavior was also decreased along with an observed decrease in stress-induced interferon-γ levels in the small intestine and stress-induced IL-18 levels in the serum of mice treated with probiotic. In another study, the same probiotic was found to attenuate neuroendocrine responses stemming from the administration of a high-fat diet in adult male mice.[59] This probiotic was also observed to affect levels of dopamine, norepinephrine, epinephrine and 5-HT in the brain and small intestine.

Besides down-regulating the activators of the HPA axis, alterations in the gut microbiota caused by psychobiotics have been shown to up-regulate the inhibitors of the HPA axis such as GABA and leptin. Furthermore, administration of a prebiotic mixture of FOS and GOS for 3 weeks in adult male mice resulted in reduced stress-induced corticosterone levels in the plasma, a reduction in anxiety and depressive behavior, an increased expression of the GABA receptor in the hippocampus, and an increase in the expression of 5-HT receptors in the prefrontal cortex.[60]

Studies involving microbiota-deficient animals, also known as germ-free (GF) animals, have provided novel insights into the role of microbiota in regulating development and function of the brain.[61] In GF mice, a mild restraint stress elicited an exaggerated release of corticosterone compared with specific-pathogen-free (SPF) control animals.[62] This stress response involving the HPA axis was completely normalized by monocolonization of GF mice with *Bifidobacterium infantis* in the neonatal period. However, when done at 6 weeks of age, monocolonization resulted in only a partial normalization of the stress response, and no response was observed at 8 weeks. These findings suggest that while the microbial composition of the gut is critical to the development of a proper stress response, the timing of these signals from gut to the brain is also critical for normal development of the HPA axis. Therefore, any abnormality in the development of the gut microbiome early in life may result in a predisposition to stress, anxiety and depressive disorders later in life. But as is evident from previous studies, there may be a ray of hope for correcting a flawed stress response by addressing the root cause, i.e. bacteria in the gut.

The gut–brain axis is a bidirectional communication between the microbiota of the gut and the brain.[63] Diet, antibiotics and probiotics can influence gut microbiome composition, and that in turn can affect mental health and behavior through the gut–brain neuroanatomical pathway, neuroendocrine-HPA axis, as well as secretion of cytokines and neurotransmitters (Fig. 8.2). In addition, in response to stress, the cortisol and other factors are released, which affect the immune function and, consequentially, the gut microbiome gene expression and signaling.[64] This alters

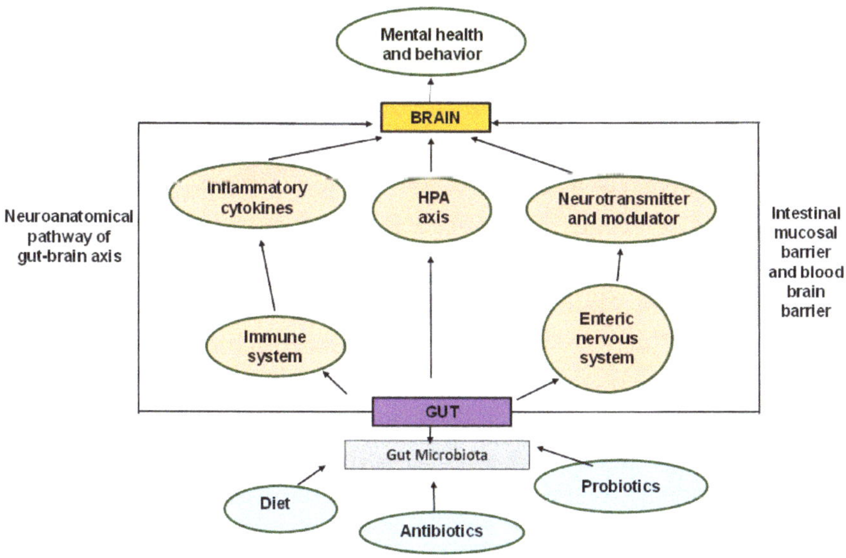

Fig. 8.2: Microbiome modulate gut–brain axis and determines predisposition to health and disease. (Adapted from Wang and Wang[65].)

the composition and activity of the microbiota in the gut. Hence, this communication between the gut and the brain is a two-way street of constant signaling and adjustment.

Vaginal Delivery versus C-Section and Healthy Gut Microbiome

As the gut–brain axis and its role in health and disease states continue to be investigated, significant leaps have been made to understand this axis. A number of clinical studies have been performed to understand the differences between babies born via either vaginal delivery or caesarean section (CS), and the effects of delivery method on the microbial composition of the baby's gut. Most of the gut microbiota develops in infancy, where vaginal delivery makes for initial exposure to maternal gut microbiota.[66–68] After this initial exposure, the baby soon develops a gut microbiome of its own. Babies born by CS develop a different gut microbiome as opposed to those born via vaginal delivery, and carry a greater long-term risk of immune disorders, obesity and mood disorders. A clinical study was performed to compare the quality of microbiota and the immune strength of infants born vaginally versus by CS.[69] Intestinal microbiota was sequenced in babies at the first week, and then at months 1, 3, 6, 12 and 24 post-delivery. The levels of Th1- and Th2-(T-helper cells 1 and 2) associated chemokines were also measured from venous blood samples at 6, 12 and 24 months. It was found that infants born vaginally had a higher overall microbiota diversity in the first 2 years of life, as well as higher levels of Th1-associated chemokines in their blood, indicating a greater immune strength compared with children born via CS and thus, unexposed to their mothers' microbiota.

The relationship of a vaginal birth with a stronger immune system and a more diverse gut microbiome leads to the likely conclusion that vaginal delivery will set up the newborn infant for a greater likelihood of health, both physically and mentally, as compared to their counterparts born via CS. While this view still needs validation with further large-scale studies, avoiding unnecessary CS may prove advantageous to newborns and, subsequently, future generations. According to the latest data from 150 countries, of all births, 18.6% occur by CS, with a range of 6%–27.2% in the least to the most developed regions, respectively.[70] This expensive surgical procedure is on the rise, especially in high-income countries, even though it had originally been used for emergency complications during delivery. There is a lack of evidence to support maternal and perinatal benefits from CS, and some studies even show a link between increasing CS rates and poor health outcomes.[71,72] While this certainly is an interesting area that warrants further research, if vaginal delivery is found to result in a stronger microbiota for the baby from day 1, and that in turn helps maintain a healthy psychological state and prevent mood disorders, this might be a good option to consider whenever possible.

Impact of Diet on Gut Microbiota

Diet, disease, infection, and environment play a major role in impacting gut health by influencing the gut microbiota.[73] The types of diet we consume such as a low fat, polysaccharide-rich diet versus high fat, high-sugar diet have a significant impact on metabolic pathways, which in turn affect the gut microbiota.

As is well known, a poor diet is associated with a multitude of health issues. The effects of a poor diet, traditionally high in fats and sugars, while low in nutrients and fiber, have been shown to alter the bacterial species in the gut.[74,75] When soluble fiber reaches the colon, it undergoes fermentation by "good bacteria" such as *Bifidobacteria* and *Lactobacillus*.[76] The by-products include short-chain fatty acids (SCFAs) that enhance the production of mucin by colonic mucosal cells, which has a protective effect on these cells.[77–79] Lactic acid, another by-product of this fermentation, aids in the reduction of the pH of the colon, maintaining conditions that discourage the growth of "bad bacteria".[80]

The protective mucin layer in the gut is crucial in the prevention of physical and mental issues.[81] A poor diet can enhance the growth of "bad bacteria," which includes gram-negative bacteria, and decrease the proliferation of "good bacteria."[82,83] While "good bacteria" aid in the production of the protective mucin layer, gram-negative bacteria along with other "bad bacteria" have been linked to an increased permeability of the colonic mucosa by affecting the function of tight junctions.[84,85] This lack of permeability of the gut lining can allow for the transfer of the immuno-stimulant lipopolysaccharides (LPS) into the circulation, resulting in chronic inflammation.[86, 87]

Gut Microbiota in Obesity and Gastrointestinal Disorders

Obesity is comorbid with many medical problems including insulin resistance, type 2 diabetes mellitus, hypertension, and cardiovascular disease. Certain psychological problems including depression are more prevalent in the obese populations.[88–92] It is possible that the composition of gut microbiota is different between obese and non-obese individuals. A study analyzed stool samples from 68 obese people and 47 controls using quantitative polymerase chain reaction (qPCR) and culture on a *Lactobacillus*-selective medium.[93] Firmicutes, Bacteroidetes, *Methanobrevibacter smithii*, *Lactococcus lactis*, *Bifidobacterium animalis*, and seven other species of *Lactobacillus* were targeted in these tests for comparison. qPCR data revealed that the gut microenvironment linked with obesity showed reduced levels of *M. smithii* as well as higher levels of *Lactobacillus reuteri*. In comparison, higher levels of *Lactobacillus plantarum*, *Lactobacillus paracases*, and *B. animalis* were observed in control subjects with normal body mass. These findings indicate that there are differences in the gut microenvironments between obese and non-obese people. The findings of this study may be useful to consider as the obesity epidemic continues to be of great concern worldwide. Alteration of an obese person's gut microbiota to resemble that of a non-obese individual could prove to be a promising avenue for potential future treatment of obesity and its associated comorbidities.

Besides obesity, inflammatory disorders such as inflammatory bowel disease (IBD) and irritable bowel syndrome (IBS) may benefit from alteration in gut microbiome. The chronic IBD known as ulcerative colitis (UC) is characterized by acute exacerbations of intestinal complications, with a relapsing and remitting course. A popular hypothesis on the cause of UC is that it is the result of a hyperactive immune response to endogenous bacteria.[94] The manipulation of this microbiota in an attempt to reduce this inflammation may present a promising treatment avenue for patients suffering from IBD. A single-center, randomized, double-blind, placebo-controlled study was conducted on patients who were in remission from UC. These patients received either probiotic therapy, which included a mixture of the strains *Streptococcus faecalis* T-110, *Clostridium butyricum* TO-A, and *Bacillus mesentericus* TO-A, or placebo for a period of 12 months.[94] The patients were evaluated for clinical symptoms at monthly intervals, and their fecal samples were collected at 3-month intervals for analysis of relapse of UC. There was observed a trend toward a lower relapse rate and higher sustained remission rate among the patients on probiotics, but the numbers were statistically significant only at 3 months, and not at 6-, 9- or 12-month time points. This data showed that probiotics may be useful in maintaining clinical remission in certain patients of UC.

Prebiotics, which are non-digestible ingredients of food, are known to promote the healthy proliferation of gut bacteria, increase anti-inflammatory action, inhibit the adherence of pathogens to the gut lining, and stimulate the growth of the gut mucosal layer. To determine the effect of prebiotics on IBS, a randomized, patient-blinded, placebo-controlled clinical study was conducted.[95] This study included 44 adult participants with IBS receiving either placebo or a prebiotic oral treatment consisting of a trans-galacto-oligosaccharide mixture. Participants were monitored for IBS symptoms, changes in fecal microflora, stool frequency and form, subjective global assessment (SGA), anxiety, and depression. It was observed that compared with the placebo, the prebiotic treatment resulted in a significant increase in fecal *Bifidobacterium*, improvements in flatulence and bloating, a change in stool consistency, and improvement in SGA and anxiety scores.[95]

The role of probiotics in the treatment of IBS has also been an area of interest for several researchers. In a randomized clinical trial involving 362 women with IBS, patients were given either placebo or a dose of *Bifidobacterium infantis* probiotics to ingest daily over a 4-week period.[96] The patients were monitored daily for IBS symptoms of abdominal pain or discomfort. A composite symptom score, which is a patient's global assessment of symptom relief, and measures of quality of life was also recorded. At the end of the 4-week study period, patients who were given *B. infantis* reported significant improvements in abdominal pain, bloating, bowel dysfunction, incomplete evacuation, straining, flatulence, and composite symptom score compared with the placebo group. These findings support the possibility of the use of probiotics in the treatment of IBS.

Synbiotics

The evidence favoring the benefit of probiotics and prebiotics for IBS has led some investigators to hypothesize that a combination of the two could have a synergistic effect.[97] The effect of supplementation with a combination of prebiotics and probiotics referred to as synbiotics (or symbiotics) in IBS patients was studied in a randomized double-blind study involving 30 IBS patients.[98] These patients were randomized in 1:1:1 ratio to receive placebo, low-dose synbiotics or high-dose synbiotics, which were administered for 8 weeks. The patients were evaluated for IBS symptoms of abdominal discomfort, bloating, frequency of formed stool, epigastric soreness, and nausea. The IBS symptom of fatigue was also assessed using the Fatigue Severity Scale (FSS). It was observed that ingestion of either low- or high-dose synbiotics resulted in significant improvements in abdominal discomfort, bloating, frequency of formed stool, and fatigue compared to the placebo group by per-protocol analysis. However, a statistically significant difference was not noted in the FSS score between the groups. Even though these results are encouraging, further large-scale studies are warranted to determine the optimum dosage for synbiotics in treating IBS.

CONCLUSIONS AND FUTURE DIRECTIONS

Prebiotics and psychobiotics have enormous potential to modulate the gut–brain axis that can determine predisposition to diseases. While the impact of certain microbiome on gastrointestinal processes, psychosis, and mood regulation has been assessed, there remains a lack of pre-clinical evaluations, which could serve to understand the mechanisms of action involved in the bidirectional communication pathways between the gut microbiome and the nervous system. Further studies are warranted that may address the interrelation of these systems, along with the exogenous factors, which influence these pathways.

Furthermore, there have been only a few large-scale placebo-controlled studies to determine the appropriate therapeutic applications of prebiotics and psychobiotics. The same is true for studies comparing the actions of commonly used probiotics and bacterial families. Certain strains of probiotics have psychotropic effects, yet it remains unknown, which mood disorder symptoms are most closely affected by which psychobiotics. Several clinical studies have measured mood and cognition, pre- and post-intervention. It may be more useful to evaluate mood and other neurological (cognitive) functions separately as opposed to generalized grouping of these functions. While psychobiotics are known to influence neurotransmission, a closer look into the relationship between ILs and specific strains of bacteria may also be useful.

There are a significant number of studies to suggest an association between the neuroendocrine system, the gut–brain axis and the immune system. However, further studies are warranted to decipher the interactions among these pathways to determine whether a particular microbe has specific relevance to certain neurotransmitters, cytokines, or other chemical messengers, and to study their specific impact on the physical and psychological processes in the human body. Understanding of these interactions and underlying molecular mechanisms can be a potential tool to develop new treatment modalities for various neurological and gastrointestinal disorders.

REFERENCES

1. Buckley MM, *et al.* (2020) Glucagon-like peptide-1 secreting l-cells coupled to sensory nerves translate microbial signals to the host rat nervous system. *Front Cell Neurosci* **14**: 95.
2. Egerod KL, *et al.* (2018) Profiling of G protein-coupled receptors in vagal afferents reveals novel gut-to-brain sensing mechanisms. *Mol Metab* **12**: 62–75.
3. Zhang J, Ma L, Chang L, Pu Y, Qu Y, Hashimoto K. (2020) A key role of the subdiaphragmatic vagus nerve in the depression-like phenotype and abnormal composition of gut microbiota in mice after lipopolysaccharide administration. *Transl Psychiatry* **10**(1): 186.
4. Haas GS, Wang W, Saffar M, Mooney-Leber SM, Brummelte S. (2020) Probiotic treatment (Bifidobacterium infantis) affects stress responsivity in male rats after chronic corticosterone exposure. *Behav Brain Res* **393**: 112718.
5. Qu Y, *et al.* (2020) Betaine supplementation is associated with the resilience in mice after chronic social defeat stress: A role of brain-gut-microbiota axis. *J Affect Disord* **272**: 66–76.
6. Kim CS, *et al.* (2021) Probiotic supplementation improves cognitive function and mood with changes in gut microbiota in community-dwelling elderly: A randomized, double-blind, placebo-controlled, multicenter trial. *J Gerontol A Biol Sci Med Sci* **76**(1): 32–40.

7. Huang WC, Hsu YJ, Huang CC, Liu HC, Lee MC. (2020) Exercise training combined with Bifidobacterium longum OLP-01 Supplementation improves exercise physiological adaption and performance. *Nutrients* **12**(4): 1145.
8. Thomann AK, *et al.* (2020) Review article: Bugs, inflammation and mood — a microbiota-based approach to psychiatric symptoms in inflammatory bowel diseases. *Aliment Pharmacol Ther* **52**(2): 247–266.
9. Gilbert JA. (2015) Our unique microbial identity. *Genome Biol* **16**(1): 97.
10. Eckburg PB, *et al.* (2005) Diversity of the human intestinal microbial flora. *Science* **308**(5728): 1635–1638.
11. Qin J, *et al.* (2010) A human gut microbial gene catalogue established by metagenomic sequencing. *Nature* **464**(7285): 59–65.
12. Human Microbiome Project Consortium. (2012) Structure, function and diversity of the healthy human microbiome. *Nature* **486**(7402): 207–214.
13. Ganci M, Suleyman E, Butt H, Ball M. (2019) The role of the brain-gut-microbiota axis in psychology: The importance of considering gut microbiota in the development, perpetuation, and treatment of psychological disorders. *Brain Behav* **9**(11): e01408.
14. Rooks MG, Garrett WS. (2016) Gut microbiota, metabolites and host immunity. *Nat Rev Immunol* **16**(6): 341–352.
15. Bermúdez-Humarán LG, Salinas E, Ortiz GG, Ramirez-Jirano LJ, Morales JA, Bitzer-Quintero OK. (2019) From probiotics to psychobiotics: Live beneficial bacteria which act on the brain-gut axis. *Nutrients* **11**(4): 890.
16. Evrensel A, Ünsalver BÖ, Ceylan ME. (2019) Psychobiotics. *Adv Exp Med Biol* **1192**: 565–581.
17. Sarkar A, Lehto S, Harty S, Dinan T, Cryan J, Burnet P. (2016) Psychobiotics and the manipulation of bacteria-gut-brain signals. *Trends Neurosci* **39**(11): 763–781.
18. Holscher H. (2017) Dietary fiber and prebiotics and the gastrointestinal microbiota. *Gut Microbes* **8**(2): 172–184.
19. Kali A. (2016) Psychobiotics: An emerging probiotic in psychiatric practice. *Biomed J* **39**(3): 223–224.
20. Desbonnet L. (2008) The probiotic *Bifidobacteria infantis*: An assessment of potential antidepressant properties in the rat. *J Psychiatr Res* **43**: 164–174

21. Collins SM, Surette M, Bercik P. (2012) The interplay between the intestinal microbiota and the brain. *Nat Rev Microbiol* **10**(11): 735–742.
22. Cryan JF, Dinan TG. (2012) Mind-altering microorganisms: The impact of the gut microbiota on brain and behaviour. *Nat Rev Neurosci* **13**(10): 701–712.
23. Bravo JA, *et al.* (2011) Ingestion of Lactobacillus strain regulates emotional behavior and central GABA receptor expression in a mouse via the vagus nerve. *Proc Natl Acad Sci USA* **108**(38): 16050–16055.
24. Liang S. (2015) Administration of Lactobacillus helveticus NS8 improves behavioral, cognitive, and biochemical aberrations caused by chronic restraint stress. *Neuroscience* **310**: 561–577.
25. Desbonnet L. (2010) Effects of the probiotic Bifidobacterium infantis in the maternal separation model of depression. *Neuroscience* **170**: 1179–1188.
26. Gareau MG. (2007) Probiotic treatment of rat pups normalises corticosterone release and ameliorates colonic dysfunction induced by maternal separation. *Gut* **56**: 1522–1528.
27. Savignac HM. (2013) Prebiotic feeding elevates central brain derived neurotrophic factor, *N*-methyl-D-aspartate receptor subunits and D-serine. *Neurochem Int* **63**: 756–764.
28. de Almeida Silva M, Mowry FE, Peaden SC, Andrade TU, Biancardi VC. (2020) Kefir ameliorates hypertension via gut-brain mechanisms in spontaneously hypertensive rats. *J Nutr Biochem* **77**: 108318.
29. Hsu YJ, *et al.* (2018) Kefir supplementation modifies gut microbiota composition, reduces physical fatigue, and improves exercise performance in mice. *Nutrients* **10**(7): 862.
30. Melo AFP, Mendonça MCP, Rosa-Castro RM. (2018) The protective effects of fermented kefir milk on azoxymethane-induced aberrant crypt formation in mice colon. *Tissue Cell* **52**: 51–56.
31. Tillisch K, *et al.* (2013) Consumption of fermented milk product with probiotic modulates brain activity. *Gastroenterology* **144**(7): 1394–1401.
32. Lee M-Y, *et al.* (2007) Anti-inflammatory and anti-allergic effects of kefir in a mouse asthma model. *Immunobiology* **212**(8): 647–654.
33. Yasar M, *et al.* (2013) The early anti-inflammatory effect of Kefir in experimental corrosive esophagitis. *Ann Ital Chir* **84**(6): 681–685.

34. Gao J, Ding G, Li Q, Gong L, Huang J, Sang Y. (2019) Tibet kefir milk decreases fat deposition by regulating the gut microbiota and gene expression of Lpl and Angptl4 in high fat diet-fed rats. *Food Res Int* **121**: 278–287.
35. Bourrie BCT, Cotter PD, Willing BP. (2018) Traditional kefir reduces weight gain and improves plasma and liver lipid profiles more successfully than a commercial equivalent in a mouse model of obesity. *J Funct Foods* **46**: 29–37.
36. Kim DH, *et al.* (2017) Kefir alleviates obesity and hepatic steatosis in high-fat diet-fed mice by modulation of gut microbiota and mycobiota: Targeted and untargeted community analysis with correlation of biomarkers. *J Nutr Biochem* **44**: 35–43.
37. Silva-Cutini MA, *et al.* (2019) Long-term treatment with kefir probiotics ameliorates cardiac function in spontaneously hypertensive rats. *J Nutr Biochem* **66**: 79–85.
38. van de Wouw M, *et al.* (2020) Distinct actions of the fermented beverage kefir on host behaviour, immunity and microbiome gut-brain modules in the mouse. *Microbiome* **8**(1): 67.
39. Yilmaz I, Dolar ME, Ozpinar H. (2019) Effect of administering kefir on the changes in fecal microbiota and symptoms of inflammatory bowel disease: A randomized controlled trial. *Turk J Gastroenterol* **30**(3): 242–253.
40. Messaoudi M. (2011) Assessment of psychotropic-like properties of a probiotic formulation (*Lactobacillus helveticus* R0052 and *Bifidobacterium longum* R0175) in rats and human subjects. *Br J Nutr* **105**: 755–764.
41. Sproston N, Ashworth J. (2018) Role of C-reactive protein at sites of inflammation and infection. *Front Immunol* **9**: 754.
42. Reigstad CS, *et al.* (2015) Gut microbes promote colonic serotonin production through an effect of short-chain fatty acids on enterochromaffin cells. *FASEB J* **29**(4): 1395–1403.
43. Yano JM, *et al.* (2015) Indigenous bacteria from the gut microbiota regulate host serotonin biosynthesis. *Cell* **161**(2): 264–276.
44. Cowen P, Browning M. (2015) What has serotonin to do with depression? *World Psychiatry* **14**(2): 158–160.
45. Carabotti M, Scirocco A, Maselli M, Severi C. (2015) The gut-brain axis: Interactions between enteric microbiota, central and enteric nervous systems. *Ann Gastroenterol* **28**(2): 203–209.

46. Dunlop BW, Wong A. (2019) The hypothalamic-pituitary-adrenal axis in PTSD: Pathophysiology and treatment interventions. *Prog Neuropsychopharmacol Biol Psychiatry* **89**: 361–379.
47. Heck AL, Handa RJ. (2019) Sex differences in the hypothalamic-pituitary-adrenal axis' response to stress: An important role for gonadal hormones. *Neuropsychopharmacology* **44**(1): 45–58.
48. Huo R, *et al.* (2017) Microbiota modulate anxiety-like behavior and endocrine abnormalities in hypothalamic-pituitary-adrenal axis. *Front Cell Infect Microbiol* **7**: 489.
49. Kallen VL, Tulen JH, Utens EM, Treffers PD, De Jong FH, Ferdinand RF. (2008) Associations between HPA axis functioning and level of anxiety in children and adolescents with an anxiety disorder. *Depress Anxiety* **25**: 131–141.
50. Erhardt A, *et al.* (2006) Regulation of the hypothalamic-pituitary-adrenocortical system in patients with panic disorder. *Neuropsychopharmacology* **31**: 2515–2522.
51. Yehuda R, Halligan SL, Golier JA, Grossman R, Bierer LM. (2004) Effects of trauma exposure on the cortisol response to dexamethasone administration in PTSD and major depressive disorder. *Psychoneuroendocrinology* **29**: 389–404.
52. Condren RM, O'Neill A, Ryan MC, Barrett P, Thakore JH. (2002) HPA axis response to a psychological stressor in generalised social phobia. *Psychoneuroendocrinology* **27**: 693–703.
53. Schreiber W, Lauer CJ, Krumrey K, Holsboer F, Krieg JC. (1996) Dysregulation of the hypothalamic-pituitary-adrenocortical system in panic disorder. *Neuropsychopharmacology* **15**: 7–15.
54. Klimes-Dougan B, Begnel E, Almy B, Thai M, Schreiner MW, Cullen KR. (2019) Hypothalamic-pituitary-adrenal axis dysregulation in depressed adolescents with non-suicidal self-injury. *Psychoneuroendocrinology* **102**: 216–224.
55. Varghese F, Brown E. (2001) The hypothalamic-pituitary-adrenal axis in major depressive disorder: A brief primer for primary care physicians. *Prim Care Companion J Clin Psychiatry* **3**(4): 151–155.
56. Farzi A, Fröhlich E, Holzer P. (2018) Gut Microbiota and the neuroendocrine system. *Neurotherapeutics* **15**(1): 5–22.

57. Muller N, Schwarz M. (2007) The immune-mediated alteration of serotonin and glutamate: Towards an integrated view of depression. *Mol Psychiatry* **12**: 988–1000.
58. Moya-Pérez A, Perez-Villalba A, Benítez-Páez A, Campillo I, Sanz Y. (2017) Bifidobacterium CECT 7765 modulates early stress-induced immune, neuroendocrine and behavioral alterations in mice. *Brain Behav Immun* **65**: 43–56.
59. Agusti A, *et al.* (2017) *Bifidobacterium pseudocatenulatum* CECT 7765 ameliorates neuroendocrine alterations associated with an exaggerated stress response and anhedonia in obese mice. *Mol Neurobiol* **55**(6): 5337–5352.
60. Burokas A, *et al.* (2017) Targeting the microbiota-gut-brain axis: Prebiotics have anxiolytic and antidepressant-like effects and reverse the impact of chronic stress in mice. *Biol Psychiatry*. **82**(7): 472–487.
61. Cenit M, Sanz Y, Codoñer-Franch P. (2017) Influence of gut microbiota on neuropsychiatric disorders. *World J Gastroenterol* **23**(30): 5486–5498.
62. Sudo N, *et al.* (2004) Postnatal microbial colonization programs the hypothalamic-pituitary-adrenal system for stress response in mice. *J Physiol* **558**(Pt 1): 263–275.
63. Mudd AT, Berding K, Wang M, Donovan SM, Dilger RN. (2017) Serum cortisol mediates the relationship between fecal Ruminococcus and brain N-acetylaspartate in the young pig. *Gut Microbes* **8**(6): 589–600.
64. Luo Y, *et al.* (2018) Gut microbiota regulates mouse behaviors through glucocorticoid receptor pathway genes in the hippocampus. *Transl Psychiatry* **8**: 187.
65. Wang H, Wang Y. (2016) Gut microbiota-brain axis and mental health. *EC Psychol Psychiatry* **1.2**: 55–60.
66. Biasucci G, Rubini M, Riboni S, Morelli L, Bessi E, Retetangos C. (2010) Mode of delivery affects the bacterial community in the newborn gut. *Early Hum Dev* **86**: 13–15.
67. Dominguez-Bello MG, *et al.* (2010) Delivery mode shapes the acquisition and structure of the initial microbiota across multiple body habitats in newborns. *Proc Natl Acad Sci USA* **107**(26): 11971–11975.
68. Li H, *et al.* (2018) The impacts of delivery mode on infant's oral microflora. *Sci Rep* **8**(1): 11938.

69. Jakobsson HE, Abrahamsson TR, Jenmalm MC, Harris K, Quince C, Jernberg C. (2014) Decreased gut microbiota diversity, delayed Bacteriodetes colonization and reduced Th1 responses in infants delivered by caesarean section. *Gut* **63**(4): 559–566.
70. Betrán AP, Ye J, Moller AB, Zhang J, Gülmezoglu AM, Torloni MR. (2016) The increasing trend in caesarean section rates: Global, regional and national estimates: 1990–2014. *PLoS One* **11**(2): e0148343.
71. Chen G, Chiang WL, Shu BC, Guo YL, Chiou ST, Chiang TL. (2017) Associations of caesarean delivery and the occurrence of neurodevelopmental disorders, asthma or obesity in childhood based on Taiwan birth cohort study. *BMJ Open* **7**(9): e017086.
72. Sandall J, *et al.* (2018) Short-term and long-term effects of caesarean section on the health of women and children. *Lancet* **392**(10155): 1349–1357.
73. De Filippis F, *et al.* (2016) High-level adherence to a Mediterranean diet beneficially impacts the gut microbiota and associated metabolome. *Gut* **65**(11): 1812–1821.
74. Wu GD, *et al.* (2011) Linking long-term dietary patterns with gut microbial enterotypes. *Science* **334**(6052): 105–108.
75. David LA, *et al.* (2014) Diet rapidly and reproducibly alters the human gut microbiome. *Nature* **505**(7484): 559–563.
76. So D, *et al.* (2018) Dietary fiber intervention on gut microbiota composition in healthy adults: A systematic review and meta-analysis. *Am J Clin Nutr* **107**(6): 965–983.
77. Gaudier E, *et al.* (2004) Butyrate specifically modulates MUC gene expression in intestinal epithelial goblet cells deprived of glucose. *Am J Physiol Gastrointest Liver Physiol* **287**(6): G1168–G1174.
78. Willemsen LE, Koetsier MA, van Deventer SJ, van Tol EA. (2003) Short chain fatty acids stimulate epithelial mucin 2 expression through differential effects on prostaglandin E(1) and E(2) production by intestinal myofibroblasts. *Gut* **52**(10): 1442–1447.
79. Shimotoyodome A, Meguro S, Hase T, Tokimitsu I, Sakata T. (2000) Short chain fatty acids but not lactate or succinate stimulate mucus release in the rat colon. *Comp Biochem Physiol A Mol Integr Physiol* **125**: 525–531.

80. Ilhan ZE, Marcus AK, Kang D-W, Rittmann BE, Krajmalnik-Brown R. (2017) pH-mediated microbial and metabolic interactions in fecal enrichment cultures. *mSphere* **2**: e00047-17.
81. Cornick S, Tawiah A, Chadee K. (2015) Roles and regulation of the mucus barrier in the gut. *Tissue Barriers* **3**(1–2): e982426.
82. Cani PD, *et al.* (2007) Metabolic endotoxemia initiates obesity and insulin resistance. *Diabetes* **56**: 1761–1772.
83. Cani PD, *et al.* (2007) Selective increases of Bifidobacteria in gut microflora improve high-fat-diet-induced diabetes in mice through a mechanism associated with endotoxaemia. *Diabetologia* **50**: 2374–2383.
84. Cani PD, *et al.* (2008) Changes in gut microbiota control metabolic endotoxemia-induced inflammation in high-fat diet-induced obesity and diabetes in mice. *Diabetes* **57**: 1470–1481.
85. Suzuki T, Hara H. (2010) Dietary fat and bile juice, but not obesity, are responsible for the increase in small intestinal permeability induced through the suppression of tight junction protein expression in LETO and OLETF rats. *Nutr Metab* **7**: 19.
86. Ghoshal S, Witta J, Zhong J, de Villiers W, Eckhardt E. (2009) Chylomicrons promote intestinal absorption of lipopolysaccharides. *J Lipid Res* **50**: 90–97.
87. Guo Y, Zhou G, He C, Yang W, He Z, Liu Z. (2015) Serum levels of lipopolysaccharide and 1,3-β-D-glucan refer to the severity in patients with Crohn's disease. *Mediators Inflamm* **2015**: 1–9.
88. Khaodhiar L, McCowen KC, Blackburn GL. (1999) Obesity and its comorbid conditions. *Clin Cornerstone* **2**(3): 17–31.
89. McIntyre RS, Konarski JZ, Wilkins K, Soczynska JK, Kennedy SH. (2006) Obesity in bipolar disorder and major depressive disorder: Results from a national community health survey on mental health and well-being. *Can J Psychiatry* **51**: 274–280.
90. Allison DB, *et al.* (2009) Obesity among those with mental disorders: A national Institute of mental health meeting report. *Am J Prev Med* **36**: 341–350.
91. Lin KP, Liang TL, Liao IC, Tsay SL. (2013) Associations among depression, obesity, and metabolic syndrome in young adult females. *Biol Res Nurs* **16**: 327–334.

92. Carey M, Small H, Yoong SL, Boyes A, Bisquera A, SansonFisher R. (2014) Prevalence of comorbid depression and obesity in general practice: A cross-sectional survey. *Br J Gen Pract* **64**: e122–e127.
93. Million M, *et al.* (2012) Obesity-associated gut microbiota is enriched in *Lactobacillus reuteri* and depleted in Bifidobacterium animalis and *Methanobrevibacter smithii*. *Int J Obes* **36**: 817–825.
94. Yoshimatsu Y, *et al.* (2015) Effectiveness of probiotic therapy for the prevention of relapse in patients with inactive ulcerative colitis. *World J Gastroenterol* **21**(19): 5985–5994.
95. Silk DB, Davis A, Vulevic J, Tzortzis G, Gibson GR. (2009) Clinical trial: The effects of a trans-galactooligosaccharide prebiotic on faecal microbiota and symptoms in irritable bowel syndrome. *Aliment Pharmacol Ther* **29**(5): 508–518.
96. Whorwell PJ, *et al.* (2006) Efficacy of an encapsulated probiotic Bifidobacterium infantis 35624 in women with irritable bowel syndrome. *Am J Gastroenterol* **101**: 1581–1590.
97. Plaza-Díaz J, Ruiz-Ojeda F, Vilchez-Padial L, Gil A. (2017) Evidence of the anti-inflammatory effects of probiotics and synbiotics in intestinal chronic diseases. *Nutrients* **9**(6): 555.
98. Lee SH, *et al.* (2019) A randomized clinical trial of synbiotics in irritable bowel syndrome: Dose-dependent effects on gastrointestinal symptoms and fatigue. *Korean J Fam Med* **40**(1): 2–8.

Chapter 9

The Gut Microbiome: Potential Clinical Applications in Disease Management

Novel Approaches Using Diet and Nutraceuticals to Reduce Manifestations of Accelerated Aging, Obesity, Type 2 Diabetes, Gastrointestinal Disorders, Neurological Disorders, Immunological Disorders, Anxiety, and Depression

Camron Davies, Jenna Bergman, Adrien A. Eshraghi, Rahul Mittal and Rebecca S. Eshraghi*

Abstract

The gut microbiome is capable of regulating physiology in many ways by altering the inflammatory response, and ultimately affecting a diverse range of illnesses and disorders. A dysbiotic gut microbiome can induce systemic inflammation and potentially lead to neuroinflammation, while a eubiotic microbiome reduces inflammation and induces immunotolerance, which

* Hearing Research and Communication Disorders Laboratory, Department of Otolaryngology, Division of Neurotology, University of Miami-Miller School of Medicine, Miami, FL, USA.

Corresponding Author: Rebecca S. Eshraghi, PhD, DNM, Hearing Research and Communication Disorders Laboratory, Department of Otolaryngology, Division of Neurotology, University of Miami-Miller School of Medicine, 1600 NW 10th Avenue, Miami, FL 33136, USA; Quantum University for Integrative Medicine, Honolulu, HI 96813, USA, E-mail: rebeccaeshraghi@gmail.com.

may either aggravate or protect against underlying pathologies of a patient. The gut microbiome has also been shown to affect neurotransmitters and gene expression in the brain, potentially allowing it to affect neurologic functioning more directly. This regulatory function implicates the gut microbiome in a wide range of disorders since many chronic health conditions are due to dysregulated inflammation leading to cellular dysfunction. Given this connection, there has been much interest recently in finding gut-based therapies to correct dysbiosis-induced inflammation, mainly via prebiotics, probiotics, and specific diets or nutrition. Gut-based interventions appear to be promising therapies for many chronic inflammatory processes as well as neurologic and psychiatric conditions, including inflammatory bowel disease, irritable bowel syndrome, Alzheimer's disease, multiple sclerosis, autism spectrum disorder, auditory processing disorders, accelerated aging, obesity, and type 2 diabetes. Despite promising preliminary results, many of the connections between the microbiome and various diseases are still unclear. More studies are needed to understand the mechanisms involved and to improve their therapeutic approach.

INTRODUCTION

The gut microbiome is a complex network of bacteria that can interact with multiple physiological systems through their diverse metabolites. One of the crucial functions of the gut microbiome and its metabolites is to help regulate inflammation and neural function. A dysbiotic microbiome tends to be pro-inflammatory, while a eubiotic microbiome tends to be anti-inflammatory and promotes immune system regulation. This regulatory ability places the gut microbiome at the center of normal physiology and a wide range of pathologies, as many common diseases and disorders are rooted in inflammation (Fig. 9.1). In this chapter, we will discuss the connections between the gut microbiome and many common clinical pathologies and present the current evidence for gut-based therapy in each case. However, to better understand the gut microbiome's role in various conditions, it is first necessary

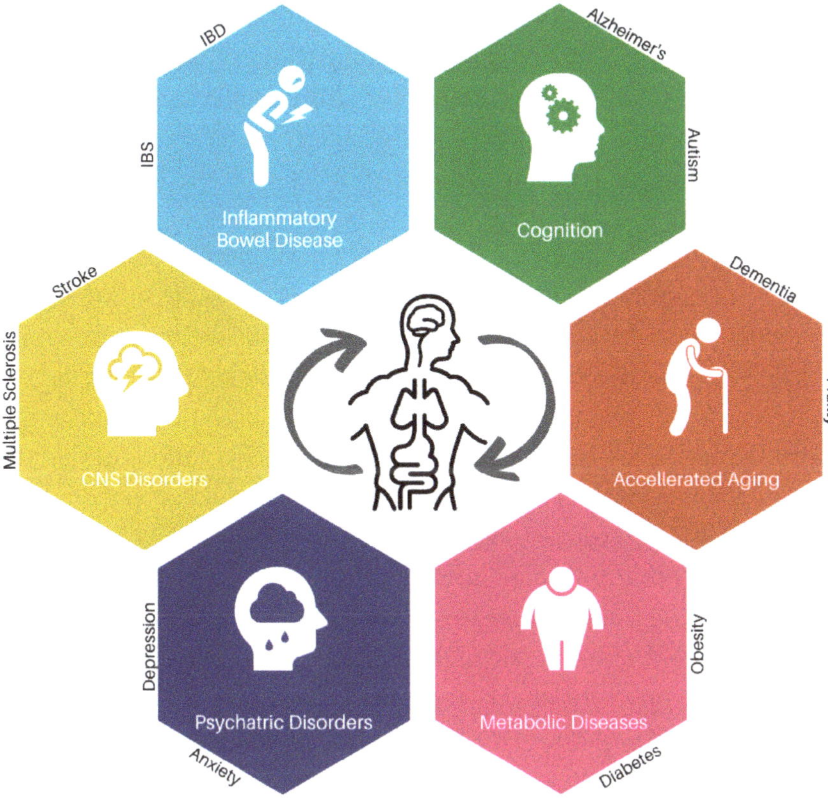

Fig. 9.1: Concept diagram illustrating the interconnectedness of the gut microbiome with a variety of conditions including cognition, accelerated aging, metabolism, gastrointestinal distress, neurologic disorders inflammatory bowel disease, irritable bowel syndrome, Alzheimer's diseases, autism, dementia, frailty, obesity, diabetes, anxiety, depression, multiple sclerosis and stroke.

to discuss the role of the microbiome in chronic disorders and chronic inflammation.

THE INFLAMMATORY PROCESS

Inflammation is the body's attempt to protect itself in response to stress or injury. Inflammation is commonly thought of in its acute phase, which is characterized by increased pain, heat, redness, swelling, and loss of function in the minutes, hours, or days following an injury. These effects are primarily mediated by the innate immune system, which increases blood flow, vascular permeability, and nutrient flow to the site of injury in an attempt to fight an infection or heal a wound. In contrast, chronic inflammation, which occurs over weeks, months, or even years, is much more subtle yet far more insidious. Chronic inflammation may be mediated by persistent exposure to pathogens, injury, foreign bodies, or autoimmune reactions. Different from the immediate effects seen in acute inflammation, persistent inflammation is characterized by tissue destruction, necrosis, calcifications, and fibrosis. Chronic inflammation may also be systemic and low grade, which is characterized by persistently high levels of pro-inflammatory cytokines like tumor necrosis factor alpha (TNF-α), C-reactive protein (CRP), and interleukin (IL)-6. Prolonged high levels of cytokines like these primes the body's immune system for autoreactivity and hyper-responsiveness, leading to conditions like atherosclerosis, insulin resistance, or autoimmunity.[1] Indeed, research has shown that low-grade chronic inflammation is a key player in many non-communicable diseases.[2]

The gut microbiome is intimately involved with these processes as gut-associated lymphoid tissue accounts for almost 70% of the entire immune system and about 80% of plasma cells.[3] An imbalance in the gut microbiome can induce alterations in the body's immune system, increasing the activation of an immune response, which thereby induces a disease state. Moreover, a dysbiotic gut microbiome may act more directly as well, since a larger number of pathogenic bacteria can increase the presence and release of pathogen-associated molecular patterns (PAMPs) like lipopolysaccharide (LPS) and enterotoxins, thereby heightening immune dysfunction, damaging the intestinal epithelium, and altering energy metabolism.[4]

MICROBIOME AND INFLAMMATORY DISORDER

The role of the intestinal microbiota is multiple and ranges from a beneficial role digesting food and providing nutrients to the host to potentially harmful role triggering disease via inflammatory processes. One major player in the proper functioning of the gastrointestinal (GI) tract are short-chain fatty acids (SCFAs). SCFAs are critical metabolites in regulating inflammatory processes. SCFAs, primarily acetate, propionate, and butyrate, are generated as by-products of bacterial fermentation of prebiotics. In simple terms, SCFAs are produced as the result of dietary fiber fermentation in the colon. Among their many functions, SCFAs are energy sources, epigenetic regulators, cell-signaling molecules, and inducers of immune tolerance in the gut.[5,6] Notably, SCFA production depends on the quantity of substrate (prebiotic/fiber) present, transit time within the gut, species, and abundance of microflora in the colon.[7,8] Once produced, SCFAs can modulate systemic inflammation in several ways, both locally within the gut and systemically in other organ systems.

SCFAs, especially butyrate, are the primary energy source for colonocytes and can epigenetically promote their function to ensure adequate barrier functionality.[9–11] Decreases in SCFAs, as found in a low-fiber diet or a dysbiotic microbiome, lead to increased translocation of bacteria and PAMPs, across the intestinal epithelial barrier, which induces *systemic inflammation*.[12–16] Moreover, SCFAs, primarily acetate, decrease the abundance of pathogenic bacteria in the gut by acidifying the lumen.[17] SCFAs also decrease the proportion of mucolytic bacteria and increase colonic mucus production, which helps preserve the first passive barrier between luminal contents and the gut epithelia.[17–20]

In addition, inflammatory and anti-inflammatory effects of the SCFAs extend beyond the gut. Systemically, SCFAs modulate the immune system by altering the body's cytokine profile and immunologic composition. Butyrate and propionate also function as histone deacetylase (HDAC) inhibitors allowing them to shift the composition of the immune system by regulating the transcription of genes contributing to the immune system and inflammatory response.[21–23] Moreover, SCFAs are agonists at free fatty acid receptors (FFAR) 2 and 3. FFAR 2/3 are G protein-coupled receptors (GPCR) found throughout the body, including on immune cells and the endothelia of the blood–brain barrier (BBB), allowing SCFAs to play a role in regulating both systemic immune responses and permeability across the BBB.[9,24–27] In the innate immune system, SCFAs can alter neutrophil chemotaxis and decrease their subsequent inflammatory response.[28–30] In the adaptive immune system, SCFAs have been shown to inhibit nuclear factor kappa B (NF-κB) release from macrophages, decrease LPS-induced pro-inflammatory markers and cytokines (IL-6 and IL-12) from monocytes as well as increase Foxp3 and IL-10 release from naive CD (Cluster of Differentiation) T cells, increase prostaglandin E_2 and decrease LPS-induced TNF-α and IFN-γ release

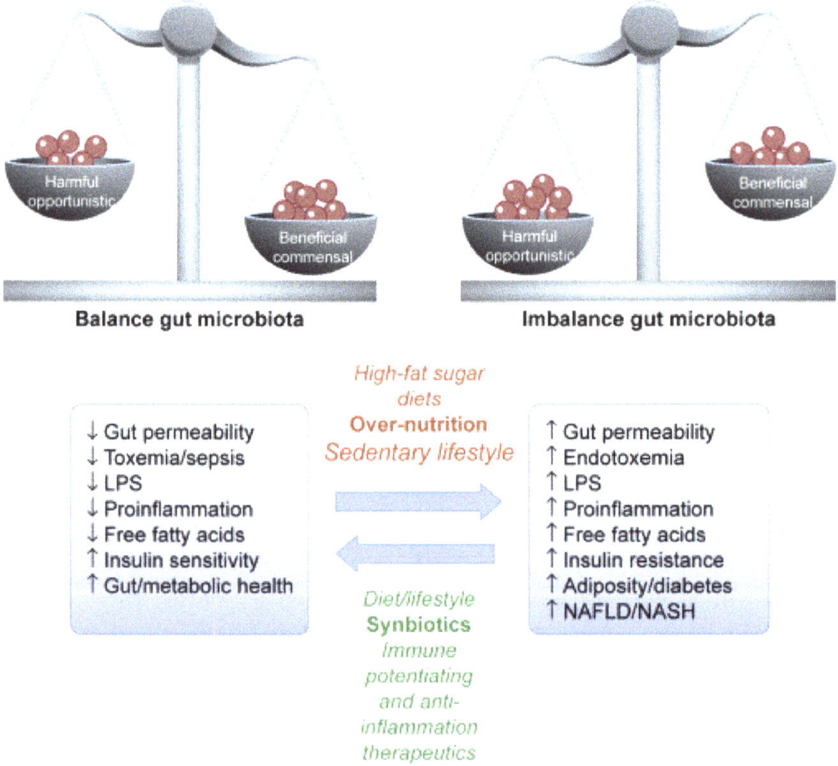

Fig. 9.2: Illustration of the inflammatory/anti-inflammatory balance regulated by the gut microbiome that is hypothesized to help induce many chronic inflammatory conditions as well as potential therapeutic modalities.

from peripheral blood mononuclear cells (PBMCs).[31–34] Finally, and perhaps most importantly, butyrate has been shown to up-regulate the differentiation of naive T cells into T regulatory (Treg) cells via induction of the Foxp3 pathway,[35,36] Tregs are critical in inhibiting mucosal immune reactions and the development of food allergies, celiac disease, inflammatory bowel disease (IBD), and even inducing systemic tolerances.[37] Given these mechanisms, many researchers have viewed the gut microbiome as a potential therapeutic modality in many diseases (Fig. 9.2).

MODULATING THE MICROBIOME

Dietary Intervention

It is increasingly clear that diet plays a significant role not only in the progression of disease directly but also indirectly via its effects on the gut microbiome. While every individual's microbiome becomes individualized and a blueprint, it can still be significantly altered via diet, drug, and antibiotics use.[38] Recent research has shown that dietary interventions can induce substantial changes in the gut microbiome as soon as 24 hours post-intervention; however, these changes returned to baseline within 48 hours of ending the diet.[39] This illustrates the gut microbiome's responsiveness to dietary intervention and demonstrates how diet might be used to create therapeutic changes in the gut microbiome. Here we will briefly discuss some of the common diets used in patient populations and their associated health outcomes, summarized in Table 9.1.

Typical Western Diet

Currently, the typical Western Diet (WD) is characterized by large quantities of highly refined and processed foods, high in sugars, salt, and fat, while almost devoid of dietary fibers and micronutrients.[40] Not only is the WD directly associated with inflammatory diseases like metabolic syndrome, but it is also associated

Table 9.1: A Summary of Primary Therapeutic Avenues with Which Clinicians Can Interact with the Gut Microbiome Including Nutraceuticals, Medications, Microbiota Transfer Therapies, and Dietary Interventions

Category	Intervention	Description
Dietary Intervention	Calorie Restriction (CR) Diet	- Fasting and CR diets limit caloric consumption while preserving micronutrient intake. - Benefits are hypothesized to derive from increasing ketosis and autophagy. - There are many ways to follow a CR diet such as intermittent fasting and limiting caloric intake, accordingly a CR diet should be planned with the assistance of a registered dietician to avoid malnutrition or deficiency of vital micronutrients or vitamins. Example diets include time-restricted feeding, alternate-day fasting, and 5:2 eating pattern.
	High-Fiber Diet	- High consumption of dietary fiber, a prebiotic, decreases intestinal inflammation by decreasing refined carbohydrate intake and promoting short-chain fatty acid (SCFA) production. - Confers the similar benefits of prebiotic supplementation. - Typical diet: contains at least 30–38 g of dietary fiber per day for men and 21–25 g for women.
	Ketogenic Diet (KD)	- The KD is characterized by extremely low levels of dietary carbohydrate intake compensated with relatively high levels of fat and moderate protein consumption. - Benefits are derived from limiting carbohydrate metabolism and promoting ketosis. - Typical diet: consuming 55%–60% of dietary calories in fat, 30%–35% in protein, and 5%–10% in carbohydrates. For a 2,000 kcal per day diet, this is equivalent to 20–50 g of carbohydrates per day, not including fiber.
	Mediterranean Diet (MD)	- Benefits derive from a combination of high fiber, moderate protein, low-refined carbohydrate, and low-saturated fat consumption. - High in whole grains, legumes, nuts, fruits, and vegetables with moderate poultry and fish consumption.

Table 9.1: (*Continued*)

Category	Intervention	Description
		- Typical diet: eating 7–10 servings a day of fruit and vegetables, eating whole-grain bread, cereal, and pasta. Consuming low-fat dairy and cheeses. The use of healthy fats like olive oil rather than butter. And replacing red meat with seafood.
Nutraceutical	Prebiotics	- Effectively equivalent to dietary fiber, these are substrates selectively metabolized by the gut microflora, which provide health benefits by fostering beneficial species and bioactive metabolites. - Sources include dietary intake of fiber, resistant starch, inulin, galacto-oligosaccharides, and fructo-oligosaccharides, which are found in foods such as avocado, broccoli, beans, and lentils.
	Probiotics	- Exogenous bacteria supplemented as either single species or multiple for a specific therapeutic effect.
	Synbiotics	- Foods containing both probiotic bacterial strains as well as the prebiotic substrate which they metabolize. - Sources include fermented foods like sauerkraut or kimchi.
	Metabiotics	- The specific bioactive molecular derivatives of bacterial metabolism, such as SCFAs.
Transfer Therapy	Microbiota Transfer Therapy (MTT)	- An oral or rectal supplement of standardized human gut microbiota, which is a purified, fecal-derived microbial extract. Benefits include quickly transferring a large number of diverse and standardized bacteria to the host's gut.
	Fecal Material Transfer (FMT)	- Transfer of fecal material from a healthy, eubiotic donor to a host's colon via colonoscopy, enema, nasogastric tube, or in capsule form. Benefits include quickly transferring a large number of diverse bacteria to the host's gut.

with bacterial overgrowth, increased pro-inflammatory species, as well as compositional and functional species-level changes in the gut microbiome.[41] Specifically, the WD has been associated with decrease quantity of total bacteria and decreased eubiotic species like *Bifidobacterium* and *Eubacterium* as well as increases in nitrosamines and reactive oxygen species (ROS).[39]

Calorie Restricted and Fasting Diet

Fasting and calorie restriction (CR) diets are thought to promote overall well-being through ketosis.[42] Rodent studies have shown that rats suddenly switched from *ad libitum* diets to CR had rapid and significant alterations in their gut microbiota, including a reduction *Firmicutes*-to-*Bacteroidetes* ratio and increased levels of *lactobacilli*.[43] Additionally, the observed increases in enzymes related to SCFA synthesis, particularly propionogenesis.[43] CR is also associated with metabolic improvements that are linked to the decreased bacterial expression of LPS, which decreases the tone of the inflammatory response.[44] CR studies in humans indicate that reduced caloric intake slowed aging and metabolism, and protected against age-related disease by decreasing oxidative stress.[45]

There are many versions and methods of implementing a CR and fasting diet, according to the National Institute on Aging, a few common regimen are time-restricted feeding, where meals are consumed within a specific time period and nothing outside of those hours; Alternate-day fasting, where no or minimal food is consumed on specific days; and 5:2 eating pattern, where unrestricted eating occurs for 5 days follow by 2 days of restricted eating. Importantly, CR and fasting diets restrict calories to minimal levels without malnutrition or deficiency of vital micronutrients or vitamins.

Ketogenic Diet

The ketogenic diet (KD) is characterized by extremely low levels of dietary carbohydrate intake with relatively high levels of fat and moderate protein consumption.[46] Uniquely, the KD has been effectively used in treating refractory epilepsy for 100 years and recently has become a very popular weight-loss diet.[47] Many of the diet's health benefits are attributed to the constant state of ketosis and the effects of β-hydroxybutyrate.[48] Interesting trends are also seen in the gut microbiota. Studies have found the KD significantly decreases total bacterial abundance in the gut as well as decreasing the *Firmicutes*-to-*Bacteroidetes* ratio, decreasing *Bifidobacteria*, as well as increasing *Akkermansia muciniphila* and *Escherichia coli.*[49,50] These changes also corresponded with a reduction in gene expression associated with carbohydrate metabolism in autism models and refractory epileptic patients. Decreases in these carbohydrate metabolism pathways are particularly concerning, given that their by-products, SCFAs, are critical to overall colonic health.[49] While some studies have found links between specific bacteria the KD's effects, more studies are needed to identify the effects of the KD on the microbiome and if these changes are linked to the diet's beneficial effects or if they are independent of microbial changes.[49,51] A KD is typically defined as consuming 55%–60% of dietary calories in fat, 30%–35% in protein, and 5%–10% in carbohydrates. For a 2,000 kcal per day diet, this is equivalent to 20–50 g of carbohydrates per day, excluding fiber.

High-Fiber Diet

Dietary fiber are prebiotics, is associated with many beneficial effects and are an important component in many of the diets discussed in this section. Many of fiber's effects are derived from its prebiotic function. Numerous studies have found that high-fiber diets increase the diversity of the intestinal microbiome, increase butyrate, and other SCFA production, as well as reduce negative GI symptoms.[52] Specifically, high-fiber diets are associated with decreased rates of colorectal cancer, IBD, as well as better outcomes in chronic obstructive pulmonary disease (COPD), asthma, obesity, and diabetes.[52] Overall, it appears that many conditions ameliorated by increased levels of SCFAs may also be aided by increased fiber intake. According to the Academy of Nutrition and Dietetics, a high-fiber diet consist of at least 30–38 g of dietary fiber per day for men and 21–25 g for women. Many foods are high in fiber, some examples include avocado, broccoli, beans, papaya, and lentils.

Mediterranean Diet

The Mediterranean diet (MD) is also exceedingly popular and associated with positive health outcomes. High in whole grains, legumes, nuts, fruits, and vegetables with moderate poultry and fish consumption, the MD has been attributed to decreased incidence of cardiovascular disease, neurodegenerative disorders, psychiatric conditions, and cancer.[9] In the gut, low adherence to MD is associated with increased *Firmicutes*-to-*Bacteroidetes* ratio and altered microbial diversity, while greater adherence was associated with higher levels of *bifidobacteria* and total SCFA.[53] Another study of non-human primates found that MD was associated with significantly higher Shannon diversity index, particularly significantly increased levels of *Clostridium*, *Lactobacillus*, *Faecalibacterium*, and *Oscillospira* and lower levels of *Ruminococcus* and *Coprococcus*.[54] These changes are hypothesized to improve metabolic functioning and reduce metabolic endotoxemia originating in the gut.[55] The typical MD consists of 7–10 servings a day of fruit and vegetables, eating whole-grain bread, cereal and pasta, low consumption of reduced-fat dairy and cheeses, the use of healthy fats like olive oil rather than butter, and replacing red meat with seafood.

Probiotics, Prebiotics, Microbiota Transfer Therapy, and Fecal Microbiota Transfer

One application that clinicians use to potentially influence the gut microbiome therapeutically is the administration of prebiotics, probiotics, and antibiotics. **Probiotics** are exogenous bacteria supplemented as either single or multiple species to provide a specific therapeutic effect. **Prebiotics** are substrates selectively metabolized by the gut microflora, which provide health benefits. In common parlance, prebiotics are equivalent to dietary fiber; within the scientific community, prebiotics are supplemented by increasing dietary intake of fiber, resistant starch, inulin, galacto-oligosaccharides (GOS), and fructo-oligosaccharides (FOS).

In addition to prebiotics and probiotics, combinations and derivatives of the two also exist called synbiotics. **Synbiotics** contain both probiotic bacterial strains as well as the prebiotic substrate, which they metabolize; these are often fermented foods like sauerkraut or kimchi.[56] **Metabiotics** are the molecular derivatives of bacterial metabolism, such as short-chain fatty acids (SCFAs); these allow clinicians and researchers to precisely deliver the biologically active molecules to provide health benefits.[57] Microbiota Transfer Therapy (MTT) is an oral or rectal supplement of standardized human gut microbiota, which is a purified, fecal-derived microbial extract.[58] Compared to probiotics, MTT can simultaneously supplement many more individual species compared to standard probiotic formulations. Similarly, fecal microbiota transfer (FMT) exogenously supplements a huge range of bacterial species by transferring fecal material from a healthy, eubiotic donor's microbiome into the host's colon typically via colonoscopy, enema, nasogastric tube, or in capsule form. Combined, probiotics and prebiotics are powerful tools that can alter the gut microbiome and its metabolites.

MICROBIOME AND VARIOUS HEALTH CONDITIONS

Accelerated Aging

Accelerated aging is characterized by inflamm-aging, a feedforward mechanism of increasing inflammation and increasing risk for chronic disease. On a cellular and molecular level, aging demonstrates increased genetic instability due to telomere shortening, epigenetic changes, and oxidative stress. In turn, these effects are seen in increased cellular senescence, stem cell exhaustion, mitochondrial dysfunction, altered intercellular communication, dysregulated immunity, and autophagy.[59,60] A large focus of research on inflammation and the aging process has been dedicated to elucidating the initiation of this process. It has been found that in addition to bacterial and viral infections, excess cellular debris, damage-associated molecular patterns (DAMPs), and denatured and oxidized proteins are significant contributors to this process.[61] The gut microbiome plays a central role in this process as it contributes to the release of inflammatory products and helps coordinate crosstalk with other systems.[62]

Interestingly, while the typical adult microbiome is stable throughout adulthood, it begins to deteriorate in the elderly, an effect exaggerated by the decreased physical activity and nutrition that often accompanies aging.[63,64] Predominantly, aging is associated with decreased microbial diversity, which is, in turn, associated with poor outcomes such as frailty, liver disease, IBD, and *Clostridium difficile* infection.[65–68] Aging is associated with a decrease in *Lactobacillus* and *Bifidobacterium*, as well as a reduced *Firmicutes*-to-*Bacteroidetes* ratio.[66,69–71] Conversely, studies have shown that the centenarians have a relatively higher *Firmicutes*-to-*Bacteroidetes* ratio as well as high levels of *Akkermansia*, *Clostridium*, *Collinsella* and *Christensenellaceae*, and decreased levels of *Faecalibacterium* and *Prevotella*.[72] These results suggest that although normal aging is associated with a distinct microbial profile, longevity is as well.

Indeed, studies in murine models have shown that loss of these bacteria impair intestinal barrier function and activate the immune system.[73] These results are supported by a similar study that found that microbiome composition correlated with frailty, comorbidity, nutritional status, inflammatory markers, and specific metabolites in fecal samples. Moreover, elderly patients residing in long-stay care facilities were significantly less diverse compared to those living in the community, and the loss of community-associated bacteria correlated with increased frailty.[66] In addition to the taxa mentioned above, *Bifidobacterium* are an especially promising genus of bacteria for supplementation in elderly populations as they are associated with colonic health during aging as well as negatively correlating with pro-inflammatory cytokine TNF-α and the anti-inflammatory cytokine IL-10.[74,75] Finally, as mentioned previously, calorie-restricted diets have been demonstrated as one of the most effective ways of improving health, increasing lifespan, and preventing metabolic- or age-related diseases.[76] As a major dietary intervention, this has a profound effect on the microbiome's composition and metabolite profile, with studies showing increased levels of eubiotic and butyrogenic bacteria and decreased levels of dysbiotic pro-inflammatory species after CR.[77] Overall, these findings suggest the therapeutic potential of gut microbiome intervention in the aging process.

Metabolic Disease: Obesity and Type 2 Diabetes

Metabolic diseases are closely related to the aging process, and studies have also demonstrated a potential role for the microbiome in their pathogenesis and therapy.[78,79] For instance, obesity, which affects 42.4% of Americans, is characterized by both an altered microbial composition and metabolite profile.[80–82] Predominantly, obesity increases the *Firmicutes*-to-*Bacteroidetes* ratio (~90% less *Bacteroidetes*), a finding which reversed after being placed on a calorie-restricted low-fat or low-carbohydrate diet.[83] It is also hypothesized that the microbiome can alter feeding behavior and appetite by modulating gut–brain signaling via the vagus nerve or gut hormones such as cholecystokinin (CCK), glucagon-like peptide-1.[84] For example, it is known that obesity is associated with leptin and CCK resistance, which results in reduced satiety and increased meal size.[85] The gut microbiome can alter this signaling, increased levels of SCFAs have been shown to increase leptin and CCK levels.[86,87] This is supported by studies examining outcomes post-bariatric surgery. By restructuring the digestive system to increase the speed of food through the stomach and proximal small intestine, bariatric surgery dumps more undigested food into the gut. While before surgery, this food would be digested and absorbed, it is now effectively a prebiotic for the gut microbiome to metabolize and ferment as evidenced by an excessive increase in flatulence experience post-operative.[88] Indeed, bariatric surgery has been demonstrated to significantly alter the gut microbiome as well as alter the hormonal and inflammatory status, reduce adiposity, and improve insulin sensitivity, and increase the bile acid pool independent of weight loss.[89–93] Specifically, the abundance of *Gammaproteobacteria* and *Verrucomicrobia*, that is, *Akkermansia*, were increased, while *Firmicutes* were decreased after bariatric surgery.[89] In studies examining post-operative outcomes of bariatric surgery, the combination of probiotics with post-operative treatment has been shown to increase bacterial diversity and enhance further weight loss while also decreasing GI side effects.[88,94] Positive anti-obesity outcomes have also been found after dietary-induced *Bifidobacterium* spp.[95] Similar results are seen in other metabolic disorders like type 2 diabetes (T2D). Recently, a study in a murine model found that insulin resistance improved, pancreatic β-cell apoptosis decreased, and inflammation decreased after FMT treatment from a healthy host. Overall, the results of these studies, combined with their microbial and metabolite changes, paint a promising picture for the role of the microbiome in metabolic syndromes like obesity and T2D.

GI Disorders: IBD and Irritable Bowel Syndrome

Inflammatory Bowel Disease

Within the gut itself, microbiome therapy has been shown to have profound effects on IBD, a chronic relapsing–remitting condition characterized by inflammation of the colon. Ulcerative colitis (UC), a form of IBD, is characterized by acute exacerbations of intestinal complications, with frequent remissions. It is hypothesized that genetics, immune disorders, intestinal dysbiosis, and environmental factors all contribute to the development of UC.[96] Naturally, the manipulation of this microbiota to reduce this inflammation may present a promising avenue of treatment for patients suffering from IBD. It has been found that probiotic administration improved the clinical symptoms, histological alterations, and mucus production in animal models.[97] In human studies, probiotic supplementation, particularly with *Bifidobacterium* spp., improved the clinical symptoms in patients with mild to moderately active UC. Additionally, it was found that *Bifidobacterium infantis* reduced plasma CRP, plasma IL-6, and LPS-stimulated TNF-α by PBMCs compared to the placebo.[98] In dietary studies of patients with IBD, it was found that the MD, which is high in fiber, is highly recommended during periods of remission due to the anti-inflammatory effect of prebiotics. However, this diet is relatively contraindicated during flares due to the acutely aggravating effects of fiber. Conversely, diets low in fiber, specifically the low fermentable oligosaccharides, disaccharides, monosaccharides, and polyols (FODMAP) diet, have also been found to improve symptoms of IBD particularly during flare-ups but also even in remission, although their effect on underlying inflammation is currently unknown.[99]

Irritable Bowel Syndrome

As opposed to the inflammation-induced symptoms of IBD, irritable bowel syndrome (IBS) is a purely functional GI disorder with no known cause, albeit a similar constellation of GI symptoms. In a recent systematic review examining the use of probiotics in IBS, it was found that multi-strain probiotics supplemented for 8 weeks or more significantly improved IBS symptoms, particularly abdominal pain.[100] However, other studies report conflicting results.

It is worth to mention here that the American Gastroenterological Association has recently released updated guidelines indicating that there is insufficient evidence to currently recommend the use of probiotics to treat Crohn's disease, UC, or IBS, rather, more well-designed randomized clinical trials are needed to evaluate their physiologic impact.[101] In dietary studies of IBS, most studies report beneficial effects of fiber on symptoms of IBS.[102] However, due to the heterogeneity in patient presentation and study design, firm conclusions are difficult to draw.

Overall, the risks and benefits of probiotic administration and high-fiber diets should be carefully assessed before starting any treatment in patients with IBD as not to exacerbate the immune response and aggravate the condition. Despite the lack of conclusions that can be drawn regarding the dietary treatment of IBS and IBD, it is important to note that a low intake of dietary fiber has also been associated with an increased incidence of IBD and IBS, thereby implicating diet in the acquisition of these gut disorders.[103] Given this, high-quality evidence from more prospective, double-blind, randomized control trials are needed to better understand the mechanisms by which probiotics and prebiotics affect IBS and IBD.

Neurologic Disorders

Another promising body of research is being developed concerning the gut microbiome and neurological function. As we discussed in-depth in Chapter 4, the gut microbiome also is connected to central nervous system (CNS) and neuroinflammation. It has been demonstrated that low-grade systemic inflammation, like that induced by a dysbiotic gut microbiome, increases the permeability across the BBB.[104] The increased permeability allows for the trafficking of PAMPs like LPS across the BBB, which activates neuroinflammation mediated by microglia, the CNS's immune cells.

Alzheimer's Disease

One active area of research within this space is Alzheimer's disease (AD). AD is characterized by a reduction in richness and diversity of gut microbiota, particularly a decreased *Firmicutes*-to-*Bacteroidetes* ratio and decreased *Bifidobacterium*. These alterations correlated strongly with a pathological load of AD markers after lumbar puncture, including phosphorylated tau.[105,106] Recent data have supported the hypothesis that probiotics can reduce the levels of amyloid-β in the brain and slow the progression of AD symptoms in mice.[107] The hypothesized mechanism behind the effect of probiotics and dysbiosis in AD is that probiotic bacteria increase the synthesis of the neurotrophic factors like brain-derived neurotrophic factor (BDNF), N-methyl-D-aspartate (NMDA) receptor, and GABA, while dysbiotic, pro-inflammatory bacteria secrete neurotoxic bacterial amyloid and LPS.[108] Despite these associations, the clinical value of microbiome intervention in AD is unclear. One recent randomized, double-blind controlled trial indicated improvements in Mini-Mental State Exam after supplementation with *Lactobacillus acidophilus*, *Lactobacillus casei*, *Bifidobacterium bifidum*, and *Lactobacillus fermentum* for 12 weeks.[109] However, another study in severe AD showed no improvement in cognitive functioning after multi-strain probiotic supplementation for 12 weeks.[110] One important consideration when analyzing these studies is that the gut microbiome is highly sensitive to lifestyle factors like diet and circadian rhythm, which are inherently altered in AD patients. Given these conflicting results and potential for confounding, more studies are needed to evaluate the relevant targets along with the gut–brain axis (GBA) in AD.

Multiple Sclerosis

Multiple sclerosis (MS) is also characterized by neuroinflammation, making it a possible target for therapeutic microbiome intervention as well. Much of the research regarding MS and the microbiome revolves around the activation of T helper 17 (Th)-17 cells. In models of MS, Th-17 cells induced a pro-inflammatory phenotype in astrocytes, critical regulators of neuronal homeostasis.[111] Indeed, ablation of the IL-17a pathway led to improvements in murine models of MS.[112] This ties back to the gut microbiome as Th-17 activation and dysfunction are increased by dysbiosis in the gut.[113] Interestingly, in experimental autoimmune encephalomyelitis (EAE), a murine model for MS, mice raised in germ-free (GF) conditions were protected against developing EAE, while the introduction of commensal bacteria into the gut restored susceptibility.[114] Additional studies have demonstrated that the microbiome itself can induce symptoms of EAE, mice undergoing FMT from a donor with EAE developed symptoms of EAE as well.[115,116] Specifically, in humans, MS is associated with increases in *Methanobrevibacter* and *Akkermansia* and decreases in *Butyricimonas* in the gut.[117] This correlates with altered expression of genes involved in dendritic cell maturation, interferon signaling, and NF-κB signaling pathways in T cells and monocytes. Moreover, patients on disease-modifying treatment show increased abundances of *Prevotella* and *Sutterella*, and decreased *Sarcina*, compared with untreated patients.[117] The protection afforded to GF models and the transferrable association of MS microbiota for developing MS raises the possibility of dietary interventions in the treatment of MS. However, more controlled and randomized studies are likely needed in human subjects to confirm this association.

Stroke

Post-stroke brain injury is another common pathology that shares many of the same pathophysiology with MS and AD (see Chapter 4 for more information). Gut permeability and microbial composition are known to be altered in murine models of stroke.[118] Murine models of ischemic stroke have found bacteria colonizing the lungs were derived from translocation across the small intestine, as opposed to aspiration from the oral cavity.[119] This increased intestinal translocation is likely to induce Th-17 dysfunction through the same mechanism as MS. Murine models have also shown that fecal material transplant (FMT) from stroke models can induce pro-inflammatory Th-17 dysfunction in GF mice.[9,118]

Finally, broad-spectrum antibiotic administration before ischemic events is associated with poorer outcomes.[9,120] Supporting these findings, inhibition of the IL-17 pathway is protective from ischemic stroke, as was the case for MS.[121] Moreover, abnormalities in Th-17 and IL-17 are also implicated in atherosclerotic plaque formation and instability.[122] Combined, the results of these implicate the Th-17 pathway and the gut microbiome in post-ischemic neuroinflammation while simultaneously opening the door for microbiome level intervention in stroke patients and as prophylaxis in those at risk for stroke.

Autism Spectrum Disorders

There is also growing evidence for the role of the gut microbiome in neurodevelopmental disorders such as autism spectrum disorder (ASD). In the United States, as many as 1 in 54 children have some form of ASD, an almost 50% increase from 2000.[123] Recent studies are pointing to the possible role of gut dysbiosis with many of the known biological mechanisms that may be associated with ASD.[124–126] The significant correlation of the GI symptoms and the Autism Treatment Evaluation Checklist (ATEC) reinforces the hypothesis that GI disorders are associated with autism severity.[127] Given this connection, several studies have found the benefit of probiotic and prebiotic administration in ASD. Probiotic supplementation has shown improvement in ASD symptoms, especially in behavior and GI symptoms. The most effective probiotics supplemented multiple species, the MTT is also very beneficial. In this study, MTT consisted of an oral or rectal supplement of standardized human gut microbiota, a purified, fecal-derived microbial extract.[58] While FMT is direct transplantation of fecal material into the hosts colon via colonoscopy, MTT is administered as a rectal or oral supplement. These interventions significantly decreased GI distress and significantly improved certain behaviors such as repetitive and opposition/defiance behaviors.[58,128,129] However, besides MTT, many of these benefits were temporary, lasting only as long as the treatment was administered, with no detectible changes after treatment.[58,128] One hypothesis for this is that by comprehensively supplementing hundreds of species, MTT — which is functionally similar to FMT — creates a more resilient microbiome compared to the supplementation of a handful of probiotics. Prebiotics have also been demonstrated to be beneficial in ASD, particularly regarding GI complaints.[130–132] Interestingly, two studies indicated that prebiotic supplementation decreased gut microbial diversity in ASD patients despite improving behavior and symptoms.[130,131] Despite the lower diversity reported, prebiotic supplementation significantly altered fecal and urine metabolites, particularly by increasing SCFAs, which are known to be significantly lower in children with autism.[127,130] Although these improvements are promising, more research is needed regarding the exact mechanisms by which modulating the gut microbiome can affect ASD symptomology.

Cognition and Central Processing

Alterations in the gut microbiome are linked to impaired central processing and emotional well-being. Gut microbiota composition has been linked with scores in speed, attention, and cognitive flexibility and alterations in neural activity in the thalamus, hypothalamus, and amygdala.[133] The microbiota composition in infants has also been associated with cognitive development as tested by the Mullen Scales of Early Learning, which found differences in brain volume and better performance in infants dominated *Bacteroides*; interestingly, this group also were more likely to be born vaginally.[134,135] In elderly populations, treatment with *Lactobacillus* spp. improved cognitive test performance compared with a placebo group.[136] Additional studies identified improvements in cognitive function after supplementation with prebiotic GOS and inulin among young healthy participants.[137,138] Many probiotics have also been associated with cognitive improvements. For example, *Lactobacillus plantarum* was found to alleviate stress and anxiety while reducing levels of pro-inflammatory cytokines in stressed adults along with enhanced social-emotional cognition and verbal learning and memory; this probiotic has also been used in studies of ASD children with positive results.[129,139] Additionally, *Bifidobacterium longum* has shown to mitigate stress-induced increases in cortisol, consequently reducing anxiety and enhancing hippocampal memory performance.[140,141] Overall, these studies emphasize the importance of gut microbiota in cognitive development and function.

It appears that the microbiome can influence these disorders by not only decreasing inflammation but also by altering neural connectivity. It has been demonstrated that GF mice have alterations in hippocampal NMDA and 5-hydroxytryptamine (5-HT$_{1A}$) receptor mRNA expression, which influence corticotropin-releasing factor (CRF) release from the hypothalamus.[9,142] Altered gene expression in GF mice is particularly apparent in the amygdala, which plays a significant part in social behavior. GF mice express increased levels of activation factors leading to neuronal hypertrophy and dendritic arborization leading to an overall increase in amygdala volume.[9,143] These changes are hypothesized to contribute to the social deficits seen in GF mice and those with antibiotic depleted or otherwise dysbiotic microbiomes as well.

Immune Disorders and Autoimmunity

Autoimmunity

Given the gut microbiome's intimate connection with the immune system and inflammatory regulation, it is increasingly apparent that it may play a role in the pathogenesis of immune disorders and autoimmune disorders. As discussed throughout this chapter, the gut microbiome may play a significant role in both the pathogenesis and treatment of many autoimmune diseases, including MS, IBD, UC, and Crohn's disease. The gut microbiome's ability to regulate these processes may be grounded in its role in inducing inflammation via abnormal microbial translocation, molecular mimicry, and dysregulation of local and systemic immunity, which primes the immune cells for autoreactivity; as well as its role in inducing immune tolerance via the differentiation of T regulatory cells (Tregs).[144] As discussed in the Microbiome and Inflammatory Disorder section, Tregs are up-regulated by SCFA activation of the Foxp3 pathway and are critical in inhibiting excessive immune reactions and inducing systemic immune tolerance. For example, in systemic lupus erythematosus (SLE), a classic example of autoimmunity, recent research has shown that SLE is characterized by a decreased *Firmicutes*-to-*Bacteroidetes* ratio and abnormal abundance of *Lactobacillus* in the gut. Additionally, increased levels of the gut bacteria *Enterococcus gallinarum* and *Lactobacillus reuteri* can be seen in systemic tissues, indicating increased bacterial translocation from the gut to the periphery.[145] These bacteria activate the body's immune system via the type 1 interferon pathway and induce autoantibody production. Moreover, increased bacteria translocation may also result in molecular mimicry of key self-epitopes like Ro60 or host DNA, two-key autoantibody targets in SLE.[72,146] These mechanisms are not unique to SLE, similar mechanisms are found in numerous autoimmune disorders, and even in the allergic response.

Allergy

The hygiene hypothesis of allergy has long been implicated in the decrease in exposure to previously common bacteria and antigens to the rapid increase in atopy since the middle of the last century.[147,148] This is closely linked to the gut microbiome as the lack of exposure to bacteria in one's environment leads to a decreased level of diversity in the gut microbiome. Indeed, other causes of decreased or altered bacterial diversity during infancy and early childhood such as C-section birth or growing up without pets have been linked to increased incidence of asthma, eczema allergies, as well as other conditions like the development of obesity and type 1 diabetes.[9,147,149,150] Additionally, experiments have shown that decreases in gut bacterial diversity as early as the first month of life precede the development of asthma, allergic rhinitis, and peripheral blood eosinophilia in school-age children. Specifically, diversity of *bifidobacterial* is thought to enhance the maturation of the mucosal immunoglobulin A (IgA) system, and early colonization with *Bacteroides fragilis* is hypothesized to down-regulate LPS responsiveness, thereby decreasing immune reactivity.[151]

Indeed, intervention within the gut to increase bacterial diversity has also been shown to decrease the allergic response and the incidence of atopy in children. Administration of prebiotic GOS and inulin pregnant and lactating mice provided a protective effect against food allergies in the offspring.[152] Additional experiments have shown that high-fiber diets therapeutically enhance the gut microbiome, increase the production of SCFAs, and increase retinal dehydrogenase activity (a key enzyme in the differentiation of Treg cells), which ultimately enhanced oral tolerance, protected against food allergy, boosted IgA production, and enhanced T follicular helper and mucosal germinal center responses. Moreover, mice lacking FFAR 2 (GPR43) or hydroxycarboxylic acid receptor 2 (GPR109A), another SCFA receptor, exhibited increased allergy responses.[153] Combined, these recent advances implicate early dysbiosis in the gut in the development of childhood atopy while also providing a potential avenue for therapy.

Anxiety and Depression

In addition to affecting cognition, the microbiome may also affect our perception of emotions, particularly stress, depression, and anxiety. Investigations into the cellular and molecular effects of depression have revealed that inflammation is bidirectionally associated with this condition. Depression is associated with increased cell-mediated immunity as well as increased oxidative and nitrosative stress, which induce chronic, low-grade inflammation and contribute to the progression of depression.[154] This hypothesis is also supported by a more recent study finding that depression was associated with increased levels of CRP, ESR, IL-1β, and IL-12.[155] Inflammatory mediators like these can alter brain signaling patterns by altering glutamate and monoamine neurotransmission, glucocorticoid receptor resistance, and hippocampal neurogenesis, affecting cognition and contributing to depression and depression-like symptoms.[156] Finally, serotonin reuptake inhibitors and serotonin–norepinephrine re-uptake inhibitors, commonly prescribed as anti-depressants, have been shown to decreased serum levels of TNF-α and increased levels of IL-10, an anti-inflammatory cytokine.[157] These findings highly implicate systemic inflammation in the bidirectional pathogenesis of depression. With this inflammatory connection in mind, decreasing the pro-inflammatory aspects of the gut microbiome and increasing its anti-inflammatory effects could potentially improve the symptoms of depression and other related disease processes such as stress and anxiety. Moreover, studies have demonstrated altered gut microbial compositions in patients with major depressive disorder (MDD) with significantly altered levels of *Firmicutes*, *Actinobacteria*, and *Bacteroidetes*; indeed, transplantation of a depression-associated microbiome into a healthy mouse resulted in depressive symptoms.[158] Moreover, neurotransmitters like neuropeptide Y, peptide YY, and pancreatic polypeptide are known to be altered by changes in microbiota composition and diet, and are associated with neuroprotection, increased neurogenesis, and decreased neuroinflammation.[159] Specifically, peripheral pancreatic polypeptide is known to be a potent activator of anxiety and fear in rodents.[159] Indeed, stress and anxiety are characterized by many of the same inflammatory processes. In the stress response, neuropeptides like substance P are released, which induces an inflammatory process. CRF is also produced, which coordinates the acute stress response and the production of acute-phase reactants, which are essential mediators of inflammation.[160] Similarly, anxiety is characterized by increased pro-inflammatory cytokines as well, predominantly IL-1β and TNF-α.[161]

CONCLUSIONS

Various inflammatory conditions are linked to the gut microbiome via its role in regulating immune responses. The gut's ability to induce systemic inflammation and neuroinflammation exacerbates many underlying pathology and predisposition to various diseases. Conversely, the gut microbiome's potent ability to induce immune-tolerance and decrease inflammation may protect against these same pathologies and predispositions. The gut's added ability to alter neurotransmitters and gene expression in the brain gives the gut added control over many neurologic functioning. Together these principles form the GBA. Intervention along the GBA is a promising avenue for future treatment in many chronic inflammatory processes like aging, obesity, type 2 diabetes, and many more conditions. Studies have also revealed promising results for microbiome level intervention with neurologic and psychiatric disorders like ASD, AD, and depression. Ultimately, the gut microbiome may have a significant role in individualized medicine in the future for many chronic conditions, tailoring the patient's treatment to both their disease and their gut microbiome and vice versa. The precise bacteria that a patient is deficient in may be supplemented to produce a specific therapeutic metabolite or outcompete a pathogenic species. Fecal material may also be transferred from a eubiotic donor to "reset" the host's microbiome. And finally, the diet as a whole may be restructured to reduce inflammation in long term or more profoundly alter physiology as in the case of a high fiber or KD. Despite these promising studies, it is essential to note that many of these connections are still unclear.

More studies are needed using the appropriate clinical-based evidence criteria to properly evaluate the risks and benefits of these novel interventions in each condition as well as elaborate the precise mechanisms by which their therapeutic benefits derive.

REFERENCES

1. Lepine JP, Briley M. (2011) The increasing burden of depression. *Neuropsychiatr Dis Treat* **7**(Suppl 1): 3–7, doi:10.2147/NDT.S19617.
2. Mathur N, Pedersen BK. (2008) Exercise as a mean to control low-grade systemic inflammation. *Mediat Inflamm* **2008**: 109502, doi:10.1155/2008/109502.
3. Vighi G, Marcucci F, Sensi L, Di Cara G, Frati F. (2008) Allergy and the gastrointestinal system. *Clin Exper Immunol* **153**(Suppl 1): 3–6, doi:10.1111/j.1365-2249.2008.03713.x.
4. Lobionda S, Sittipo P, Kwon HY, Lee YK. (2019) The role of gut microbiota in intestinal inflammation with respect to diet and extrinsic stressors. *Microorganisms* **7**(8): 271, doi:10.3390/microorganisms7080271.
5. Baxter NT, Schmidt AW, Venkataraman A, Kim KS, Waldron C, Schmidt TM. (2019) Dynamics of human gut microbiota and short-chain fatty acids in response to dietary interventions with three fermentable fibers. *mBio* **10**(1): e02566–18, doi:10.1128/mBio.02566-18.
6. Macfarlane S, Macfarlane GT. (2003) Regulation of short-chain fatty acid production. *Proc Nutr Soc* **62**(1): 67–72, doi:10.1079/PNS2002207.
7. Flint HJ. (2012) The impact of nutrition on the human microbiome. *Nutr Rev* **70**(Suppl 1): S10–13, doi:10.1111/j.1753-4887.2012.00499.x.
8. Wong JM, de Souza R, Kendall CW, Emam A, Jenkins DJ. (2006) Colonic health: Fermentation and short chain fatty acids. *J Clin Gastroenterol* **40**(3): 235–243, doi:10.1097/00004836-200603000-00015.
9. Cryan JF, *et al.* (2019) The microbiota-gut-brain axis. *Physiol Rev* **99**(4): 1877–2013, doi:10.1152/physrev.00018.2018.
10. Hamer HM, Jonkers D, Venema K, Vanhoutvin S, Troost FJ, Brummer RJ. (2008) Review article: The role of butyrate on colonic function. *Aliment Pharmacol Ther* **27**(2): 104–119, doi:10.1111/j.1365-2036.2007.03562.x.
11. Kelly CJ, *et al.* (2015) Crosstalk between microbiota-derived short-chain fatty acids and intestinal epithelial HIF augments tissue barrier function. *Cell Host Microbe* **17**(5): 662–671, doi:10.1016/j.chom.2015.03.005.
12. Miao W, *et al.* (2016) Sodium butyrate promotes reassembly of tight junctions in caco-2 monolayers involving inhibition of MLCK/MLC2 pathway and phosphorylation of PKCβ2. *Int J Mol Sci* **17**: 1696.

13. Peng L, Li ZR, Green RS, Holzman IR, Lin J. (2009) Butyrate enhances the intestinal barrier by facilitating tight junction assembly via activation of AMP-activated protein kinase in Caco-2 cell monolayers. *J Nutr* **139**(9): 1619–1625, doi:10.3945/jn.109.104638.
14. Tong LC, *et al.* (2016) Propionate ameliorates dextran sodium sulfate-induced colitis by improving intestinal barrier function and reducing inflammation and oxidative stress. *Front Pharmacol* **7**: 253, doi:10.3389/fphar.2016.00253.
15. Yan H, Ajuwon KM. (2017) Butyrate modifies intestinal barrier function in IPEC-J2 cells through a selective upregulation of tight junction proteins and activation of the Akt signaling pathway. *PloS One* **12**: e0179586.
16. Zheng L, *et al.* (2017) Microbial-derived butyrate promotes epithelial barrier function through IL-10 receptor-dependent repression of claudin-2. *J Immunol* **199**(8): 2976–2984, doi:10.4049/jimmunol.1700105.
17. Iacob S, Iacob DG. (2019) Infectious threats, the intestinal barrier, and its trojan horse: Dysbiosis. *Front Microbiol* **10**: 1676, doi:10.3389/fmicb.2019.01676.
18. Barcelo A, Claustre J, Moro F, Chayvialle JA, Cuber JC, Plaisancie P. (2000) Mucin secretion is modulated by luminal factors in the isolated vascularly perfused rat colon. *Gut* **46**(2): 218–224, doi:10.1136/gut.46.2.218.
19. Jung TH, Park JH, Jeon WM, Han KS. (2015) Butyrate modulates bacterial adherence on LS174T human colorectal cells by stimulating mucin secretion and MAPK signaling pathway. *Nutr Res Pract* **9**(4): 343–349, doi:10.4162/nrp.2015.9.4.343.
20. Willemsen LE, Koetsier MA, van Deventer SJ, van Tol EA. (2003) Short chain fatty acids stimulate epithelial mucin 2 expression through differential effects on prostaglandin E(1) and E(2) production by intestinal myofibroblasts. *Gut* **52**(10): 1442–1447, doi:10.1136/gut.52.10.1442.
21. Chang PV, Hao L, Offermanns S, Medzhitov R. (2014) The microbial metabolite butyrate regulates intestinal macrophage function via histone deacetylase inhibition. *Proc Natl Acad Sci USA* **111**(6): 2247–2252, doi:10.1073/pnas.1322269111.
22. Parada Venegas D, *et al.* (2019) Short Chain Fatty Acids (SCFAs)-Mediated gut epithelial and immune regulation and its relevance for inflammatory bowel diseases. *Front Immunol* **10**: 277, doi:10.3389/fimmu.2019.00277.

23. Shakespear MR, Halili MA, Irvine KM, Fairlie DP, Sweet MJ. (2011) Histone deacetylases as regulators of inflammation and immunity. *Trends Immunol* **32**(7): 335–343, doi:10.1016/j.it.2011.04.001.
24. Brown AJ, *et al.* (2003) The orphan G protein-coupled receptors GPR41 and GPR43 are activated by propionate and other short chain carboxylic acids. *J Biol Chem* **278**(13): 11312–11319, doi:10.1074/jbc.M211609200.
25. Le Poul E, *et al.* (2003) Functional characterization of human receptors for short chain fatty acids and their role in polymorphonuclear cell activation. *J Biol Chem* **278**(28): 25481–25489, doi:10.1074/jbc.M301403200.
26. Nilsson NE, Kotarsky K, Owman C, Olde B. (2003) Identification of a free fatty acid receptor, FFA2R, expressed on leukocytes and activated by short-chain fatty acids. *Biochem Biophys Res Commun* **303**(4): 1047–1052, doi:10.1016/s0006-291x(03)00488-1.
27. Nohr MK, *et al.* (2015) Expression of the short chain fatty acid receptor GPR41/FFAR3 in autonomic and somatic sensory ganglia. *Neuroscience* **290**: 126–137, doi:10.1016/j.neuroscience.2015.01.040.
28. La Sala MS, *et al.* (2013) Modulation of taste responsiveness by the satiation hormone peptide YY. *FASEB J* **27**(12): 5022–5033, doi:10.1096/fj.13-228064.
29. Vieira AT, *et al.* (2017) Dietary fiber and the short-chain fatty acid acetate promote resolution of neutrophilic inflammation in a model of gout in mice. *J Leuk Biol* **101**(1): 275–284, doi:10.1189/jlb.3A1015-453RRR.
30. Vinolo MA, *et al.* (2011) SCFAs induce mouse neutrophil chemotaxis through the GPR43 receptor. *PloS One* **6**(6): e21205, doi:10.1371/journal.pone.0021205.
31. Cox MA, *et al.* (2009) Short-chain fatty acids act as antiinflammatory mediators by regulating prostaglandin E(2) and cytokines. *World J Gastroenterol* **15**(44): 5549–5557, doi:10.3748/wjg.15.5549.
32. Luhrs H, *et al.* (2002) Butyrate inhibits NF-kappaB activation in lamina propria macrophages of patients with ulcerative colitis. *Scand J Gastroenterol* **37**(4): 458–466, doi:10.1080/003655202317316105.
33. Nastasi C, *et al.* (2015) The effect of short-chain fatty acids on human monocyte-derived dendritic cells. *Sci Rep* **5**: 16148, doi:10.1038/srep16148.
34. Smith PM, *et al.* (2013) The microbial metabolites, short-chain fatty acids, regulate colonic Treg cell homeostasis. *Science* **341**(6145): 569–573, doi:10.1126/science.1241165.

35. Arpaia N, *et al.* (2013) Metabolites produced by commensal bacteria promote peripheral regulatory T-cell generation. *Nature* **504**(7480): 451–455, doi:10.1038/nature12726.
36. Furusawa Y, *et al.* (2013) Commensal microbe-derived butyrate induces the differentiation of colonic regulatory T cells. *Nature* **504**(7480): 446–450, doi:10.1038/nature12721.
37. Luu M, Steinhoff U, Visekruna A. (2017) Functional heterogeneity of gut-resident regulatory T cells. *Clin Transl Immunol* **6**(9): e156, doi:10.1038/cti.2017.39.
38. Aleman FDD, Valenzano DR. (2019) Microbiome evolution during host aging. *PLoS Pathogens* **15**(7): e1007727, doi:10.1371/journal.ppat.1007727.
39. Singh RK, *et al.* (2017) Influence of diet on the gut microbiome and implications for human health. *J Transl Med* **15**(1): 73, doi:10.1186/s12967-017-1175-y.
40. Statovci D, Aguilera M, MacSharry J, Melgar S. (2017) The impact of Western diet and nutrients on the microbiota and immune response at mucosal interfaces. *Front Immunol* **8**: 838, doi:10.3389/fimmu.2017.00838.
41. Zinocker MK, Lindseth IA. (2018) The Western diet-microbiome-host interaction and its role in metabolic disease. *Nutrients* **10**(3): 365, doi:10.3390/nu10030365.
42. Maalouf M, Rho JM, Mattson MP. (2009) The neuroprotective properties of calorie restriction, the ketogenic diet, and ketone bodies. *Brain Res Rev* **59**(2): 293–315, doi:10.1016/j.brainresrev.2008.09.002.
43. Tanca A, *et al.* (2018) Caloric restriction promotes functional changes involving short-chain fatty acid biosynthesis in the rat gut microbiota. *Sci Rep* **8**(1): 14778, doi:10.1038/s41598-018-33100-y.
44. Fabbiano S, *et al.* (2018) Functional gut microbiota remodeling contributes to the caloric restriction-induced metabolic improvements. *Cell Metab* **28**(6): 907–921.e7, doi:10.1016/j.cmet.2018.08.005.
45. Redman LM, Smith SR, Burton JH, Martin CK, Il'yasova D, Ravussin E. (2018) Metabolic slowing and reduced oxidative damage with sustained caloric restriction support the rate of living and oxidative damage theories of aging. *Cell Metab* **27**(4): 805–815.e4, doi:10.1016/j.cmet.2018.02.019.
46. Paoli A, Bosco G. (2015) The ketogenic mediterranean diet, in *The Mediterranean Diet.* San Diego, CA, Academic Press, pp. 271–280.

47. Neal EG, *et al.* (2008) The ketogenic diet for the treatment of childhood epilepsy: A randomised controlled trial. *Lancet Neurol* **7**(6): 500–506, doi:10.1016/S1474-4422(08)70092-9.
48. Veech RL. (2004) The therapeutic implications of ketone bodies: The effects of ketone bodies in pathological conditions: Ketosis, ketogenic diet, redox states, insulin resistance, and mitochondrial metabolism. *Prostaglandins Leukot Essential Fatty Acids* **70**(3): 309–319, doi:10.1016/j.plefa.2003.09.007.
49. Lindefeldt M, *et al.* (2019) The ketogenic diet influences taxonomic and functional composition of the gut microbiota in children with severe epilepsy. *NPJ Biofilms Microbi* **5**(1): 5, doi:10.1038/s41522-018-0073-2.
50. Newell C, Bomhof MR, Reimer RA, Hittel DS, Rho JM, Shearer J. (2016) Ketogenic diet modifies the gut microbiota in a murine model of autism spectrum disorder. *Mol Autism* **7**(1): 37, doi:10.1186/s13229-016-0099-3.
51. Paoli A, Mancin L, Bianco A, Thomas E, Mota JF, Piccini F. (2019) Ketogenic diet and microbiota: Friends or enemies? *Genes* **10**(7): 534, doi:10.3390/genes10070534.
52. Makki K, Deehan EC, Walter J, Backhed F. (2018) The impact of dietary fiber on gut microbiota in host health and disease. *Cell Host Microbe* **23**(6): 705–715, doi:10.1016/j.chom.2018.05.012.
53. Garcia-Mantrana I, Selma-Royo M, Alcantara C, Collado MC. (2018) Shifts on gut microbiota associated to mediterranean diet adherence and specific dietary intakes on general adult population. *Front Microbiol* **9**: 890, doi:10.3389/fmicb.2018.00890.
54. Nagpal R, *et al.* (2018) Gut microbiome composition in non-human primates consuming a Western or Mediterranean diet. *Front Nutr* **5**: 28, doi:10.3389/fnut.2018.00028.
55. Bailey MA, Holscher HD. (2018) Microbiome-mediated effects of the mediterranean diet on inflammation. *Adv Nutr* **9**(3): 193–206, doi:10.1093/advances/nmy013.
56. Pandey KR, Naik SR, Vakil BV. (2015) Probiotics, prebiotics and synbiotics — A review. *J Food Sci Technol* **52**(12): 7577–7587, doi:10.1007/s13197-015-1921-1.
57. Shenderov BA. (2013) Metabiotics: Novel idea or natural development of probiotic conception. *Microb Ecol Health Dis* **24**(1): 20399, doi:10.3402/mehd.v24i0.20399.

58. Kang DW, *et al.* (2017) Microbiota Transfer Therapy alters gut ecosystem and improves gastrointestinal and autism symptoms: An open-label study. *Microbiome* **5**(1): 10, doi:10.1186/s40168-016-0225-7.
59. Lopez-Otin C, Blasco MA, Partridge L, Serrano M, Kroemer G. (2013) The hallmarks of aging. *Cell* **153**(6): 1194–1217, doi:10.1016/j.cell.2013.05.039.
60. Ren R, Ocampo A, Liu GH, Izpisua Belmonte JC. (2017) Regulation of stem cell aging by metabolism and epigenetics. *Cell Metab* **26**(3): 460–474, doi:10.1016/j.cmet.2017.07.019.
61. Franceschi C, Campisi J. (2014) Chronic inflammation (inflammaging) and its potential contribution to age-associated diseases. *J Gerontol Ser A Biol Sci Med Sci* **69**(Suppl 1): S4–S9, doi:10.1093/gerona/glu057.
62. Franceschi C, Garagnani P, Parini P, Giuliani C, Santoro A. (2018) Inflammaging: A new immune-metabolic viewpoint for age-related diseases. *Nat Rev Endocrinol* **14**(10): 576–590, doi:10.1038/s41574-018-0059-4.
63. Claesson MJ, *et al.* (2011) Composition, variability, and temporal stability of the intestinal microbiota of the elderly. *Proc Natl Acad Sci USA* **108**(Suppl 1): 4586–4591, doi:10.1073/pnas.1000097107.
64. Vauzour D, *et al.* (2017) Nutrition for the ageing brain: Towards evidence for an optimal diet. *Ageing Res Rev* **35**: 222–240, doi:10.1016/j.arr.2016.09.010.
65. Biagi E, *et al.* (2010) Through ageing, and beyond: Gut microbiota and inflammatory status in seniors and centenarians. *PloS One* **5**(5): e10667, doi:10.1371/journal.pone.0010667.
66. Claesson MJ, *et al.* (2012) Gut microbiota composition correlates with diet and health in the elderly. *Nature* **488**(7410): 178–184, doi:10.1038/nature11319.
67. Jackson MA, *et al.* (2016) Signatures of early frailty in the gut microbiota. *Genome Med* **8**(1): 8, doi:10.1186/s13073-016-0262-7.
68. Kriss M, Hazleton KZ, Nusbacher NM, Martin CG, Lozupone CA. (2018) Low diversity gut microbiota dysbiosis: Drivers, functional implications and recovery. *Curr Opin Microbiol* **44**: 34–40, doi:10.1016/j.mib.2018.07.003.
69. Mariat D, *et al.* (2009) The firmicutes/bacteroidetes ratio of the human microbiota changes with age. *BMC Microbiol* **9**(1): 123, doi:10.1186/1471-2180-9-123.

70. Nagpal R, *et al.* (2018) Gut microbiome and aging: Physiological and mechanistic insights. *Nutr Health Aging* **4**(4): 267–285, doi:10.3233/NHA-170030.
71. Odamaki T, *et al.* (2016) Age-related changes in gut microbiota composition from newborn to centenarian: A cross-sectional study. *BMC Microbiol* **16**(1): 90, doi:10.1186/s12866-016-0708-5.
72. Kim BS, *et al.* (2019) Comparison of the gut microbiota of centenarians in longevity villages of South Korea with those of other age groups. *J Microbiol Biotechnol* **29**(3): 429–440, doi:10.4014/jmb.1811.11023.
73. Bodogai M, *et al.* (2018) Commensal bacteria contribute to insulin resistance in aging by activating innate B1a cells. *Sci Transl Med* **10**(467): eaat4271, doi:10.1126/scitranslmed.aat4271.
74. Ouwehand AC, *et al.* (2008) Bifidobacterium microbiota and parameters of immune function in elderly subjects. *FEMS Immunol Med Microbiol* **53**(1): 18–25, doi:10.1111/j.1574-695X.2008.00392.x.
75. van der Lugt B, *et al.* (2018) Integrative analysis of gut microbiota composition, host colonic gene expression and intraluminal metabolites in aging C57BL/6J mice. *Aging* **10**(5): 930–950, doi:10.18632/aging.101439.
76. National Institute on Aging. (2020) *Calorie Restriction and Fasting Diets: What Do We Know?* https://www.ncbi.nlm.nih.gov/pubmed/.
77. Zheng X, Wang S, Jia W. (2018) Calorie restriction and its impact on gut microbial composition and global metabolism. *Front Med* **12**(6): 634–644, doi:10.1007/s11684-018-0670-8.
78. Bonomini F, Rodella LF, Rezzani R. (2015) Metabolic syndrome, aging and involvement of oxidative stress. *Aging Dis* **6**(2): 109–120, doi:10.14336/AD.2014.0305.
79. Dominguez LJ, Barbagallo M. (2016) The biology of the metabolic syndrome and aging. *Curr Opin Clin Nutr Metab Care* **19**(1): 5–11, doi:10.1097/MCO.0000000000000243.
80. Castaner O, *et al.* (2018) The gut microbiome profile in obesity: A systematic review. *Int J Endocrinol* **2018**: 4095789, doi:10.1155/2018/4095789.
81. CDC. (2020) Adult obesity facts | overweight & obesity | CDC. *Centers for Disease Controll and Prevention*. https://www.cdc.gov/obesity/data/adult.html.
82. Rangel-Huerta OD, Pastor-Villaescusa B, Gil A. (2019) Are we close to defining a metabolomic signature of human obesity? A systematic review

of metabolomics studies. *Metabolomics* **15**(6): 93, doi:10.1007/s11306-019-1553-y.

83. Davis CD. (2016) The gut microbiome and its role in obesity. *Nutr Today* **51**(4): 167–174, doi:10.1097/NT.0000000000000167.
84. Bonaz B, Bazin T, Pellissier S. (2018) The vagus nerve at the interface of the microbiota-gut-brain axis. *Front Neurosci* **12**: 49–49, doi:10.3389/fnins.2018.00049.
85. de Lartigue G, Barbier de la Serre C, Espero E, Lee J, Raybould HE. (2012) Leptin resistance in vagal afferent neurons inhibits cholecystokinin signaling and satiation in diet induced obese rats. *PloS One* **7**(3): e32967, doi:10.1371/journal.pone.0032967.
86. Byrne CS, Chambers ES, Morrison DJ, Frost G. (2015) The role of short chain fatty acids in appetite regulation and energy homeostasis. *Int J Obes* **39**(9): 1331–1338. doi:10.1038/ijo.2015.84.
87. Martin AM, Sun EW, Rogers GB, Keating DJ. (2019) The influence of the gut microbiome on host metabolism through the regulation of gut hormone release. *Front Physiol* **10**: 428, doi:10.3389/fphys.2019.00428.
88. Chen JC, Lee WJ, Tsou JJ, Liu TP, Tsai PL. (2016) Effect of probiotics on postoperative quality of gastric bypass surgeries: A prospective randomized trial. *Surg Obes Related Dis* **12**(1): 57–61, doi:10.1016/j.soard.2015.07.010.
89. Aron-Wisnewsky J, Clement K. (2014) The effects of gastrointestinal surgery on gut microbiota: Potential contribution to improved insulin sensitivity. *Curr Atheroscler Rep* **16**(11): 454, doi:10.1007/s11883-014-0454-9.
90. Hafida S, Mirshahi T, Nikolajczyk BS. (2016) The impact of bariatric surgery on inflammation: Quenching the fire of obesity? *Curr Opin Endocrinol Diabetes Obes* **23**(5): 373–378, doi:10.1097/MED.0000000000000277.
91. Kong LC, *et al.* (2013) Gut microbiota after gastric bypass in human obesity: Increased richness and associations of bacterial genera with adipose tissue genes. *Am J Clin Nutr* **98**(1): 16–24, doi:10.3945/ajcn.113.058743.
92. Luijten J, Vugts G, Nieuwenhuijzen GAP, Luyer MDP. (2019) The importance of the microbiome in bariatric surgery: A systematic review. *Obesity Surgery* **29**(7): 2338–2349, doi:10.1007/s11695-019-03863-y.
93. Maruvada P, Leone V, Kaplan LM, Chang EB. (2017) The human microbiome and obesity: Moving beyond associations. *Cell Host Microbe* **22**(5): 589–599, doi:10.1016/j.chom.2017.10.005.

94. Woodard GA, *et al.* (2009) Probiotics improve outcomes after Roux-en-Y gastric bypass surgery: A prospective randomized trial. *J Gastrointest Surg* **13**(7): 1198–1204, doi:10.1007/s11605-009-0891-x.
95. Kovatcheva-Datchary P, Arora T. (2013) Nutrition, the gut microbiome and the metabolic syndrome. *Best Pract Res Clin Gastroenterol* **27**(1): 59–72, doi:10.1016/j.bpg.2013.03.017.
96. Shen ZH, *et al.* (2018) Relationship between intestinal microbiota and ulcerative colitis: Mechanisms and clinical application of probiotics and fecal microbiota transplantation. *World J Gastroenterol* **24**(1): 5–14, doi:10.3748/wjg.v24.i1.5.
97. Plaza-Diaz J, Ruiz-Ojeda FJ, Vilchez-Padial LM, Gil A. (2017) Evidence of the anti-inflammatory effects of probiotics and synbiotics in intestinal chronic diseases. *Nutrients* **9**(6): 555. doi:10.3390/nu9060555.
98. Groeger D, *et al.* (2013) Bifidobacterium infantis 35624 modulates host inflammatory processes beyond the gut. *Gut Microbes* **4**(4): 325–339, doi:10.4161/gmic.25487.
99. Mentella MC, Scaldaferri F, Pizzoferrato M, Gasbarrini A, Miggiano GAD. (2020) Nutrition, IBD and gut microbiota: A review. *Nutrients* **12**(4): 944, doi:10.3390/nu12040944.
100. Dale HF, Rasmussen SH, Asiller OO, Lied GA. (2019) Probiotics in irritable bowel syndrome: an up-to-date systematic review. *Nutrients* **11**(9): 2048, doi:10.3390/nu11092048.
101. Su GL, *et al.* (2020) AGA clinical practice guidelines on the role of probiotics in the management of gastrointestinal disorders. *Gastroenterology* **159**: 697–705,
102. Axelrod CH, Saps M. (2018) The role of fiber in the treatment of functional gastrointestinal disorders in children. *Nutrients* **10**(11): 1650, doi:10.3390/nu10111650.
103. Hou JK, Abraham B, El-Serag H. (2011) Dietary intake and risk of developing inflammatory bowel disease: A systematic review of the literature. *Am J Gastroenterol* **106**(4): 563–573, doi:10.1038/ajg.2011.44.
104. Varatharaj A, Galea I. (2017) The blood-brain barrier in systemic inflammation. *Brain Behav Immun* **60**: 1–12, doi:10.1016/j.bbi.2016.03.010.
105. Vogt NM, *et al.* (2017) Gut microbiome alterations in Alzheimer's disease. *Sci Rep* **7**(1): 13537, doi:10.1038/s41598-017-13601-y.

106. Zhuang ZQ, *et al.* (2018) Gut microbiota is altered in patients with Alzheimer's disease. *J Alzheimer's Dis* **63**(4): 1337–1346, doi:10.3233/JAD-180176.
107. Abraham D, *et al.* (2019) Exercise and probiotics attenuate the development of Alzheimer's disease in transgenic mice: Role of microbiome. *Exp Gerontol* **115**: 122–131, doi:10.1016/j.exger.2018.12.005.
108. Askarova S, *et al.* (2020) The links between the gut microbiome, aging, modern lifestyle and Alzheimer's disease. *Front Cell Infect Microbiol* **10**: 104, doi:10.3389/fcimb.2020.00104.
109. Akbari E, *et al.* (2016) Effect of probiotic supplementation on cognitive function and metabolic status in Alzheimer's disease: A randomized, double-blind and controlled trial. *Front Aging Neurosci* **8**: 256, doi:10.3389/fnagi.2016.00256.
110. Agahi A, *et al.* (2018) Does severity of Alzheimer's disease contribute to its responsiveness to modifying gut microbiota? A double blind clinical trial. *Front Neurol* **9**: 662, doi:10.3389/fneur.2018.00662.
111. Prajeeth CK, *et al.* (2017) Effectors of Th1 and Th17 cells act on astrocytes and augment their neuroinflammatory properties. *J Neuroinflammation* **14**(1): 204, doi:10.1186/s12974-017-0978-3.
112. Kang Z, *et al.* (2010) Astrocyte-restricted ablation of interleukin-17-induced Act1-mediated signaling ameliorates autoimmune encephalomyelitis. *Immunity* **32**(3): 414–425, doi:10.1016/j.immuni.2010.03.004.
113. Pandiyan P, Bhaskaran N, Zou M, Schneider E, Jayaraman S, Huehn J. (2019) Microbiome dependent regulation of T_{regs} and Th17 cells in mucosa. *Front Immunol* **10**: 426, doi:10.3389/fimmu.2019.00426.
114. Berer K, *et al.* (2011) Commensal microbiota and myelin autoantigen cooperate to trigger autoimmune demyelination. *Nature* **479**(7374): 538–541, doi:10.1038/nature10554.
115. Berer K, *et al.* (2017) Gut microbiota from multiple sclerosis patients enables spontaneous autoimmune encephalomyelitis in mice. *Proc Natl Acad Sci USA* **114**(40): 10719–10724, doi:10.1073/pnas.1711233114.
116. Cekanaviciute E, *et al.* (2017) Gut bacteria from multiple sclerosis patients modulate human T cells and exacerbate symptoms in mouse models. *Proc Natl Acad Sci USA* **114**(40): 10713–10718, doi:10.1073/pnas.1711235114.

117. Jangi S, *et al.* (2016) Alterations of the human gut microbiome in multiple sclerosis. *Nat Commun* **7**(1): 12015, doi:10.1038/ncomms12015.
118. Singh V, *et al.* (2016) Microbiota dysbiosis controls the neuroinflammatory response after stroke. *J Neurosci* **36**(28): 7428–7440, doi:10.1523/JNEUROSCI.1114-16.2016.
119. Stanley D, *et al.* (2016) Translocation and dissemination of commensal bacteria in post-stroke infection. *Nat Med* **22**(11): 1277–1284, doi:10.1038/nm.4194.
120. Winek K, Dirnagl U, Meisel A. (2016) The gut microbiome as therapeutic target in central nervous system diseases: Implications for stroke. *Neurotherapeutics* **13**(4): 762–774, doi:10.1007/s13311-016-0475-x.
121. Gelderblom M, *et al.* (2012) Neutralization of the IL-17 axis diminishes neutrophil invasion and protects from ischemic stroke. *Blood* **120**(18): 3793–3802, doi:10.1182/blood-2012-02-412726.
122. Cipollini V, Anrather J, Orzi F, Iadecola C. (2019) Th17 and cognitive impairment: Possible mechanisms of action. *Front Neuroanat* **13**: 95, doi:10.3389/fnana.2019.00095.
123. National Center on Birth Defects and Developmental Disabilities, C. f. D. C. a. P. (2020) *Data & Statistics on Autism Spectrum Disorder | CDC*. https://www.cdc.gov/ncbddd/autism/data.html.
124. Altves S, Yildiz HK, Vural HC. (2020) Interaction of the microbiota with the human body in health and diseases. *Biosci Microbiota Food Health* **39**(2): 23–32, doi:10.12938/bmfh.19-023.
125. Mathee K, Cickovski T, Deoraj A, Stollstorff M, Narasimhan G. (2020) The gut microbiome and neuropsychiatric disorders: Implications for Attention Deficit Hyperactivity Disorder (ADHD). *J Med Microbiol* **69**(1): 14–24, doi:10.1099/jmm.0.001112.
126. Tyler Patterson T, Grandhi R. (2020) Gut microbiota and neurologic diseases and injuries, in *Gut Microbiota and Pathogenesis of Organ Injury*, vol. 1238. Singapore, Springer Singapore, pp. 73–91.
127. Adams JB, Johansen LJ, Powell LD, Quig D, Rubin RA. (2011) Gastrointestinal flora and gastrointestinal status in children with autism — Comparisons to typical children and correlation with autism severity. *BMC Gastroenterol* **11**(1): 22, doi:10.1186/1471-230X-11-22.

128. Arnold LE, *et al.* (2019) Probiotics for gastrointestinal symptoms and quality of life in autism: A placebo-controlled pilot trial. *J Child Adol Psychopharmacol* **29**(9): 659–669, doi:10.1089/cap.2018.0156.
129. Liu YW, *et al.* (2019) Effects of lactobacillus plantarum PS128 on children with autism spectrum disorder in Taiwan: A randomized, double-blind, placebo-controlled trial. *Nutrients* **11**(4): 820, doi:10.3390/nu11040820.
130. Grimaldi R, *et al.* (2018) A prebiotic intervention study in children with autism spectrum disorders (ASDs). *Microbiome* **6**(1): 133, doi:10.1186/s40168-018-0523-3.
131. Inoue R, *et al.* (2019) Dietary supplementation with partially hydrolyzed guar gum helps improve constipation and gut dysbiosis symptoms and behavioral irritability in children with autism spectrum disorder. *J Clin Biochem Nutr* **64**(3): 217–223, doi:10.3164/jcbn.18-105.
132. Sanctuary MR, *et al.* (2019) Pilot study of probiotic/colostrum supplementation on gut function in children with autism and gastrointestinal symptoms. *PloS One* **14**(1): e0210064, doi:10.1371/journal.pone.0210064.
133. Fernandez-Real JM, *et al.* (2015) Gut microbiota interacts with brain microstructure and function. *J Clin Endocrinol Metabol* **100**(12): 4505–4513, doi:10.1210/jc.2015-3076.
134. Carlson AL, *et al.* (2018) Infant gut microbiome associated with cognitive development. *Biol Psychiatry* **83**(2): 148–159, doi:10.1016/j.biopsych.2017.06.021.
135. Polidano C, Zhu A, Bornstein JC. (2017) The relation between cesarean birth and child cognitive development. *Sci Rep* **7**(1): 11483, doi:10.1038/s41598-017-10831-y.
136. Chung Y-C, *et al.* (2014) Fermented milk of Lactobacillus helveticus IDCC3801 improves cognitive functioning during cognitive fatigue tests in healthy older adults. *J Funct Foods* **10**: 465–474, doi:10.1016/j.jff.2014.07.007.
137. Smith AP, Sutherland D, Hewlett P. (2015) An investigation of the acute effects of oligofructose-enriched inulin on subjective wellbeing, mood and cognitive performance. *Nutrients* **7**(11): 8887–8896, doi:10.3390/nu7115441.

138. Tillisch K, *et al.* (2013) Consumption of fermented milk product with probiotic modulates brain activity. *Gastroenterology* **144**(7): 1394–1401.e4, doi:10.1053/j.gastro.2013.02.043.
139. Lew LC, *et al.* (2019) Probiotic lactobacillus plantarum P8 alleviated stress and anxiety while enhancing memory and cognition in stressed adults: A randomised, double-blind, placebo-controlled study. *Clin Nutr* **38**(5): 2053–2064, doi:10.1016/j.clnu.2018.09.010.
140. Allen AP, *et al.* (2016) Bifidobacterium longum 1714 as a translational psychobiotic: Modulation of stress, electrophysiology and neurocognition in healthy volunteers. *Transl Psychiatry* **6**(11): e939, doi:10.1038/tp.2016.191.
141. Savignac HM, Kiely B, Dinan TG, Cryan JF. (2014) Bifidobacteria exert strain-specific effects on stress-related behavior and physiology in BALB/c mice. *Neurogastroenterol Motil* **26**(11): 1615–1627, doi:10.1111/nmo.12427.
142. Neufeld KM, Kang N, Bienenstock J, Foster JA. (2011) Reduced anxiety-like behavior and central neurochemical change in germ-free mice. *Neurogastroenterol Motil* **23**(3): 255–264, e119, doi:10.1111/j.1365-2982.2010.01620.x.
143. Luczynski P, McVey Neufeld KA, Oriach CS, Clarke G, Dinan TG, Cryan JF. (2016) Growing up in a bubble: Using germ-free animals to assess the influence of the gut microbiota on brain and behavior. *Int J Neuropsychopharmacol* **19**(8): pyw020, doi:10.1093/ijnp/pyw020.
144. Zhang X, Chen BD, Zhao LD, Li H. (2020) The gut microbiota: Emerging evidence in autoimmune diseases. *Trends Mol Med* **26**: 862–873, doi:10.1016/j.molmed.2020.04.001.
145. Kim JW, Kwok SK, Choe JY, Park SH. (2019) Recent advances in our understanding of the link between the intestinal microbiota and systemic lupus erythematosus. *Int J Mol Sci* **20**(19): 4871, doi:10.3390/ijms20194871.
146. Greiling TM, *et al.* (2018) Commensal orthologs of the human autoantigen Ro60 as triggers of autoimmunity in lupus. *Sci Transl Med* **10**(434): eaan2306, doi:10.1126/scitranslmed.aan2306.
147. Azad MB, *et al.* (2013) Infant gut microbiota and the hygiene hypothesis of allergic disease: Impact of household pets and siblings on microbiota composition and diversity. *Allergy Asthma Clin Immunol* **9**(1): 15, doi:10.1186/1710-1492-9-15.

148. Strachan DP. (2000) Family size, infection and atopy: The first decade of the "hygiene hypothesis." *Thorax* **55**(Suppl 1): S2–S10, doi:10.1136/thorax.55.suppl_1.s2.
149. Abrahamsson TR, Jakobsson HE, Andersson AF, Bjorksten B, Engstrand L, Jenmalm MC. (2014) Low gut microbiota diversity in early infancy precedes asthma at school age. *Clin Exp Allergy* **44**(6): 842–850, doi:10.1111/cea.12253.
150. Penders J, *et al.* (2007) Gut microbiota composition and development of atopic manifestations in infancy: The KOALA birth cohort study. *Gut* **56**(5): 661–667, doi:10.1136/gut.2006.100164.
151. Sjogren YM, Tomicic S, Lundberg A, Bottcher MF, Bjorksten B, Sverremark-Ekstrom E, Jenmalm MC. (2009) Influence of early gut microbiota on the maturation of childhood mucosal and systemic immune responses. *Clin Exper Allergy* **39**(12): 1842–1851, doi:10.1111/j.1365-2222.2009.03326.x.
152. Bouchaud G, *et al.* (2016) Maternal exposure to GOS/inulin mixture prevents food allergies and promotes tolerance in offspring in mice. *Allergy* **71**(1): 68–76, doi:10.1111/all.12777.
153. Tan J, *et al.* (2016) Dietary fiber and bacterial SCFA enhance oral tolerance and protect against food allergy through diverse cellular pathways. *Cell Rep* **15**(12): 2809–2824, doi:10.1016/j.celrep.2016.05.047.
154. Berk M, *et al.* (2013) So depression is an inflammatory disease, but where does the inflammation come from? *BMC Med* **11**: 200, doi:10.1186/1741-7015-11-200.
155. Beydoun MA, *et al.* (2019) Systemic inflammation is associated with depressive symptoms differentially by sex and race: A longitudinal study of urban adults. *Mol Psychiatry* **25**: 1286–1300, doi:10.1038/s41380-019-0408-2.
156. Amodeo G, Trusso MA, Fagiolini A. (2017) Depression and inflammation: Disentangling a clear yet complex and multifaceted link. *Neuropsychiatry* **7**(4): 448–457.
157. Ohgi Y, Futamura T, Kikuchi T, Hashimoto K. (2013) Effects of antidepressants on alternations in serum cytokines and depressive-like behavior in mice after lipopolysaccharide administration. *Pharmacol Biochem Behav* **103**(4): 853–859, doi:10.1016/j.pbb.2012.12.003.

158. Zheng P, *et al.* (2016) Gut microbiome remodeling induces depressive-like behaviors through a pathway mediated by the host's metabolism. *Mol Psychiatry* **21**(6): 786–796, doi:10.1038/mp.2016.44.
159. Lach G, Schellekens H, Dinan TG, Cryan JF. (2018) Anxiety, depression, and the microbiome: A role for gut peptides. *Neurotherapeutics* **15**(1): 36–59, doi:10.1007/s13311-017-0585-0.
160. Black PH. (2002) Stress and the inflammatory response: A review of neurogenic inflammation. *Brain Behav Immun* **16**(6): 622–653, doi:10.1016/s0889-1591(02)00021-1.
161. Renna ME, O'Toole MS, Spaeth PE, Lekander M, Mennin DS. (2018) The association between anxiety, traumatic stress, and obsessive-compulsive disorders and chronic inflammation: A systematic review and meta-analysis. *Depress Anxiety* **35**(11): 1081–1094, doi:10.1002/da.22790.

Chapter 10

Role of Gut–Brain Axis in HIV-1 Infection

Tanya Chhibber[1,*], Behnaz Lahooti[1,*], Sree-Pooja Varahachalam[1,*], Sounak Bagchi[1], Abhijeet Joshi[2] and Rahul Dev Jayant[1]

Abstract

The gut–brain axis (GBA) represents the network that connects the gut and brain, and plays a crucial role in maintaining homeostasis. It is now gaining ever more attention in fields exploring the biological and physiological basis of infectious, age-related, psychiatric, neurodevelopmental, and neurodegenerative disorders. The brain and gut are connected both physically and biochemically in several different ways such as through the immune system, vagus nerve, or enteric nervous system, involving microbial metabolites. Gut microbiota plays an important role in regulating the GBA. Published data have suggested the role of gut microbiota in many neurological conditions, including autism, anxiety, schizophrenia, Parkinson's disease, Alzheimer's disease, and human immunodeficiency virus-1 (HIV-1) infection.

[1]Department of Pharmaceutical Sciences, JH School of Pharmacy, Texas Tech University, Health Sciences Center (TTUHSC), Amarillo, TX, USA.

[2]Centre for Biosciences and Bio-medical Engineering, Indian Institute of Technology Indore, Indore, India.

* These authors contributed equally to this work.

Corresponding Author: Rahul Dev Jayant, PhD, Department of Pharmaceutical Sciences, School of Pharmacy, Texas Tech University Health Sciences Center, 1406 S. Coulter Street, Suite 1104, Amarillo, TX 79106, USA, E-mail: rahuldev.jayant@gmail.com. Ph no: +17864090516.

The crosstalk between gut microbiota and GBA may have a crucial impact on neurogenerative processes during HIV-1 infection. In this chapter, we discuss our current understanding about gut microbiota composition, its role in HIV-1 infection, and promising therapeutic approaches to treat the gut dysbiosis in HIV-1 infection.

Keywords: HIV-1, Gut microbiome, Microbiota, Dysbiosis, Mucosal barrier, Mucosal immunity, Anti-retroviral (ARV) drugs.

INTRODUCTION

The human immunodeficiency virus (HIV) infects cells of the immune system (CD4+ T cells) and subsequently destroys or disrupts the function of the immune system. In the more advanced stages of HIV infection, an acquired immunodeficiency syndrome (AIDS) develops. The AIDS-associated neurological disorder can be instigated by HIV infection of the brain cells, by opportunistic infections or other viruses, or by the cytotoxic properties of anti-retroviral (ARV) drugs. One of the critical neurological problems associated with AIDS is HIV-associated dementia (HAD).[1,2] HIV enters the brain through mononuclear phagocytes utilizing the "Trojan Horse" mechanism.[3] Brain cells (e.g., macrophages, astrocytes, and microglia) are activated by HIV infection through communication with viral proteins (Tat, gp120, or Nef) or via other environmental factors. Neurons may also be affected indirectly by releasing viral proteins and leading to neurodegeneration.[4,5]

The HIV infection of the brain eventually leads to motor and cognitive dysfunction. The inflammation and circulating monocytes (specifically CD16 and CD69) play an important role in damaging the brain microvasculature as well as blood–brain barrier (BBB) integrity, and are significant in neuroHIV etiology.[4] There is also a growing indication that gut microbial composition can intensely affect numerous neurological disorders such as anxiety, stroke, depression, autism, and HIV. The microbiota–gut–brain axis is a bidirectional communication system between the brain and gut microbiota, and is a leading factor in preserving neuroendocrine and neuroimmune equilibrium. Gut microbiota yield metabolites such as short-chain fatty acids (SCFAs), and neurotransmitters that can enter the central nervous system (CNS), alter neuronal activity, or affect brain function via the immune system.[6,7]

INFLUENCE OF ANTI-RETROVIRAL THERAPY ON MICROBIOME COMPOSITION

Anti-retroviral therapy (ART) has been used to control HIV viral replication (via reducing viral loads) in plasma to undetectable levels that can increase CD4+ T cell counts. Yet it still fails to eliminate HIV reservoirs,[5,8] unable to restore the normal gut microbiome, and resolve gut inflammation.[9–11] Undoubtedly, ART has transformed the face of HIV infection from a lethal infection to a chronic controllable disease condition. Numerous studies have shown that ART (short-term or long-term use) is incapable of reinstating gut well-being regardless of active suppression of plasma HIV-1 viral load. Undeniably, the resolution of residual inflammation and gut microbial translocation is partially under ART.[12] Regardless of these developments, HIV-associated morbidities continue to be a substantial concern in the post-ART era.

Studies have shown that ARV drugs could lead to gut dysbiosis.[13] Different ARV combinations could have vivid effects on the gut microbiome as ARVs themselves have anti-microbial properties. In addition, gut microbiota (mainly bacteria) could metabolize ARV drugs. The augmentation or exhaustion of some essential microbes within the gut microbiota in HIV-infected individuals has been associated with HIV disease progression. For example, the reduction of butyrate-producing bacteria (BPB) has been linked with higher microbial translocation and immune activation.[14] Bacteria maintain epithelial health and immune homeostasis through delivering short-chain fatty acids (SCFAs) such as butyrate, which have been steadily found in lesser amounts in HIV-infected individuals on ART compared with healthy individuals.[15,16] One study conducted by Villanueva-Millán *et al.* revealed that ART combinations of protease inhibitors (PIs), non-nucleoside reverse transcriptase inhibitors (NNRTIs), and integrase strand transfer inhibitors (INSTIs) with nucleoside reverse transcriptase inhibitors (NRTIs) lead to increased plasma levels of endothelial damage markers compared to non-HIV-infected controls. Of the three, the PI-based combination showed the most endothelial damage compared with both NNRTIs and INSTIs.[16]

Similarly, Pinto-Cardoso *et al.* showed that a ritonavir-boosted PI-based combination leads to increased endothelial damage compared with both NNRTIs and non-HIV-infected controls. In contrast, NNRTI-based ART damage was augmented compared with HIV-uninfected controls.[15] Both these studies have used soluble CD14 as a marker to envisage mortality in HIV infection. Thus, these results offer a strong indication that ARV combinations endorse disparity dysbiosis, leading to inflammation and immune activation. One likely explanation for the different efficacies of ARV drugs on the gut microbiome (specifically in terms of markers of microbial translocation, immune activation, and gut inflammation) would be the degree of ARV drug penetration on the gastrointestinal (GI) tissue and their pharmacokinetic-pharmacodynamic properties. For example, Raltegravir (integrase inhibitor) has been shown to penetrate quicker in the GI tissue compared to other drugs from the same class.[12,17] The newer studies have started to offer relevant information suggesting that gut microbes can affect the efficacy of some cancer drugs[18] and ART.[19] Klatt *et al.* have shown that the microbiota in the female reproductive tract can directly metabolize Tenofovir (NRTI), and that the presence of *Gardnerella vaginalis* bacteria is associated with decreased efficacy of topical pre-exposure prophylaxis (PrEP) in women. Thus, we need to refine our understanding of the influence of ART and different ARV combinations to create a comprehensive view, as new HIV-infected individuals will have access to ART and as more people will have ART access to PrEP in the absence of HIV-1 infection.

MICROBIOTA COMPOSITION AND DIVERSITY DURING HIV-1 INFECTION

The human CNS and the GI tract interconnect via the GBA. The composition of gut microbiota affects the intestinal barrier permeability. During dysbiosis (microbial imbalance or impaired microbiota), the activation of gut-associated lymphoid tissue (GALT) as well as inflammation may lead to higher intestinal permeability.[20–22] Communication between the intestinal barrier elements has a selective ability to absorb and secrete specific substances, while preventing the translocation of microorganisms and the diffusion of toxins and other harmful antigens.[20,23] Due to this "leakiness" of the intestinal barrier, the GI tract becomes susceptible to infection by bacteria or viruses, leading to increased predisposition to inflammatory, autoimmune, or neurological disorders. The gut barrier bears a resemblance in structure (being composed of epithelial and endothelial cells) and function (cellular tight junctions) to the BBB. The biochemical information between the gut and CNS is carried out by cytokines, chemokines, neurotransmitters, and micro-vesicles via the neural and endocrine pathways present in the gut.[24]

It is well established that HIV-infection leads to extensive damage to the GI tract causing mucosal inflammation, abnormal cytokine production, immunological changes, structural damage of the epithelial barrier, and disruption of intestinal homeostasis as shown in Fig. 10.1. HIV-1 replicates in the GALT in the preliminary phases of infection, resulting in a massive reduction of mucosal effector CD4+ T cells, including T helper (Th) 17 and Th22 cells, which are essential for preserving the integrity of the gut.[25] Even in the presence of a highly active anti-retroviral therapy (HAART), due to the presence of comorbidities, HIV infection

(a)

(b)

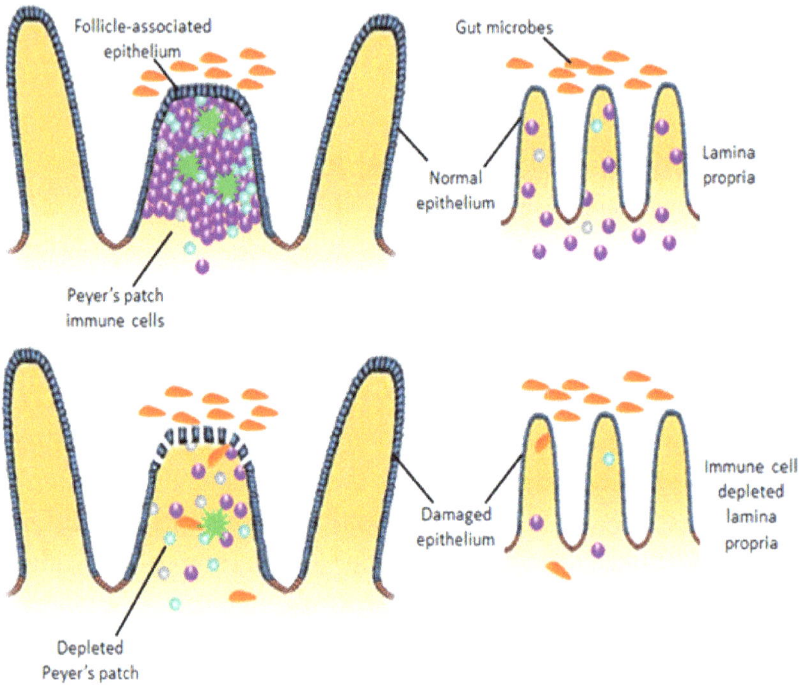

Fig. 10.1: Role of microbial translocation and immune cell activation in human immunodeficiency virus-1 (HIV-1) infection. (a) The normal small intestine epithelium contains a follicle-associated epithelium (FAE) that specializes in the transport of antigen into underlying lymphoid cells in Peyer's patches. Immune cell traffic from inductive sites such as Peyer's patches through the lymph and systemic circulations to effector sites in the small intestine lamina propria. (b) During immune depletion in HIV-1 infection, a breakdown of FAE tight junctions occurs, resulting in microbial products in plasma. These microbial products may prompt systemic immune activation. (Adapted from Haynes[26] and created with BioRender.com)

leads to chronic immune activation, neuroinflammation, and neurocognitive dementia. The association between HIV-1-associated mucosal pathogenesis and the microbiome is probably reciprocal, with changes in mucosa leading to dysbiosis and consequently, playing a pivotal role in the disruption of intestinal homeostasis as well as contributing to HIV-1-related immune activation and inflammation. Therefore, it is essential to study the interplay between the microbiome and HIV-1 pathogenesis to understand the role of ART drugs in dysbiosis, viral suppression, and restoration of GI functions in HIV-infected individuals.

Studies exploring the role of microbiome in the setting of HIV-1 infection have used various sequence-based technologies to systematically recognize and determine the composition of gut microbiome.[6,9,10,27,28] Alterations in gut microbiome have been detected in the three most dominant bacterial species found in the GI tract, Proteobacteria, Bacteroidetes, and Firmicutes.[29] *In situ* fluorescence hybridization and quantitative Polymerase Chain Reaction (PCR) (16S RNA gene expression) techniques also have been widely employed to study the composition (such as Enterobacteriales, Bacteroidales, and Clostridiales) and diversity of microbiota using fecal samples. In most of the studies, it has been observed that HIV-1 infection is related to increased Enterobacteriales that link with lower frequencies of CD4+ T cells.[25,30] For microbial diversity studies (i.e., comparison of relative taxa), an enormous difference has been observed in HIV-infected subjects compared to uninfected control subjects in both mucosal[28] and fecal samples.[28,31,32] Diversity is further impacted by ART-treated versus non-treated subjects, and HIV-associated variability by men who have sex with men (MSM) versus non-MSM, more for the mucosal samples than the fecal samples.[33] Therefore, understanding critical gut microbiome functional pathways that are transformed during HIV-1 infection will provide novel insights regarding the association between the host microbiota and immune system. The information derived from these studies will help in developing novel treatment modalities to improve the health of HIV-1 infected patients by modulating the GBA.[25]

APPROACHES TO RESCUE GUT MUCOSAL INTEGRITY AND IMMUNITY

Studies have shown that both abnormal environmental factors and host immune dysregulation contribute to GI imbalance and inflammatory diseases. The specific gut microbiota and metabolites mainly influence the host's innate and adaptive immunity in the GI tract via inducing immune cell function and differentiation or through the expression of numerous molecules involved in gut barrier function. A similar situation has been observed in the relationship between gut dysbiosis and HIV-1 pathogenesis. Several studies have started to better understand the possible mechanisms that lead to microbiota-associated immune activation and inflammation during HIV-1 infection.[34,35] Even though they are not conclusive, the majority of indications show that changes in gut microbiota secondary to HIV-1 infection play a crucial role in the severity of HIV-1-linked chronic inflammation. Studies have attempted to alter the gut microbiome to diminish HIV-1-associated inflammation, immune activation, and poor clinical outcomes.[36,37] Probiotics have shown beneficial effects in maintaining the gut microbiome environment via stimulating the growth or activity of helpful bacteria in the host, or through consumption of products that comprise a blend of both prebiotics and probiotics (synbiotics)[38,39] Diet has a vital connection to the intestinal microbiota and the host immunity. So far, very few studies have been conducted to study the impact of diet on the intestinal microbiome in regard to HIV-1 infection.

Therefore, by assessing the configuration of the intestinal microbiome at the taxonomic level, several technologies now allow the validation of microbial community function. Re-establishing the gut microbiome in HIV-1 infection using probiotics and prebiotics have an assortment of consequences for intestinal barrier functions, resistance to pathogenic colonization, and restoration of the mucosal immune function[11,40] as depicted in Fig. 10.2. Some studies have shown the modest improvement of CD4+ T cell counts and GI health.[41,42] Another study showed a reduction in the activation of circulating CD4+ T cells, production of inflammatory cytokines and chemokines, and increased natural killer (NK) cell activity, thus emphasizing the clinical potential to sustain immune rebuilding and ease pro-virus removal.[43,44]

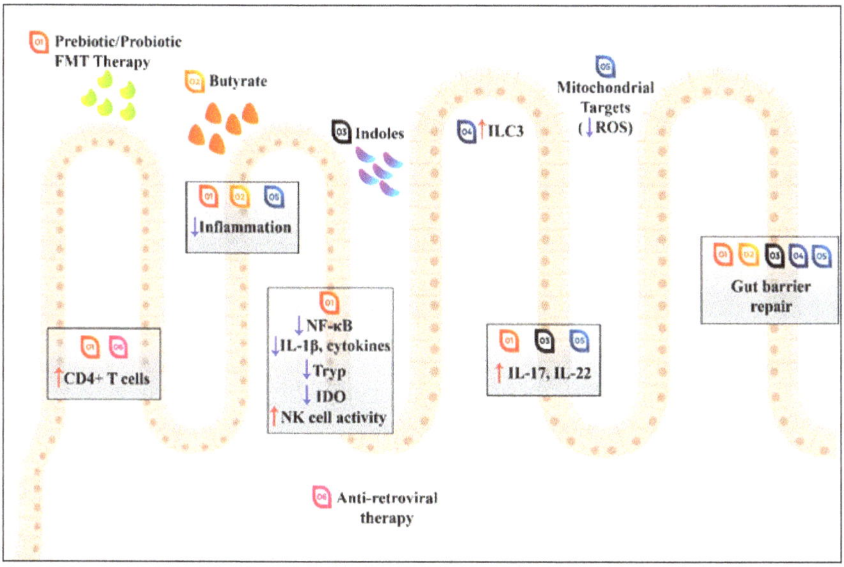

Fig. 10.2: Therapeutic approaches used to restore mucosal injury during human immunodeficiency virus-1 (HIV-1) infection. Strategies comprise (01) retrieval of gut microbial communities, (02) prebiotic stimulation using butyrate, (03) adding of bacterial metabolites such as indoles, (04) augmenting host immune cell function through innate lymphoid cells, (05) contesting cellular metabolism via mitochondrial targets such as reactive oxygen species (ROS), and (06) use of anti-retroviral therapy (ART). (Created with BioRender.com.)

A similar observation was also reported in an (Simian immunodeficiency virus) SIV-infected macaque model.[45]

Studies have shown that probiotics/prebiotics supplementation during ART treatment results in partial improvement of the GI tract function, healthier reconstitution and functionality of intestinal CD4+ T cells, and reduced fibrosis in colonic lymphoid follicles. Table 10.1 highlights the therapeutic approaches used to improve the dysbiosis resulting from the HIV-1 infection. Alternatively, fecal microbiota transplantation has shown excellent results in maintaining epithelial homeostasis in SIV-1-infected individuals via enhancing the number of Th17 and Th22 cells.[46] Research efforts to advance gut mucosal immunity have also concentrated on the refinement of host cellular signaling pathways that are transformed during HIV infection.

Table 10.1: Therapeutic Approaches Used to Modulate the Gut Microbiome During HIV-1 Infection

Type of Treatment	Microorganism Species and Products Used
Probiotics	Single (e.g., *Lactobacillus rhamnosus*) and multiple strains of bacteria (e.g., *L. rhamnosus*, *Lactobacillus reuteri*) supplemented in yogurt.
	Single (e.g., *Lactobacillus casei*) and multiple strains of bacteria (e.g., *Lactobacillus* spp., *Bifidobacteria* spp.) supplemented with fermented milk.
	Single (e.g., *Saccharomyces boulardii*; *Bacillus coagulans*) and multiple strains of bacteria (e.g., *Streptococcus* spp., *Bifidobacteria* spp., *Lactobacillus* spp.) delivered through capsules.
Prebiotics	Single compounds (e.g., oligosaccharides: short-chain galacto-oligosaccharides, long-chain fructo-oligosaccharides (inulin), pectin hydrolysate-derived acidic oligosaccharides).
	Multiple compounds (e.g., oligosaccharides, bovine colostrum protein, cysteine, lipids, lactose, and glucose).
Synbiotics	Strains of bacteria (e.g., *Pediococcus pentosaceus*, *Leuconostoc mesenteroides*, *Lactobacillus paracasei*, *Lactobacillus plantarum*) + non-digestible, fermentable dietary fibers (e.g., betaglucan, inulin, pectin, and resistant starch).

CONCLUSION AND FUTURE PERSPECTIVES

In conclusion, all the studies performed so far (*in vitro* or *in vivo*) have shown that host immune cells, intestinal epithelial cells and gut microbiota are essential in the preservation of gut homeostasis by interconnecting with each other. Together, abnormal environmental factors and gut immune system dysfunction leads to inflammation in the GI tract. In HIV-1 infection, how gut microbiota is involved in the host inflammatory progression that affects HIV replication and its associated mechanism is in very infantile (preclinical or clinical) stages. Non-human primate studies have advanced our understanding about how gut-associated inflammation, viral transmission, primary immune reactions, and prevention and/or treatment strategies can be developed. Recent technological innovations allow the evaluation of the intestinal microbiome composition and diversity. These technologies have facilitated the disentangling of the multi-faceted interactions among the intestinal microbial taxa and the host in the context of HIV-1 infection. Studies have revealed that HIV-1 infection and HIV-1-related gut dysbiosis is connected to gut inflammation and disease pathogenesis.[35] However, HIV has different modes of transmission and the infection rate varies based on gender, the span of infection, ART commencement periods, and the role of comorbidities. Therefore, it is challenging to draw a universal conclusion, based on any specific type study conducted so far, to comprehend the influence of HIV-1 on the intestinal microbiome and how PrEP or ART has induced the dysbiosis profile. Figure 10.3 represents future strategies that can be used to better understand

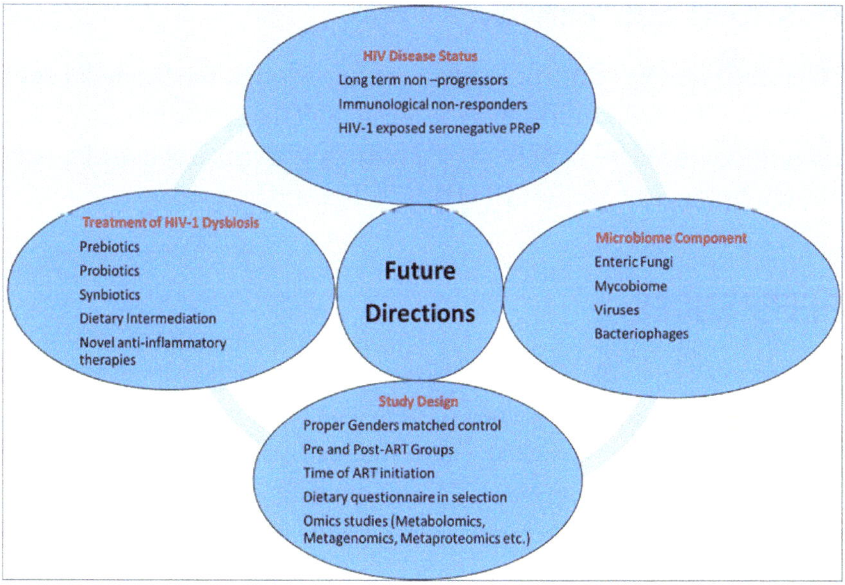

Fig. 10.3: Future strategies to be considered for studying the relationship among gut microbiota and human immunodeficiency virus-1 (HIV-1) pathogenesis.

the gut microbiome and HIV-1 pathogenesis relationship for possible treatment modalities.

HIV-1 microbiome studies are in their initial stages, and as with further studies examining human dysbiosis, there is a need to better understand the biological and clinical effects of gut dysbiosis. If disparities in gut microbiota aggravate HIV-1 pathogenesis, then reinstating "normal" gut microbiota composition could decrease pathogenesis. Otherwise, if dysbiosis arises exclusively because of HIV-1 infection but does not add to pathogenesis, then treating dysbiosis would be unsuccessful. Therefore, determining the etiology of dysbiosis through appropriately designed studies, and with additional broad use of *in vitro* models (two-dimesional [2D], three-dimesional [3D], or organoid models) and *in vivo* (animal) studies, will have a fundamental influence on the appropriate clinical interventions necessary to reinstate equilibrium to the microbe/host connection in the context of HIV-1 infection.

ACKNOWLEDGMENTS

Dr. Rahul Dev Jayant would like to acknowledge the financial support from Start-up funds and the Basic Science Pilot funding from the Office of Science, School of Pharmacy, Texas Tech University Health Sciences Center (TTUHSC), Texas, USA.

REFERENCES

1. Eggers C, *et al.* (2017) HIV-1-associated neurocognitive disorder: Epidemiology, pathogenesis, diagnosis, and treatment. *J Neurol* **264**: 1715–1727.
2. McArthur JC, *et al.* (2003) Human immunodeficiency virus-associated dementia: An evolving disease. *J Neurovirol* **9**(2): 205–221.
3. González-Scarano F, Martín-García J. (2005) The neuropathogenesis of AIDS. *Nat Rev Immunol* **5**(1): 69–81.
4. Kaul M, Garden GA, Lipton SA. (2001) Pathways to neuronal injury and apoptosis in HIV-associated dementia. *Nature* **410**(6831): 988–994.

5. Nair M, Jayant RD, Kaushik A, Sagar V. (2016) Getting into the brain: Potential of nanotechnology in the management of NeuroAIDS. *Adv Drug Deliv Rev* **103**: 202–217.
6. Monaco CL, *et al.* (2016) Altered virome and bacterial microbiome in human immunodeficiency virus-associated acquired immunodeficiency syndrome. *Cell Host Microbe* **19**(3): 311–322.
7. Gootenberg DB, Paer JM, Luevano J-M, Kwon DS. (2017) HIV-associated changes in the enteric microbial community: Potential role in loss of homeostasis and development of systemic inflammation. *Curr Opin Infect Dis* **30**(1): 31.
8. Jayant RD, Atluri VS, Agudelo M, Sagar V, Kaushik A, Nair M. (2015) Sustained-release nanoART formulation for the treatment of neuroAIDS. *Int J Nanomed* **10**: 1077.
9. Dinh DM, *et al.* (2015) Intestinal microbiota, microbial translocation, and systemic inflammation in chronic HIV infection. *J Infect Dis* **211**(1): 19–27.
10. Nowak P, *et al.* (2015) Gut microbiota diversity predicts immune status in HIV-1 infection. *Aids* **29**(18): 2409–2418.
11. Crakes KR, Jiang G. (2019) Gut microbiome alterations during HIV/SIV infection: Implications for HIV Cure. *Front Microbiol* **10**: 1104.
12. Pinto-Cardoso S, Klatt NR, Reyes-Terán G. (2018) Impact of antiretroviral drugs on the microbiome: Unknown answers to important questions. *Curr Opin HIV AIDS* **13**(1): 53.
13. Li SX, Armstrong AJ, Neff CP, Shaffer M, Lozupone CA, Palmer BE. (2016) Complexities of gut microbiome dysbiosis in the context of HIV infection and antiretroviral therapy. *Clin Pharmacol Ther* **99**(6): 600–611.
14. Dillon SM, *et al.* (2017) Low abundance of colonic butyrate-producing bacteria in HIV infection is associated with microbial translocation and immune activation. *AIDS (London, England)* **31**(4): 511.
15. Pinto-Cardoso S, *et al.* (2017) Fecal bacterial communities in treated HIV infected individuals on two antiretroviral regimens. *Sci Rep* **7**: 43741.
16. Villanueva-Millán MJ, Pérez-Matute P, Recio-Fernández E, Lezana Rosales JM, Oteo JA. (2017) Differential effects of antiretrovirals on microbial translocation and gut microbiota composition of HIV-infected patients. *J Int AIDS Soc* **20**(1): 21526.

17. Patterson KB, *et al*. (2013) Differential penetration of raltegravir throughout gastrointestinal tissue: Implications for eradication and cure. *AIDS (London, England)* **27**(9): 1413.
18. García-González AP, Ritter AD, Shrestha S, Andersen EC, Yilmaz LS, Walhout AJ. (2017) Bacterial metabolism affects the C. elegans response to cancer chemotherapeutics. *Cell* **169**(3): 431–441.e8.
19. Klatt NR, *et al*. (2017) Vaginal bacteria modify HIV tenofovir microbicide efficacy in African women. *Science* **356**(6341): 938–945.
20. Groschwitz KR, Hogan SP. (2009) Intestinal barrier function: Molecular regulation and disease pathogenesis. *J Allergy Clin Immunol* **124**(1): 3–20.
21. Spadoni I, *et al*. (2015) A gut-vascular barrier controls the systemic dissemination of bacteria. *Science* **350**(6262): 830–834.
22. Skonieczna-Żydecka K, Marlicz W, Misera A, Koulaouzidis A, Łoniewski I. (2018) Microbiome — The missing link in the gut-brain axis: Focus on its role in gastrointestinal and mental health. *J Clin Med* **7**(12): 521.
23. Brown EM, Sadarangani M, Finlay BB. (2013) The role of the immune system in governing host-microbe interactions in the intestine. *Nat Immunol* **14**(7): 660.
24. Paolicelli RC, Bergamini G, Rajendran L. (2019) Cell-to-cell communication by extracellular vesicles: Focus on microglia. *Neuroscience* **405**: 148–157.
25. Dillon SM, Frank DN, Wilson CC. (2016) The gut microbiome and HIV-1 pathogenesis: A two way street. *AIDS (London, England)* **30**(18): 2737.
26. Haynes BF. (2006) Gut microbes out of control in HIV infection. *Nat Med* **12**(12): 1351.
27. Dillon S, *et al*. (2014) An altered intestinal mucosal microbiome in HIV-1 infection is associated with mucosal and systemic immune activation and endotoxemia. *Mucosal Immunol* **7**(4): 983–994.
28. Mutlu EA, *et al*. (2014) A compositional look at the human gastrointestinal microbiome and immune activation parameters in HIV infected subjects. *PLoS Pathogens* **10**(2): e1003829.
29. Di Bella JM, Bao Y, Gloor GB, Burton JP, Reid G. (2013) High throughput sequencing methods and analysis for microbiome research. *J Microbiol Methods* **95**(3): 401–414.
30. Lozupone CA, *et al*. (2013) Alterations in the gut microbiota associated with HIV-1 infection. *Cell Host Microbe* **14**(3): 329–339.

31. McHardy IH, *et al.* (2013) HIV Infection is associated with compositional and functional shifts in the rectal mucosal microbiota. *Microbiome* **1**(1): 26.
32. Lozupone CA, Rhodes ME, Neff CP, Fontenot AP, Campbell TB, Palmer BE. (2014) HIV-induced alteration in gut microbiota: Driving factors, consequences, and effects of antiretroviral therapy. *Gut Microbes* **5**(4): 562–570.
33. Noguera-Julian M, *et al.* (2016) Gut microbiota linked to sexual preference and HIV infection. *EBioMedicine* **5**: 35–146.
34. Favre D, *et al.* (2010) Tryptophan catabolism by indoleamine 2,3-dioxygenase 1 alters the balance of TH17 to regulatory T cells in HIV disease. *Sci Transl Med.* **2**: 32ra36.
35. Vujkovic-Cvijin I, *et al.* (2013) Dysbiosis of the gut microbiota is associated with HIV disease progression and tryptophan catabolism. *Sci Transl Med* **5**: 193ra91.
36. Kim CJ, *et al.* (2016) Can probiotics reduce inflammation and enhance gut immune health in people living with HIV: Study designs for the Probiotic Visbiome for Inflammation and Translocation (PROOV IT) pilot trials. *HIV Clin Trials* **17**: 147–157.
37. Hummelen R, Vos AP, van't Land B, van Norren K, Reid G. (2010) Altered host-microbe interaction in HIV: A target for intervention with pro- and pre-biotics. *Int Rev Immunol* **29**: 485–513.
38. Santos ASEAC, Silveira EAD, Falco MO, Nery MW, Turchi MD. (2017) Effectiveness of nutritional treatment and synbiotic use on gastrointestinal symptoms reduction in HIV-infected patients: Randomized clinical trial. *Clin Nutr* **36**: 680–685.
39. Slavin J. (2013) Fiber and prebiotics: Mechanisms and health benefits. *Nutrients* **5**(4): 1417–1435.
40. D'Angelo C, Reale M, Costantini E. (2017) Microbiota and probiotics in health and HIV infection. *Nutrients* **9**(6): 615.
41. Cunningham-Rundles S, *et al.* (2011) Effect of probiotic bacteria on microbial host defense, growth, and immune function in human immunodeficiency virus type-1 infection. *Nutrients* **3**(12):1042–1070.
42. Miller H, Ferris R, Phelps B. (2016) The effect of probiotics on CD4 counts among people living with HIV: A systematic review. *Benef Microbes* **7**(3): 345–351.

43. Gori A, *et al.* (2011) Specific prebiotics modulate gut microbiota and immune activation in HAART-naive HIV-infected adults: Results of the "COPA" pilot randomized trial. *Mucosal Immunol* **4**(5): 554–563.
44. Falasca K, Vecchiet J, Ucciferri C, Di Nicola M, D'Angelo C, Reale M. (2015) Effect of probiotic supplement on cytokine levels in HIV-infected individuals: A preliminary study. *Nutrients* **7**(10): 8335–8347.
45. Klatt NR, *et al.* (2013) Probiotic/prebiotic supplementation of antiretrovirals improves gastrointestinal immunity in SIV-infected macaques. *J Clin Invest* **123**(2): 903–907.
46. Hensley-McBain T, *et al.* (2016) Effects of fecal microbial transplantation on microbiome and immunity in simian immunodeficiency virus-infected macaques. *J Virol* **90**(10): 4981–4989.

Chapter 11

Functional Integration of Gut–Brain Axis and Fecal Microbiota Transplantation Therapy

Rahul Mittal*, Rahul Iyengar, Richard A. Skelton, Abdul Aljohani and Priyanka Sinha

Abstract

The gut–brain axis (GBA) is the bidirectional communication system between the central nervous system (CNS) and the enteric nervous system (ENS). The functions of the GBA include connection of intestinal functions with emotional and cognitive brain regions in the CNS, and integration with the gut microbiome. The GBA also ensures maintenance of physiological function in the gastrointestinal (GI) tract. Research into the GBA has suggested an influence of gut microbiota on mood and higher cognitive functioning via an integrated signaling pathway composed of the CNS and ENS. The gut microbiome is crucial in influencing the GBA communication and recent studies have investigated the clinical applications of the manipulation of

* Hearing Research and Communication Disorders Laboratory, Department of Otolaryngology, University of Miami-Miller School of Medicine, Miami, FL, USA.

Corresponding Author: Rahul Mittal, PhD, Hearing Research and Communication Disorders Laboratory, Department of Otolaryngology, University of Miami-Miller School of Medicine, 1600 NW 10th Avenue, Miami, FL 33136, USA, Tel: 305-243-7117, E-mail: r.mittal11@med.miami.edu.

the gut microbiome for treatment of depression, autism spectrum disorder, Parkinson's disease, Alzheimer's disease, and other neurodevelopmental disorders. These treatments range from probiotics to vitamins and prebiotics to fecal microbiome transplants to treat dysbiosis of the gut microbiome and restore balance to the GBA.

INTRODUCTION

The gut–brain axis (GBA) refers to the complex bidirectional communication system between the central nervous system (CNS) and the enteric nervous systems (ENS), with gut microbiota implicated in the cross-communication of neural, endocrine, immune, and humoral signaling.[1–7] The multi-faceted interactions of the GBA include functional linkage of peripheral intestinal functions with emotional and cognitive brain regions.[8,9] Physiologically, the GBA ensures maintenance of GI homeostasis and mediates neuroimmune–endocrine communication such as immune activation, intestinal permeability, enteroendocrine signaling, and enteric reflexes.[10–12]

Insights into the GBA have demonstrated the broader influence of gut microbiota on motivation, and higher cognitive functioning via an integrated signaling pathway composed of the CNS, ENS, autonomic nervous system (ANS), and hypothalamic–pituitary–adrenal (HPA) axis (Fig. 11.1).[5,13–15] The brain is able to influence the function of several effector cells including immune cells, enteric neurons, smooth muscle cells, epithelial cells, and enterochromaffin cells.[16–18] The sympathetic and parasympathetic branches of the ANS drive both afferent and efferent signaling between the CNS and the intestinal wall, while the HPA axis is implicated in coordination and execution of adaptive responses to stressors.[14] As an integral component of the limbic system, the HPA axis ultimately responds to environmental or emotional stress by regulating the release of cortisol, the body's major stress hormone from the adrenal glands, which has multiple systemic effects

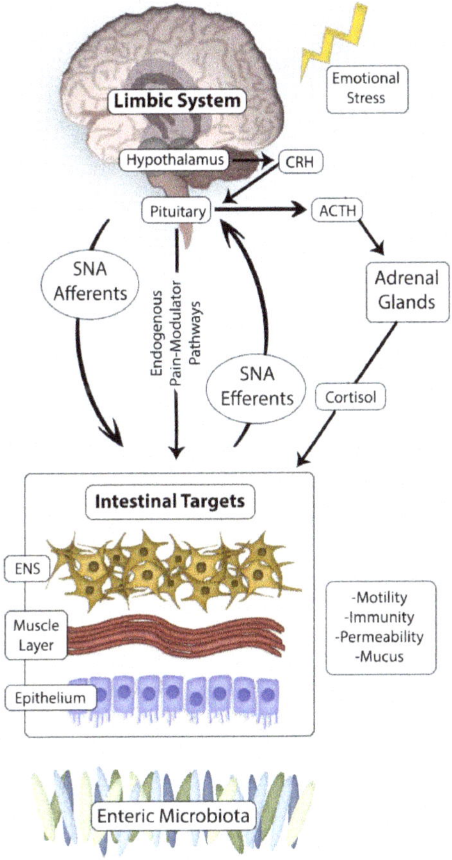

Fig. 11.1: Functional integration between gut–brain axis. There is a bidirectional complex communication between gut microbiome and brain through an integrated signaling pathway composed of enteric nervous system (ENS), autonomic nervous system (ANS), and hypothalamic–pituitary–adrenal (HPA) axis. (Adapted from Carabotti *et al.*[23])

(Fig. 11.1).[14,19] Reciprocal neuroendocrine hormonal signaling allows for both the brain and endogenous gut microbiota to modulate enteric function via parallel signaling pathways.[20–22]

The importance of the gut microbiome in influencing GBA communication has been a topic of recent interest, especially concerning the clinical applications of fecal microbiome transplant (FMT) therapy for the treatment of GBA-related disorders. In this review article, we aim to outline the important features of the gut microbiome and the ENS as well as describe their integrations with the CNS through a variety of mechanisms including vagal afferent and efferent nerves, microbiome production of neurotransmitters, hormones, and peptides as a result of metabolism as well as signaling and neuroinflammation from the gut epithelium as a result of microbiome and inflammation. We will then discuss the manipulation of this GBA through various mechanisms such as fecal transplant therapy for the treatment of a variety of disorders, including autism spectrum disorder (ASD), Parkinson's disorder (PD), and other neurodevelopmental disorder.

GUT MICROBIOME

The gut microbiome refers to the community of all microorganisms that inhabit the digestive system and GI tract.[24,25] The gut of humans is inhabited by trillions of microbes, the majority of which are bacteria that exist as commensal organisms with the human hosts. However, it is very likely that fungi, viruses and phages also exist in the human gut.[24,26–28] This commensal relationship between the gut microbiome and the human host plays an important role in homeostasis throughout life. The development of next-generation high-throughput sequencing techniques for genetic material has allowed for the classification and characterization of the taxonomic diversity of the human gut microbiome.[29–31] The gut microbiome is very individual specific and varies greatly among different human populations, as the majority of humans only share a third of gut microbe genes.[24] Changes in host physiology, environment, and diet affect the composition of gut microbiome and this dysbiosis of the gut microbiome may be associated with many diseases, such as cancer, diabetes, and even obesity.[32–34]

The establishment of the gut microbiome is influenced by the vaginal microbiome prenatally.[18,35,36] Studies have shown that prenatal stress-induced changes to the maternal microbiome can alter the microbiome of the offspring, which can create a niche of microbes that are predisposed to disease risk.[36–38] Initial colonization of the microbiome occurs *in utero*, and the microbial community is greatly influenced by exposures at birth as well.[24,39] Breast milk, the placenta, and nutrition are also important for the early establishment of a healthy microbiome, which prevents opportunistic microbes and infections.[40–42] Studies have shown that children suffering from malnutrition have a decreased diversity of gut flora, which is associated with increased risk of certain diseases such as type 1 diabetes and other autoimmune diseases.[43,44] While these factors play a role in the early gut microbiome composition, around 1–2 years after birth, the gut microbiome is fully established.[24,44,45] A healthy gut microbiome provides many beneficial effects to the host, such as nutrient metabolization, defense against pathogens, and contributing to the epithelial barrier.[46–49] Additionally, commensal gut bacteria play an essential role in immune system development, as demonstrated in germ-free mice who develop immune defects.[18,46,49,50] The flora in the gut microbiome also have a direct effect on the CNS as they produce a number of neuroactive signaling molecules, such as catecholamines, histamine, melatonin, acetylcholine, gamma-aminobutyric acid, and serotonin (Fig. 11.2).[17,18,51]

Gut enterotypes have been determined using 16S ribosomal RNA, and it has been reported that there are species diversity between individuals.[53,54] Healthy adult gut microbiomes consist largely of the anaerobic bacteria *Bacteroidetes*, *Ruminococcus*, and *Firmicutes* phyla, and are also composed of *Actinobacteria*, *Proteobacteria*, and *Verrucomicrobia* at lower levels.[55,56] In studies of populations, there seem to be three differing enterotypes, which all appears to employ relatively different routes of energy generation and nutrient metabolization based on the substrates available

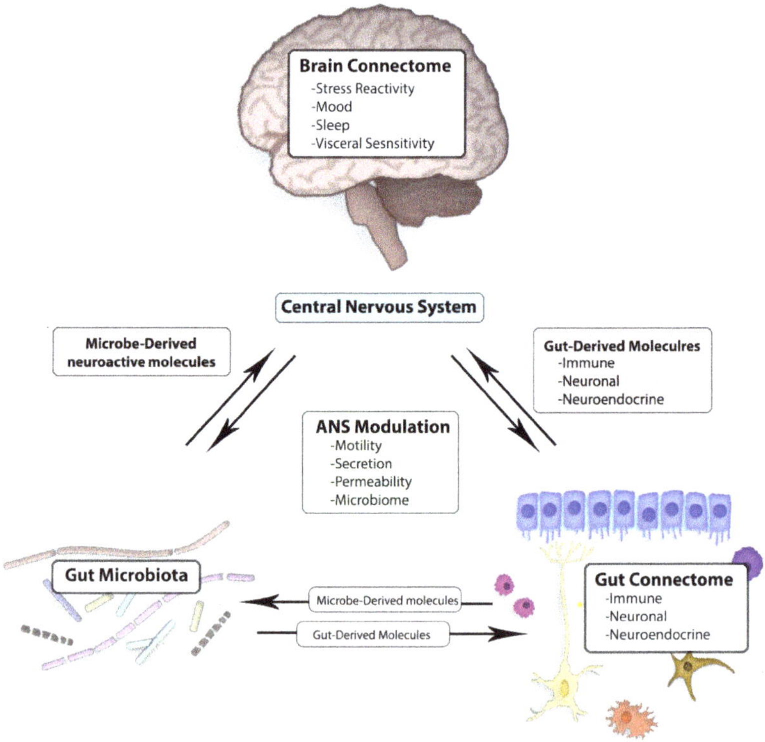

Fig. 11.2: A schematic representation of gut–microbiome–brain interaction. Gut microbiome produces a number of neuroactive signaling molecules that can have a direct effect on the central nervous system and influence brain function. (Adapted from Martin *et al.*[52])

in the colon.[56] This energy generation is mostly based on metabolization of carbohydrates and vitamin production, which differ between enterotypes.[56,57] The microorganisms of the gut benefit the human body by fermenting fiber into short-chain fatty acids, such as butyric and acetic acids, which the body is able to absorb through the gut wall.[55–57] Both vitamins B and K are synthesized by the intestinal bacteria, and microbiome also metabolize bile acids, sterols, and xenobiotics.[58] Examples of the differing digestion capabilities of the enterotypes include one enterotype that seems to be enriched in *Bacteroides* and *Parabacteroides*, which derive energy primarily from carbohydrates and through fermentation.[56] Another enterotype is composed of mostly *Ruminococcus*, which is characterized by many simple sugar transport proteins and specializes in mucin degradation and absorption of vitamins like folate and thiamine. The third enterotype contains mostly *Prevotella* and *Desulfovibrio*, which act to degrade mucin glycoproteins in the mucosal layer of the gut.[55,56]

Disruption of a balanced gut microbiome is implicated in a number of disease processes, such as immunological dysregulation and autoimmunity leading to irritable bowel syndrome and even some cancers.[18,59] One of the bacterium that has been correlated with GI disorders is *Helicobacter pylori*, which is associated with 80%–90% of all ulcers present in the GI tract and its presence is linked with the development of gastric cancer.[18,60,61] *Helicobacter pylori* creates ulcers through the secretion of urease enzyme that converts urea to ammonia to create a more habitable environment.[62] Another microbe associated with dysbiosis of the gut is *Clostridium difficile*, which is a gram-positive bacterium that causes diarrhea in the host. *Clostridium difficile* is an opportunistic infection that secretes toxins and becomes pathogenic when competing commensal gut bacteria are eliminated by antibiotic treatment or by another mechanism.[63–66] Other diseases such as irritable bowel syndrome exhibit a linkage between the gut dysbiosis that leads to not only GI symptoms but also CNS effects, and it has also been linked to early life trauma and regional brain volumes during development.[65,67] There have also been observed effects of depletion of the gut microbiome via broad-spectrum antibiotics early in life, leading to inflammatory disorders as well as metabolic disorders.[38,68,69]

ENTERIC NERVOUS SYSTEM

One of the mechanisms through which gut microbiome interacts with CNS is through ENS.[70] ENS is a division of the ANS that is composed of a mesh-like system of neurons that function to regulate the secretion, motility, mucosal maintenance, and immunological defenses of the GI system.[71–73] The function of ENS is to communicate between the gut and the CNS, as well as initiate reflexes and modulate function of the GI system.[18,74] It is a subset of the peripheral nervous system that allows it to govern the functions of the GI tract independently of the CNS, although the CNS can also influence the ENS via the vagus nerve (parasympathetic nervous system) and the prevertebral ganglia (sympathetic nervous system).[18,75] The ENS is unique from other parts of the ANS in that it can both respond to stimuli and initiate motor output.[71,76] Over 90% of the vagal innervation of the ENS is afferent, meaning that the brain has more sensory neurons in the GI system than motor and can transmit the senses of bloating, nausea, and pain to the brain.[75,76] The vagus nerve is also involved in affecting mood, brain development, and behavior, as well as may provide a pathway for disease transmission from the gut to the brain.[76]

The ENS contains hundreds of millions of neurons embedded in the wall of the GI tract, which is more than all other peripheral ganglia.[76] Nerve cells and the supporting glial cells are grouped in clusters called enteric ganglia, which are small in size but large in quantity, and have processes that connect with the processes of neurons from outside of the ENS.[77] The sensation and modulation of luminal conditions is accomplished through mechanoreceptors and chemoreceptors such as enteroendocrine cells and mucosal nerve fibers, ascending and descending interneurons, intrinsic primary afferent neurons, and motor neurons.[76,78] There is a vast array of neurotransmitters that are used in the ENS, many of which are similar to the CNS such as acetylcholine, dopamine, and serotonin.[17,18] Of note, the 5-hydroxytryptamine (5-HT) receptor, a serotonin G protein-coupled receptor (GPCR) that stimulates cyclic adenosine monophosphate (cAMP) production, is a gut neurotransmitter receptor that has been targeted for developing therapeutic modalities and improves gut motility.[79–81] This means that many pathophysiological processes that underlie CNS disease also have a negative effect on the ENS. Enteric neuronal function is compromised in Varicella Zoster virus, PD, and ASD.[18] Gut pathology related to the sensing of pathological consequences of disease are involved in Alzheimer's disease (AD), and amyotrophic lateral sclerosis (ALS).[18] These are all disorders that have neurological and GI pathologies. Hirschsprung's disease is a genetic disorder in which the myenteric neurons are absent in the colon, which results in a loss of parasympathetic innervation of the bowel.[82] It leads to a functional bowel obstruction due to the loss of motility.[83]

The ENS neurons are embedded in the wall of the GI tract in the form of plexuses, such as the myenteric and submucosal plexuses. The myenteric (Auerbach's) plexus is located in between the longitudinal and the circular layers of smooth muscle, whereas the submucosal (Meissner's) plexus is located in the dense connective tissue of the submucosa.[78] The myenteric plexus supplies motor innervation to both the circular and longitudinal layers of smooth muscle in the GI tract and has input from both the parasympathetic and sympathetic nervous systems.[84] It controls GI tract motility, peristalsis and churning of intestinal content.[78] The submucosal plexus is composed of myenteric plexus branches that perforate the circular muscle layer and has only parasympathetic innervation. It provides motor innervation to the gut and carry reflexes through intrinsic primary interneurons, which have mechanoreceptive endings. These endings can sense the distension of the gut wall and can also respond to serotonin release by enterochromaffin cells in the gut epithelium.[78] The activation of reflexes in this system results in a proximal to distal travel of intestinal contents.

GUT–BRAIN INTEGRATION AND SIGNALING

The gut–brain integration is essential to maintain homeostasis. It is a bidirectional neurohormonal interaction pathway that is regulated by the ENS and the CNS, as well as the immune system, endocrine system, and metabolic pathways such as the HPA axis.[18] The microbiome in the gut is directly involved in signaling between the gut and the brain through its interactions with the

ENS as well as the CNS. The molecular mechanisms underlying this signaling involve vagal and spinal afferent neurons, immune mediators (cytokine release), gut hormones, and gut microbe-derived signaling molecules (Table 11.1).[17,18,75] The CNS communicates with and has effect on the gut microbiome through four other mechanisms: sympathetic neurons, parasympathetic neurons, and neuroendocrine factors released from both the adrenal medulla and the adrenal cortex.[16,75] Bacterial colonization of the gut plays a crucial role in the development of the ENS and the CNS, and altered gut microbiome composition can affect the development of the CNS via their integration.[18,85–88]

The vagus nerve is the main connection between the gut viscera and the CNS.[18,89] Vagal sensory neurons and afferent spinal neurons are responsible for carrying information to the brain from the intestinal end.[92,93] The vagus nerve consists of 60,000 fibers in humans, which has both afferent sensory and efferent fibers, and is the only nerve that directly connects the GI system to the brain.[91,93,94] The vagus nerve is involved in the control of mood, immune response, digestion, and heart rate.[95] It interfaces with monoaminergic systems in the brain stem, which interacts with the hypothalamus and the limbic system.[93] The hypothalamic–pituitary–adrenal axis is a hormonal network that is activated by stressful situations, which can in turn cause changes in both the intestinal epithelium and the gut flora through vagal nerve efferent fibers.[18] This makes the vagus nerve a potent target for treatment of psychiatric and GI disorders.[96,97]

The vagus nerve effects the contraction of smooth muscle and glandular secretion in the GI wall. There are afferent sensory fibers in the intestines, which include mucosal mechanoreceptors, chemoreceptors, and tension receptors.[98,99] Activation of the vagus nerve leads to the release of acetylcholine at synaptic junctions, which binds to nicotinic and muscarinic receptors, stimulating muscle contractions as part of the parasympathetic nervous system.[89,99] These sensory receptors send afferent signals to the

Table 11.1: Molecular Mechanisms of Gut–Brain Axis Integration

Method of Integration	Mechanism	References
Direct activation of neuronal pathways	The vagus nerve connects the gut viscera to the CNS. Sensory neurons, such as mucosal mechanoreceptors, chemoreceptors and tension receptors, and afferent spinal neurons carry information regarding gut epithelial function, contents, motility, and function, to the brain from the intestines; providing a mechanism for gut interaction with the brain. Autonomic efferent fibers carry information from the brain to the gut, providing a mechanism for the brain to affect the gut. The vagus nerve interfaces with the hypothalamus and can manipulate the HPA axis to change a variety of systems in the body.	89
Microbial production of metabolites and neurotransmitters	Catecholamines, histamine, melatonin, acetylcholine, gamma-aminobutyric acid, brain derived neurotrophic factor and serotonin are neuroactive compounds that can be secreted by host microbes. These can cross the blood–brain barrier to affect the brain.	18
Gut epithelium production of peptides and hormones	Enteroendocrine cells, due to their proximity to the gut microbiota, are able to create signals based on gut microbe characteristics. Cells at the gut epithelium do this through the production of neuropeptides in response to gut microbes, such as peptide YY, neuropeptide Y, substance P, glucagon-like peptide, and cholecystokinin. These enter the blood stream and cause CNS and ENS effects.	90
Immune modulation	Microbes in the gut generate an inflammatory response which leads to the release of cytokines, which up-regulate neuroinflammation, either by acting directly on the brain or by signaling vagal and spinal afferent neurons. Pathogen associated molecular patterns, such as lipopolysaccharide, are able to activate immune receptors, such as toll-like receptors, in the CNS as well as throughout the body. The commensal gut microbiome contributes to intestinal epithelium integrity, however, when pathogenic bacteria outcompete the commensal bacteria, this integrity can be compromised. This change in permeability can allow an inflammatory response to become systemic, potentially affecting the blood–brain barrier.	18, 91

nucleus tractus solitarii in the brain, which projects these signals to the locus coeruleus, the rostral ventrolateral medulla, the amygdala, and the thalamus.[89]

As stated previously, a number of neurotransmitters and hormones produced in the gut are also used in neurohormonal signaling, and, therefore, these compounds are neuroactive.[18] Such compounds include catecholamines, histamine, melatonin, acetylcholine, gamma-aminobutyric acid, brain-derived neurotrophic factor (BDNF), and serotonin.[100] Alterations in the contents of GI microbiota due to drugs, disease or diet, can cause changes in the levels of cytokines produced, which can in turn affect the CNS.[18,100] Hormones and neurotransmitters produced by the ENS can be released into the blood and can cross the blood–brain barrier.[18,100]

Enteroendocrine cells are stimulated by these products of the gut bacteria, which leads to the production of neuropeptides that include peptide YY, neuropeptide Y, substance P, glucagon-like peptide, and cholecystokinin.[101–103] All these neuropeptides are involved in inflammation, defense, stress, and repair.[90,104] These all influence the ENS after their entry into the bloodstream, and even lead to effects in the CNS.[90,105] Peptide YY is released from the ileum and the colon in response to food consumption, specifically metabolism of dietary fibers by gut microbes, and plays a role in appetite reduction.[106–108] Neuropeptide Y controls the gut microbe impact on inflammatory processes, pain, brain function, and behavior.[89,109–111] Peptide YY is released from enteroendocrine cells in response to short-chain fatty acids produced by gut microbes from fiber digestion.[89] These short-chain fatty acids include butyric and acetic acids, which stimulate the nervous system and can also influence memory and learning in the brain.[89] Substance P is known as a vasodilator and is involved in inflammatory processes, pain sensation, as well as regulation of mood and anxiety.[112,113] Cholecystokinin is secreted from the duodenum and triggers release of bile and digestive enzymes from the pancreas.[114,115] Enterochromaffin cells in the gut also produce 95% of serotonin (5-HT) in the body through tryptophan metabolism.[100,116] Serotonin controls

GI secretion, motility, and pain perception in the gut, and mood and cognition in the brain.[100,116] Serotonin is also produced by gut microbes of the genus *Candida* and *Enterococcus*.[116] The main inhibitory neurotransmitter, gamma-aminobutyric acid, is produced by gut microbes of the genus *Lactobacillus* and *Bifidobacterium*.[90,116,117] Dopamine is also produced by microbes in the gut, specifically from the *Bacillus* species.[5,90]

The immune system is also a mechanism of linkage between the gut microbiome and the brain. The lymphoid tissue associated with the gut comprises about 70% of the immune system in the body.[18,118] Microbes generate an inflammatory response and the release of cytokines, which upregulate neuroinflammation by acting directly on the brain or by signaling vagal and spinal afferent neurons.[118] Lipopolysaccharide (LPS) from microbes can travel through the gut through the mucosal barrier and can transmit a message from the microbes to cells around the body.[119] These LPS molecules can even travel through circulation to interact with the CNS directly, via toll-like receptor 4 (TLR 4), which is expressed at high levels in the CNS.[119,120] Through stimulation of TLR 4 in the brain, LPS can trigger pro-inflammatory cytokine production, which can be implicated in neurodegeneration, repair, and even the pathologies of some psychiatric disorders.[78,119,120]

Increased peripheral inflammatory markers have been identified in individuals suffering from major depressive disorder (MDD) compared to individuals who do not suffer from MDD.[120–122] It is thought that this inflammation may be found in a particular subset of individuals suffering from MDD.[120] It has been shown that certain types of probiotic treatments, specifically *Bifidobacterium longum* and *Lactobacillus* spp., have decreased the severity of depression in pre-clinical experimental animal models and human patients.[18,123–125]

The gut microbiome can also integrate with the CNS through altered intestinal permeability, which is associated with chronic stress that can lead to the development of psychiatric disorders.[78] High levels of bacterial endotoxins and LPS are a risk factor for dysbiosis of gut microbiota.[78,126] The gut microbiome contributes to tight junction integrity and the intestinal barrier, which allows it to moderate the HPA axis and certain characteristics of the ANS.[47,48,78] For example, cortisol is released by the HPA axis, which affects immune functions as well as alters the composition of gut microbes and the permeability of the intestinal epithelium.[78] The presence of pathogenic bacteria in the gut environment can also outcompete commensal bacteria and activate mucosal innate responses, which lead to a decrease in epithelial permeability and inflammatory response.[91,127] This change in gut permeability can allow the inflammatory response to become systemic, affecting and impairing the blood–brain barrier, which will promote neuroinflammation resulting in neuronal injury and degradation, which is postulated to be the mechanism through which the gut microbiome can lead to neurodegenerative disorders.[91]

Disease manifestations in the gut, such as transmissible spongiform encephalopathies, ASD, PD, AD, ALS, and Varicella Zoster infection all involve the CNS as well as the ENS and have effects on the GI system.[18] Experiments have shown that brain functions are highly involved in the control of the gut–brain microbiome signaling and various forms of social stressors do impact the secretion of mucus and the composition and biomass of the gut microbiota.[88] Through signaling from the brain, peptides and neurotransmitters are released by enterochromaffin cells and directly affect the composition of the gut microbiome.[88] Vagus nerve stimulation, which produces effects similar to afferent signaling from the ENS, has been used to treat depression.[88] Treatments aimed at the vagus nerve influence vagal tone by reducing cytokine production.[23]

Neurodevelopment

Due to the similarities between the ENS and the CNS, it suggests that the mechanisms through which the gut microbiome influences the ENS, via hormones, peptides, and signaling processes, can in turn affect the development of the CNS as well.

In studies of germ-free mice, the animals that lacked BDNF in the cortex, hippocampus, and amygdala, showed altered development of the brain and behavior abnormalities.[18,69,128,129] Germ-free mice were observed to have increased anxiety-related behaviors and increased plasma corticosterone levels, indicating a dysregulation of the HPA axis in these mice.[128] Germ-free mice had increased expression of BDNF mRNA and a decrease in serotonin receptor expression in the dentate gyrus of the hippocampus.[128] Decreased serotonergic signaling has been implicated in increased anxiety-related behaviors in mice.[129] The up-regulation of BDNF has also been linked to increased anxiety-related behaviors in mouse models and thus could also play a role in the germ-free mice behavior studied here.[129] These results have been replicated in multiple studies of germ-free mice having developmental delays and atrophy in the hippocampus and the amygdala.[128,129]

These observations have also been made with respect to treatment with antibiotics early in life.[128] Administration of antibiotics early in life has been shown to lead to changes in the microbiome and can predispose the child to inflammatory and metabolic disorders later on in life.[38,128,130] It is even hypothesized that sub-therapeutic doses of antibiotics from food can result in population level changes in the gut microbiome and its metabolic properties. This is strengthened by studies showing that sub-therapeutic doses of antibiotics in mice led to increased adiposity, changes in metabolic enzyme profiles, and a dysbiosis and change of microbiome taxonomy.[130]

FECAL TRANSPLANT THERAPY

Recently, there has been an increased interest in exploring the efficacy of administrating vitamin A as well as the use of probiotics and prebiotics for the treatment of GI and neurological disorders.[18] Probiotics are living organisms that are prescribed for many GI conditions, such as irritable bowel syndrome, and normally contain gram-positive bacteria from the genus *Lactobacillus* and *Bifidobacterium*. *Bifidobacterium longum* has been shown to reduce depression in patients with irritable bowel syndrome and improve their quality of life.[18,131,132] On par with these findings, it has been shown that *Lactobacillus casei* reduces the response to stress, leading to decreased cortisol and other stress hormones.[133] However, other studies have shown that probiotics have no benefit over the placebo group with regard to mood, anxiety, stress, or sleep quality.[134] On the other hand, prebiotics are non-digestible carbohydrates that provide resources for metabolism in the gut microbiome and select for beneficial microbiota and have been shown to reduce stress behaviors in mouse models.[18]

Besides probiotics and prebiotics, FMT has started to be explored for neurological disorders (Table 11.2). FMT is a procedure where healthy donor fecal microbes are transferred into the recipient patient, either orally, through an enema, or through a colonoscopy.[18] FMT is a method that has been developed to treat disorders involving the GBA, which include numerous diseases resulting from neurodevelopmental disorders such as autism, anxiety, mood disorders, and PD (Table 11.2). It has even been used to treat refractory *C. difficile* infections with success.[18] It has also been used for obesity-related illnesses and insulin insensitivity, where FMTs from healthy donors showed improved insulin insensitivity compared to control FMTs (autotransplants of the patients own microbiome).[135] PD, depression and autism are neurological disorders that often have associated comorbidities in the GI tract as well as the CNS, and thus potentially candidates for therapy using FMT and microbiome manipulation.

Parkinson's Disease

PD is a progressive neurodegenerative disease, affecting between 100 and 200 per 100,000 people over age 40, which results in damage to the dopaminergic neurons in the CNS.[141–143] It often presents clinically with a typical tremor, bradykinesia, rigidity and postural

Table 11.2: Effect of Modulating Gut Microbiome on Neurological and Gastrointestinal Disorders

Disorder	Study Design	Findings/Results	References
Autism spectrum disorder	Microbiota transfer therapy was conducted in 18 patients who were 7–16 years old. Treatment consisted of 2 weeks of antibiotics treatment, followed by 8 weeks of a high initial dose and low maintenance dose of FMT. The participants were evaluated using GI symptoms scales as well as ASD behavior ratings.	GI rating scale showed an 80% reduction in symptoms related to the GI tract, and these symptoms persisted after maintenance doses of microbiome transplant had ended. There were significant improvements in ASD symptoms and behaviors. Analysis of gut microbes showed that there was an increase in the amount of *Bifidobacterium*, *Prevotella*, and *Desulfovibrio*, as well as a general increase in microbiome diversity, which are beneficial changes to the gut microbiome.	136
	Sixty-four 1–8-year-old children with autism received vitamin A supplementation. Children were followed for 6 months and evaluated using a series of autism behavior scales to assess autism symptoms, and blood draws were completed to detect CD38 and RORA mRNA levels (biochemical markers of autism) to assess autism-related biochemical changes, and high-throughput gene sequencing was used to identify changes in the gut microbiome.	Improvements in metabolic modifications, ability to focus, listening to instructions, and an increase in and normalization of *Bacteroidetes/Bacteroidales* levels, and a decrease of *Bifidobacterium*. No other behavioral changes were noticed, and further research is required.	132

(*Continued*)

Table 11.2: (*Continued*)

Disorder	Study Design	Findings/Results	References
	The fecal microbiome of 10 children with autism, their siblings, and 10 children without autism were investigated with polymerase chain reaction (PCR). Participants were given a probiotics supplement with *Lactobacillus*, *Bifidumbacteria* and *Streptococcus*. Fecal specimens were obtained, GI symptom score was tracked, and ASD severity was evaluated.	There was a strong positive correlation of ASD severity associated with increased GI symptoms. The probiotics treatment was able to better normalize the gut microbiome and increase diversity. This led to an increase in *Firmicutes* and *Bacteroidetes*. However, changes in behavior related to ASD were not measured in this study.	137
	100 preschoolers with ASD were classified into GI symptom group or non-GI symptom group. Subjects in both groups were randomized to control regular diet with placebo probiotic or regular diet with probiotic. Participants are evaluated for GI symptoms, ASD symptoms, comorbidities, and other markers.	Under investigation currently (ongoing clinical trial)	138
Parkinson's disease	Patients with PD on levodopa therapy were given standard breath test to detect *H. pylori*. The *H. pylori* positive patients were given eradication therapy for the infection and monitored for progression of symptoms and response to levodopa therapy.	The patients who were *H. pylori* positive generally had a more severe disease, based on the unified Parkinson's disease rating scale. Treatment and eradication of infection led to improved motor function and quality of life, and improved levodopa effectiveness and onset time.	139

Table 11.2: (*Continued*)

Disorder	Study Design	Findings/Results	References
	The gut microbiome, from PD model mice was analyzed and then FMT was performed. Behavioral tests were used to analyze mouse PD pathology and severity. Samples were also collected to measure dysbiosis and gut microbe levels as well as metabolite levels in the blood and nervous system.	Gut dysbiosis was reduced by the FMT, and there was also reduced inflammation and microglial activation in the CNS. There was reduced TLR 4 expression and signalling in the CNS as well. Physical impairment was decreased with FMT by the mouse behavioral tests.	140
Irritable Bowel Syndrome — Depression and Anxiety	Randomized, double-blinded, study of 44 individuals with IBS and diarrhea with mild to moderate anxiety and/or depression. Patients were randomized and given placebo or *Bifidobacterium longum* probiotic for 6 weeks. Depression and anxiety were measured at 0, 6 and 10 weeks, and microbiome samples were collected.	At 6 weeks, 14 of 22 patients in treatment arm had a reduced depression score of 2 points, compared with only 7 of 22 in the control group. Reductions continued through week 10. Probiotic treatment had no effect on anxiety or IBS symptoms. The treatment arm had a significant increase in quality of life scores, and fMRI studies showed that the probiotic reduces responses to negative stimuli in multiple parts of the brain.	130

instability.[144,145] Patients' movements become slow even when doing simple tasks, their steps become shorter while walking, and they also suffer from muscular pain and rigidity.[143] It can also present with mood disorders such as anxiety, depression, and disrupted sleep, as well as autonomic dysfunction resulting in constipation, urinary difficulties, dysphagia and diaphoresis.[143,146] GI dysfunction is a common symptom in PD and emerges in early stages of the disease as well.[147,148] Extensive research focuses on how the GI tract and the ENS influence the development of PD. PD is believed to be caused by both genetic and environmental factors, and family history is a major risk factor as well as exposure to certain environmental pesticides, and traumatic brain injury.[141,143,149] PD occurs mostly in people over 60 and affects males more so than females.[142] PD can result in the death of neurons in the midbrain, called the substantia nigra, and in a decreased production of dopamine.[143] Levodopa therapy is conducted on PD patients in order to restore dopaminergic neurotransmission, which seems to help alleviate motor symptoms while leaving non-motor symptoms untreated.[132]

There is a well-studied link between *H. pylori* and PD, as *H. pylori* reduces the absorption of levodopa and is associated with increased severity and disease progression.[150,151] The treatment and eradication of *H. pylori* in persons with PD has resulted in clinical improvement of symptoms.[152–154] Studies involving high-throughput sequencing and metagenomics have demonstrated a significant dysbiosis of the gut microbiome in patients with PD and AD. This gut dysbiosis involves suppression of *Prevotellacae* and anti-inflammatory gut genera such as *Blautia* and *Coprococcus*, with an increase in pro-inflammatory *Proteobacteria* and *Enterococcaceae*.[153,155–157] These genera have also been linked to decreased gut wall function and increased permeability.[128] This suppression of anti-inflammatory genera with an up-regulation of pro-inflammatory genera is also seen in studies of AD, and both of these diseases have also been associated with increased prior exposures to bacterial infections.[128] The physical movement disorder symptoms and pathology of neurodegeneration in PD is exacerbated by the presence of gut microbiota versus germ-free mice.[146] It was observed that antibiotic treatment ameliorated this deficit, while giving microbiome-related metabolites could exacerbate it, and that the FMT of healthy human patients reduced symptoms, whereas the transplant from symptomatic PD patients exacerbated symptoms.[146] However, due to the multiple confounding variables and comorbidities with these diseases, it is difficult to isolate one single microbiome signature in PD or AD. It is even possible that the presence of PD or AD drives the changes in the gut microbiome and not the other way around, due to the bidirectional nature of the connection between the CNS and the gut microbiome.

Another area of study regarding PD is the presence of a dysregulated bile acid profile.[18,158] Bile acids are considered a type of gut hormone because they are deconjugated by gut microbiota, which can then signal through the nuclear farnesoid X receptor and the bile acid GPCR.[159] Bile acid is related to weight regulation and altered bile acid profiles can be associated with various systemic diseases.[159,160] Bile acids such as Tauroursodeoxycholic acid, produced only by GI bacteria, have neuroprotective and anti-inflammatory properties and were shown to be protective against AD and PD.[161,162] Bile acids are also linked to circadian rhythm, which becomes dysregulated in animals with neurodegenerative diseases.[18,163] This is another mechanism through which it is thought that there is a link between the gut microbiome, CNS and neurodegeneration that is pathological in PD.

Autism Spectrum Disorder

ASD comprises a group of neurodevelopmental disorders that are mainly characterized by social and communication impairments, as well as restrictive patterns of behavior, interests and activities.[164,165] Individuals with ASD often have GI complications, such as abdominal pain, reflux, vomiting, and alterations in bowel habits. The presence of GI symptoms is a marker of ASD severity and has been correlated with more pronounced neurodevelopmental deficits.[138] The pathophysiology of ASD is unclear and likely multi-factorial, however, it involves genetic interplay and environmental factors.[166–170] Several studies have noted the dysbiosis in the gut microbiome seen in children with ASD, with predominance of *Clostridiales*, *Suterrella*, and *Ruminococcus* populations.[38,138,171–173] Children having ASD that received antibiotic treatment within the first 3 years of their lives after birth can also lead to further dysbiosis of the gut microbiome.[38,138,173]

The gut microbiome can have an effect on the CNS development through the innate-immune system, vagus nerve signaling via peptide production of the microbes themselves as well as the epithelium.[18] Altered gut microbe metabolites affect GI function and lead to alterations in neurobiological conditions through the production of short-chain fatty acids, production of neuroactive metabolites, and the production of p-cresol (a molecule produced by microbes in the gut).[174,175] Inflammation in the brain related to ASD development seems to be associated with interleukin 6, tumor necrosis factor alpha, and a class of cytokines called adipokines.[138] All these mechanisms can contribute to interference with CNS development through interruption of synaptic transmission. One hypothesis is that children with a genetic predisposition to autism can become more prone to the autism phenotype due to the presence of altered gut microbes as seen in many children with ASD.[18,138]

Treatment using FMT, vitamin A supplementation, and probiotics has been studied in populations of children with ASD with mixed results. Kaluzna-Czaplinska and Blaszczyk gave oral probiotic supplements consisting of *Lactobacillus acidophilus* twice daily for 2 months to children with ASD and found significant metabolic modifications and an improved ability to focus and listen to instructions; however, no other improvements in behavior or emotional affects were noted.[176] Children with ASD treated with *Lactobacillus plantarum* probiotics had significantly improved behavior compared to the control treatment group.[177] Another study revealed a strong positive correlation of autism severity with increased GI symptoms, and found that probiotic treatment with *Lactobacillus*, *Bifidobacter* and *Streptococci* three times a day for 4 months normalized gut microbiota; however, the change in behavior was not measured in this study.[137] It was observed that the level of *Desulfovibrio* decreased to the level of healthy individuals, and this microbe is thought to play an important role in the pathophysiology of autism.[137] Vitamin A, which promotes mucosal immunity and improved epithelial integrity, has been used as a supplement for children with ASD, and it has shown beneficial properties and improvement in gut microbiome composition to the level of healthy individuals as autism is a spectrum.[132] It could be that there is a subgroup of children diagnosed with ASD who experience GI issues, which could benefit therapeutically from these treatments, leading to improved GI symptoms, improved behavior, and improved cognition and adaptation functioning.[132,138]

FMT can be carried out in the treatment of ASD to rebalance gut microbiota dysbiosis and mitigate symptoms of ASD.[178,179] Treatment with FMT for *C. difficile* infection requires only a single dose of FMT in order to see therapeutic benefits.[18,136] A study evaluated safety, tolerability, and efficacy of FMT from healthy donors for amelioration of some of GI and behavior symptoms affecting children with ASD, and it was observed that there were notable improvement in their symptoms during the 8-week FMT treatment.[136] This was the first clinical trial for the use of FMT to treat patients with autism and it provided important feedback and demonstrated the importance of focusing on the gut microbiome in the treatment of ASD.[136] However, there were various adverse effects in the FMT treatment group, although it was generally well-tolerated. These adverse effects included mild to moderate hyperactivity as well as tantrums and aggression.[136] Further studies are warranted to continue evaluating the safety, efficacy, and procedures of FMT, with particular focus on analyzing the bacterial species responsible for improving the clinical manifestations of ASD.

CONCLUSIONS AND FUTURE DIRECTIONS

GBA serves as a connection between the brain and the gut as well as the CNS and the ENS. Brain disorders involving neurodevelopment and neurodegeneration, such as ASD and PD respectively, affect this GBA connection between the CNS and the gut.[18] Research has shown that neurodevelopment is influenced by the gut microbiome through the CNS's connection with the ENS, hormones, peptides, and immune signaling processes.[128] These same processes are thought to be part of the pathology involved in ASD. There have been a number of studies that have found promising benefits with the treatment of ASD with probiotics, vitamin A supplementation, and FMTs.[136–138,176] However, these studies have looked at different endpoints, such as microbiome composition, behavior, and GI-related symptoms, and the data are somewhat mixed on outcomes. Further research is necessary to elucidate the safety and efficacy of these treatments with regard to specific outcomes. PD is a neurodegenerative disorder that is often associated with GI symptoms, which implicates dysbiosis of the gut and is tightly associated with pathogenic microbes in the gut, such as *H. pylori*. Treatment of these infections has demonstrated improved clinical symptoms with regard to PD.[18,139]

An important step and research question moving forward will be to understand when gut dysbiosis induces disease versus when it is a result, or symptom, of a disease. Further research into the composition of healthy versus pathogenic microbes in the gut, as well as research into adverse effects, is necessary to improve the efficacy and effectiveness of FMT for the treatment of neurodevelopmental disorders.

ACKNOWLEDGMENT

We are thankful to Dr. Valerie Gramling for critical reading of the manuscript.

REFERENCES

1. Li XJ, *et al.* (2020) Bidirectional brain-gut-microbiota axis in increased intestinal permeability induced by central nervous system injury. *CNS Neurosci Ther.* 28(6): 783–790.
2. Wijdeveld M, Nieuwdorp M, IJzerman R. (2020) The interaction between microbiome and host central nervous system: The gut-brain axis as a potential new therapeutic target in the treatment of obesity and cardiometabolic disease. *Expert Opin Ther Targets* **22**: 1–15.
3. Mörkl S, Butler MI, Holl A, Cyran JF, Dinan TG. (2020) Probiotics and the microbiota-gut-brain axis: Focus on psychiatry. *Curr Nutr Rep.* 9(3): 171–182.
4. Adelman MW, *et al.* (2020) The gut microbiome's role in the development, maintenance, and outcomes of sepsis. *Crit Care* **24**(1): 278.
5. Baj A, Moro E, Bistoletti M, Orlandi V, Crema F, Giaroni C. (2019) Glutamatergic signaling along the microbiota-gut-brain axis. *Int J Mol Sci* **20**(6): 1482.
6. Heym N, *et al.* (2019) The role of microbiota and inflammation in self-judgement and empathy: Implications for understanding the brain-gut-microbiome axis in depression. *Psychopharmacology (Berl)* **236**(5): 1459–1470.
7. Pan JX, *et al.* (2019) Absence of gut microbiota during early life affects anxiolytic Behaviors and monoamine neurotransmitters system in the hippocampal of mice. *J Neurol Sci* **400**: 160–168.
8. Davidson GL, Cooke AC, Johnson CN, Quinn JL. (2018) The gut microbiome as a driver of individual variation in cognition and functional behaviour. *Philos Trans R Soc Lond B Biol Sci* **373**(1756): 20170286.
9. Wekerle H. (2019) Secondary progressive multiple sclerosis and the gut-brain axis. *Brain* **142**(4): 838–840.

10. Serra D, Almeida LM, Dinis TCP. (2019) The impact of chronic intestinal inflammation on brain disorders: The microbiota-gut-brain axis. *Mol Neurobiol* **56**(10): 6941–6951.
11. Kolacz J, Kovacic KK, Porges SW. (2019) Traumatic stress and the autonomic brain-gut connection in development: Polyvagal theory as an integrative framework for psychosocial and gastrointestinal pathology. *Dev Psychobiol* **61**(5): 796–809.
12. Roubalová R, Procházková P, Papežová H, Smitka K, Bilej M, Tlaskalová-Hogenová H. (2019) Anorexia nervosa: Gut microbiota-immune-brain interactions. *Clin Nutr* **39**(3): 676–684.
13. Drews E, Fertuck EA, Koenig J, Kaess M, Arntz A. (2019) Hypothalamic-pituitary-adrenal axis functioning in borderline personality disorder: A meta-analysis. *Neurosci Biobehav Rev* **96**: 316–334.
14. Huo R, *et al*. (2017) Microbiota modulate anxiety-like behavior and endocrine abnormalities in hypothalamic-pituitary-adrenal axis. *Front Cell Infect Microbiol* **7**: 489.
15. Tsigos C, Chrousos GP. (2002) Hypothalamic-pituitary-adrenal axis, neuroendocrine factors and stress. *J Psychosom Res* **53**: 865–871.
16. Mohajeri MH, La Fata G, Steinert RE, Weber P. (2018) Relationship between the gut microbiome and brain function. *Nutr Rev* **76**(7): 481–496.
17. Mittal R, *et al*. (2017) Neurotransmitters: The critical modulators regulating gut-brain axis. *J Cell Physiol* **232**(9): 2359–2372.
18. Bruce-Keller AJ, Salbaum JM, Berthoud HR. (2017) Harnessing gut microbes for mental health: Getting from here to there. *Biol Psychiatry* **83**(3): 214–223.
19. DeMorrow S. (2018) Role of the hypothalamic-pituitary-adrenal axis in health and disease. *Int J Mol Sci* **19**(4): 986.
20. Dinan TG, Cryan JF. (2017) The microbiome-gut-brain axis in health and disease. *Gastroenterol Clin North Am* **46**(1): 77–89.
21. Wang HX, Wang YP. (2016) Gut microbiota-brain axis. *Chin Med J (Engl)* **129**(19): 2373–2380.
22. Brinks V, van der Mark M, de Kloet R, Oitzl M. (2007) Emotion and cognition in high and low stress sensitive mouse strains: A combined neuroendocrine and behavioral study in BALB/c and C57BL/6J mice. *Front Behav Neurosci* **1**: 8.

23. Carabotti M, Scirocco A, Maselli MA, Severi C. (2015) The gut-brain axis: Interactions between enteric microbiota, central and enteric nervous systems. *Ann Gastroenterol* **28**(2): 203–209.
24. Lloyd-Price J, Abu-Ali G, Huttenhower C. (2016) The healthy human microbiome. *Genome Med* **8**: 5.
25. Shafquat A, Joice R, Simmons SL, Huttenhower C. (2014) Functional and phylogenetic assembly of microbial communities in the human microbiome. *Trends Microbiol* **22**(5): 261–266.
26. Khan Mirzaei M, *et al.* (2020) Bacteriophages isolated from stunted children can regulate gut bacterial communities in an age-specific manner. *Cell Host Microbe* **27**(2): 199–212.e5.
27. Shkoporov AN, *et al.* (2019) The human gut virome is highly diverse, stable, and individual specific. *Cell Host Microbe* **26**(4): 527–541.e5.
28. Robles Alonso V, Guarner F. (2013) Linking the gut microbiota to human health. *Br J Nutr* **109**(Suppl 2): S21–S26.
29. Mas-Lloret J, *et al.* (2020) Gut microbiome diversity detected by high-coverage 16S and shotgun sequencing of paired stool and colon sample. *Sci Data* **7**(1): 92.
30. Laudadio I, Fulci V, Palone F, Stronati L, Cucchiara S, Carissimi C. (2018) Quantitative assessment of shotgun metagenomics and 16S rDNA amplicon sequencing in the study of human gut microbiome. *OMICS* **22**(4): 248–254.
31. ENCODE Project Consortium, *et al.* (2007) Identification and analysis of functional elements in 1% of the human genome by the ENCODE pilot project. *Nature* **447**: 799–816.
32. Karhu E, *et al.* (2020) Nutritional interventions for autism spectrum disorder. *Nutr Rev* **78**(7): 515–531.
33. Dahl WJ, Rivero Mendoza D, Lambert JM. (2020) Diet, nutrients and the microbiome. *Prog Mol Biol Transl Sci* **171**: 237–263.
34. Zheng J, *et al.* (2020) Dietary inflammatory potential in relation to the gut microbiome: Results from a cross-sectional study. *Br J Nutr* **1**: 1–31.
35. Rutayisire E, Huang K, Liu Y, Tao F. (2016) The mode of delivery affects the diversity and colonization pattern of the gut microbiota during the first year of infants' life: A systematic review. *BMC Gastroenterol* **16**(1): 86.

36. Aagaard K, Stewart CJ, Chu D. (2016) Una destinatio, viae diversae: Does exposure to the vaginal microbiota confer health benefits to the infant, and does lack of exposure confer disease risk? *EMBO Rep* **17**: 1679–1684.
37. Kang LJ, *et al.* (2020) Maternal psychological distress before birth influences gut immunity in mid-infancy. *Clin Exp Allergy* **50**(2): 178–188.
38. Eshraghi RS, *et al.* (2018) Early disruption of the microbiome leading to decreased antioxidant capacity and epigenetic changes: Implications for the rise in autism. *Front Cell Neurosci* **12**: 256.
39. Kumbhare SV, Patangia DVV, Patil RH, Shouche YS, Patil NP. (2019) Factors influencing the gut microbiome in children: From infancy to childhood. *J Biosci* **44**(2): 49.
40. Angelopoulou A, Field D, Pérez-Ibarreche M, Warda AK, Hill C, Ross RP. (2020) Vancomycin and nisin A are effective against biofilms of multi-drug resistant Staphylococcus aureus isolates from human milk. *PLoS One* **15**(5): e0233284.
41. Vandenplas Y, *et al.* (2020) Factors affecting early-life intestinal microbiota development. *Nutrition* **78**: 110812.
42. Kostic AD, *et al.* (2015) The dynamics of the human infant gut microbiome in development and in progression toward type 1 diabetes. *Cell Host Microbe* **17**(2): 260–273.
43. Million M, Diallo A, Raoult D. (2017) Gut microbiota and malnutrition. *Microb Pathog* **106**: 127–138.
44. Subramanian S, *et al.* (2014) Persistent gut microbiota immaturity in malnourished Bangladeshi children. *Nature* **510**: 417–421.
45. Bäckhed F, *et al.* (2015) Dynamics and stabilization of the human gut microbiome during the first year of life. *Cell Host Microbe* **17**(5): 690–703.
46. Kuczma MP, *et al.* (2020) Commensal epitopes drive differentiation of colonic T_{regs}. *Sci Adv* **6**(16): eaaz3186.
47. Yadav M, Verma MK, Chauhan NS. (2018) A review of metabolic potential of human gut microbiome in human nutrition. *Arch Microbiol* **200**(2): 203–217.
48. Hiippala K, *et al.* (2018) The potential of gut commensals in reinforcing intestinal barrier function and alleviating inflammation. *Nutrients* **10**(8): 988.
49. Tillisch K. (2014) The effects of gut microbiota on CNS function in humans. *Gut Microbes* **5**: 404–410.

50. Tang TWH, *et al.* (2019) Loss of gut microbiota alters immune system composition and cripples postinfarction cardiac repair. *Circulation* **139**(5): 647–659.
51. Lyte JM, *et al.* (2020) Gut-brain axis serotonergic responses to acute stress exposure are microbiome-dependent. *Neurogastroenterol Motil* 32(11): 13881.
52. Martin CR, Osadchiy V, Kalani A, Mayer EA. (2018) The brain-gut-microbiome axis. *Cell Mol Gastroenterol Hepatol* **6**: 133–148.
53. Kuczynski J, *et al.* (2011) Experimental and analytical tools for studying the human microbiome. *Nat Rev Genet* **13**(1): 47–58.
54. Zoetendal EG, Rajilic-Stojanovic M, de Vos WM. (2008) High-throughput diversity and functionality analysis of the gastrointestinal tract microbiota. *Gut* **57**: 1605–1615.
55. Christensen L, Roager HM, Astrup A, Hjorth MF. (2018) Microbial enterotypes in personalized nutrition and obesity management. *Am J Clin Nutr* **108**(4): 645–651.
56. Arumugam M, *et al.* (2011) Enterotypes of the human gut microbiome. *Nature* **473**: 174–180.
57. Martens EC, Koropatkin NM, Smith TJ, Gordon JI. (2009) Complex glycan catabolism by the human gut microbiota: The Bacteroidetes Sus-like paradigm. *J Biol Chem* **284**: 24673–24677.
58. Kamada N, Chen GY, Inohara N, Núñez G. (2013) Control of pathogens and pathobionts by the gut microbiota. *Nat Immunol* **14**: 685–690.
59. Singh A, *et al.* (2020) Microbiome and host crosstalk: A new paradigm to cancer therapy. *Semin Cancer Biol* **70**. 71–84.
60. Sukri A, Hanafiah A, Mohamad Zin N, Kosai NR. (2020) Epidemiology and role of Helicobacter pylori virulence factors in gastric cancer carcinogenesis. *APMIS* **128**(2): 150–161.
61. Crowe SE. (2019) Helicobacter pylori Infection. *N Engl J Med* **380**(12): 1158–1165.
62. Abadi ATB. (2017) Strategies used by Helicobacter pylori to establish persistent infection. *World J Gastroenterol* **23**(16): 2870–2882.
63. De Roo AC, Regenbogen SE. (2020) Clostridium difficile Infection: An epidemiology update. *Clin Colon Rectal Surg* **33**(2): 49–57.
64. Czepiel J, *et al.* (2019) Clostridium difficile infection: Review. *Eur J Clin Microbiol Infect Dis* **38**(7): 1211–1221.

65. Borody TJ, Peattie D, Mitchell SW. (2015) Fecal microbiota transplantation: Expanding horizons for clostridium difficile infections and beyond. *Antibiotics (Basel)* **4**: 254–266.
66. Orenstein. (2013).
67. Leclercq S, Forsythe P, Bienenstock J. (2016) Posttraumatic stress disorder: Does the gut microbiome hold the key? *Can J Psychiatry* **61**(4): 204–213.
68. Mayer EA, Savidge T, Shulman RJ. (2014) Brain-gut microbiome interactions and functional bowel disorders. *Gastroenterology* **146**: 1500–1512.
69. Mayer EA, Tillisch K, Gupta A. (2015) Gut/brain axis and the microbiota. *J Clin Invest* **125**: 926–938.
70. Spencer NJ, Hu H. (2020) Enteric nervous system: Sensory transduction, neural circuits and gastrointestinal motility. *Nat Rev Gastroenterol Hepatol* **17**(6): 338–351.
71. Fung C, Vanden Berghe P. (2020) Functional circuits and signal processing in the enteric nervous system. *Cell Mol Life Sci.* **77**: 4505–4522.
72. Chalazonitis A, Rao M. (2018) Enteric nervous system manifestations of neurodegenerative disease. *Brain Res* **1693**(Pt B): 207–213.
73. Furness JB. (2012) The enteric nervous system and neurogastroenterology. *Nat Rev Gastroenterol Hepatol* **9**: 286–294.
74. Chng SH, Pachnis V. (2020) Enteric nervous system: Lessons from neurogenesis for reverse engineering and disease modelling and treatment. *Curr Opin Pharmacol* **50**: 100–106.
75. Furness JB, Callaghan BP, Rivera LR, Cho HJ. (2014) The enteric nervous system and gastrointestinal innervation: Integrated local and central control. *Adv Exp Med Biol* **817**: 39–71.
76. Rao M, Gershon MD. (2016) The bowel and beyond: The enteric nervous system in neurological disorders. Nature reviews. *Gastroenterol hepatol* **13**: 517–528.
77. Neunlist M, Van Landeghem L, Mahé MM, Derkinderen P, des Varannes SB, Rolli-Derkinderen M. (2013) The digestive neuronal–glial–epithelial unit: A new actor in gut health and disease. *Nat Rev Gastroenterol Hepatol* **10**: 90–100.
78. Forsythe P, Bienenstock J, Kunze WA. (2014) Vagal pathways for microbiome-brain-gut axis communication. *Adv Exp Med Biol* **817**: 115–133.

79. Wan M, Ding L, Wang D, Han J, Gao P. (2020) Serotonin: A potent immune cell modulator in autoimmune diseases. *Front Immunol* **11**: 186.
80. Liu QQ, *et al.* (2020) Role of 5-HT receptors in neuropathic pain: Potential therapeutic implications. *Pharmacol Res* **159**: 104949.
81. Rush AJ, *et al.* (2005) Vagus nerve stimulation for treatment-resistant depression: A randomized, controlled acute phase trial. *Biol Psychiatry* **58**(5): 347–354.
82. Arshad A, Powell C, Tighe MP. (2012) Hirschsprung's disease. *BMJ* **345**: e5521.
83. George MS, *et al.* (2000) Vagus nerve stimulation: A new tool for brain research and therapy. *Biol Psychiatry* **47**(4): 287–295.
84. Shahrestani J, Das JM. (2020) Neuroanatomy, auerbach plexus, in *StatPearls*. Treasure Island, FL, StatPearls Publishing.
85. Cerdó T, Diéguez E, Campoy C. (2019) Impact of gut microbiota on neurogenesis and neurological diseases during infancy. *Curr Opin Pharmacol* **50**: 33–37.
86. Long-Smith C, O'Riordan KJ, Clarke G, Stanton C, Dinan TG, Cryan JF. (2020) Microbiota-Gut-Brain axis: New therapeutic opportunities. *Annu Rev Pharmacol Toxicol* **60**: 477–502.
87. Ma Q, Xing C, Long W, Wang HY, Liu Q, Wang RF. (2019) Impact of microbiota on central nervous system and neurological diseases: The gut-brain axis. *J Neuroinflammation* **16**(1): 53.
88. Yang Y, Tian J, Yang B. (2018) Targeting gut microbiome: A novel and potential therapy for autism. *Life Sci* **194**: 111–119.
89. Breit S, Kupferberg A, Rogler G, Hasler G. (2018) Vagus nerve as modulator of the Brain-Gut axis in psychiatric and inflammatory disorders. *Front Psychiatry* **9**: 44.
90. Holzer P, Farzi A. (2014) Neuropeptides and the microbiota-gut-brain axis. *Adv Exp Med Biol* **817**: 195–219.
91. Berthoud HR, Kressel M, Raybould HE, Neuhuber WL. (1995) Vagal sensors in the rat duodenal mucosa: Distribution and structure as revealed by in vivo DiI-tracing. *Anat Embryol* **191**: 203–212.
92. Fülling C, Dinan TG, Cryan JF. (2019) Gut microbe to brain signaling: What happens in vagus…. *Neuron* **101**: 998–1002.
93. Berthoud HR, Blackshaw LA, Brookes SJ, Grundy D. (2004) Neuroanatomy of extrinsic afferents supplying the gastrointestinal tract. *Neurogastroenterol Motil* **16**(Suppl 1): 28–33.

94. Suarez AN, *et al.* (2018) Gut vagal sensory signaling regulates hippocampus function through multi-order pathways. *Nat Commun* **9**(1): 2181.
95. Browning KN, Verheijden S, Boeckxstaens GE. (2017) The vagus nerve in appetite regulation, mood, and intestinal inflammation. *Gastroenterology* **152**(4): 730–744.
96. Bonaz B, Sinniger V, Pellissier S. (2017) Vagus nerve stimulation: A new promising therapeutic tool in inflammatory bowel disease. *J Intern Med* **282**(1): 46–63.
97. Bonaz B, Sinniger V, Pellissier S. (2019) Vagus nerve stimulation at the interface of brain-gut interactions. *Cold Spring Harb Perspect Med* **9**(8): a034199.
98. Waise TMZ, Dranse HJ, Lam TKT. (2018) The metabolic role of vagal afferent innervation. *Nat Rev Gastroenterol Hepatol* **15**(10): 625–636.
99. Berthoud HR, Neuhuber WL. (2000) Functional and chemical anatomy of the afferent vagal system. *Auton Neurosci* **85**: 1–17.
100. O'Mahony SM, Clarke G, Borre YE, Dinan TG, Cryan JF. (2015) Serotonin, tryptophan metabolism and the brain-gut-microbiome axis. *Behav Brain Res* **277**: 32–48.
101. El-Salhy M, Hatlebakk JG, Hausken T. (2020) Possible role of peptide YY (PYY) in the pathophysiology of irritable bowel syndrome (IBS). *Neuropeptides* **79**: 101973.
102. Wei P, Keller C, Li L. (2020) Neuropeptides in gut-brain axis and their influence on host immunity and stress. *Comput Struct Biotechnol J* **18**: 843–851.
103. Kaelberer MM, Bohórquez DV. (2018) The now and then of gut-brain signaling. *Brain Res* **1693**(Pt B): 192–196.
104. Alviña K, Jodeiri Farshbaf M, Mondal AK. (2020) Long term effects of stress on hippocampal function: Emphasis on early life stress paradigms and potential involvement of neuropeptide Y. *J Neurosci Res* **99**(1): 57–66.
105. Li C, Wu X, Liu S, Zhao Y, Zhu J, Liu K. (2019) Roles of neuropeptide Y in neurodegenerative and neuroimmune diseases. *Front Neurosci* **13**: 869.
106. Alexiadou K, Tan TM. (2020) Gastrointestinal peptides as therapeutic targets to mitigate obesity and metabolic syndrome. *Curr Diab Rep* **20**(7): 26.
107. Persaud SJ, Bewick GA. (2014) Peptide YY: More than just an appetite regulator. *Diabetologia* **57**(9): 1762–1769.

108. Chandarana K, Batterham R. (2008) Peptide YY. *Curr Opin Endocrinol Diabetes Obes* **15**(1): 65–72.
109. Comeras LB, Herzog H, Tasan RO. (2019) Neuropeptides at the crossroad of fear and hunger: A special focus on neuropeptide Y. *Ann N Y Acad Sci* **1455**(1): 59–80.
110. Wu Y, He H, Cheng Z, Bai Y, Ma X. (2019) The role of neuropeptide Y and peptide YY in the development of obesity via gut-brain axis. *Curr Protein Pept Sci* **20**(7): 750–758.
111. Carniglia L, *et al.* (2017) Neuropeptides and microglial activation in inflammation, pain, and neurodegenerative diseases. *Mediators Inflamm* **2017**: 5048616.
112. Graefe S, Mohiuddin SS. (2020) Biochemistry, Substance P, in *StatPearls*. Treasure Island, FL, StatPearls Publishing.
113. Schank JR, Heilig M. (2017) Substance P and the neurokinin-1 receptor: The new CRF. *Int Rev Neurobiol* **136**: 151–175.
114. Williams JA. (2019) Cholecystokinin (CCK) regulation of pancreatic acinar cells: physiological actions and signal transduction mechanisms. *Compr Physiol* **9**(2): 535–564.
115. Ebner K, Singewald N. (2006) The role of substance P in stress and anxiety responses. *Amino Acids* **31**: 251–272.
116. Rege S. (2017) The Simplified Guide to the Gut-Brain Axis — How the gut and the brain talk to each other. *Psych Scene Hub*.
117. Ngo DH, Vo TS. (2019) An updated review on pharmaceutical properties of gamma-aminobutyric acid. *Molecules* **24**(15): 2678.
118. Tomkovich S, Jobin C. (2016) Microbiota and host immune responses: A love-hate relationship. *Immunology* **147**: 1–10.
119. Calatayud M, Dezutter O, Hernandez-Sanabria E, Hidalgo-Martinez S, Meysman FJR, Van de Wiele T. (2019) Development of a host-microbiome model of the small intestine. *FASEB J* **33**(3): 3985–3996.
120. Raison CL, Miller AH. (2011) Is depression an inflammatory disorder? *Curr Psychiatry Rep* **13**: 467–475.
121. He Y, *et al.* (2020) Major depression accompanied with inflammation and multiple cytokines alterations: Evidences from clinical patients to macaca fascicularis and LPS-induced depressive mice model. *J Affect Disord* **271**: 262–271.

122. Osimo EF, Pillinger T, Rodriguez IM, Khandaker GM, Pariante CM, Howes OD. (2020) Inflammatory markers in depression: A meta-analysis of mean differences and variability in 5,166 patients and 5,083 controls. *Brain Behav Immun* **87**: 901–909.
123. Karakula-Juchnowicz H, *et al.* (2019) The study evaluating the effect of probiotic supplementation on the mental status, inflammation, and intestinal barrier in major depressive disorder patients using gluten-free or gluten-containing diet (SANGUT study): A 12-week, randomized, double-blind, and placebo-controlled clinical study protocol. *Nutr J* **18**(1): 50.
124. Tian P, *et al.* (2019) Ingestion of Bifidobacterium longum subspecies infantis strain CCFM687 regulated emotional behavior and the central BDNF pathway in chronic stress-induced depressive mice through reshaping the gut microbiota. *Food Funct* **10**(11): 7588–7598.
125. Han SK, Kim DH. (2019) Lactobacillus mucosae and Bifidobacterium longum synergistically alleviate immobilization stress-induced anxiety/depression in mice by suppressing gut dysbiosis. *J Microbiol Biotechnol* **29**: 1369–1374.
126. Wang C, Li Q, Ren J. (2019) Microbiota-immune interaction in the pathogenesis of gut-derived infection. *Front Immunol* **10**: 1873.
127. Pédron T, Sansonetti P. (2008) Commensals, bacterial pathogens and intestinal inflammation: An intriguing ménage à trois. *Cell Host Microbe* **3**: 344–347.
128. Quigley EMM. (2017) Microbiota-brain-gut axis and neurodegenerative diseases. *Curr Neurol Neurosci Rep* **17**: 94.
129. Neufeld KM, Kang N, Bienenstock J, Foster JA. (2010) Reduced anxiety-like behavior and central neurochemical change in germ-free mice. *Neurogastroenterol Motil* **23**: 255–264.
130. Cho I, *et al.* (2012) Antibiotics in early life alter the murine colonic microbiome and adiposity. *Nature* **488**: 621–626.
131. Pinto-Sanchez MI, *et al.* (2017) Probiotic bifidobacterium longum NCC3001 reduces depression scores and alters brain activity: A pilot study in patients with irritable bowel syndrome. *Gastroenterology* **153**: 448–459.
132. Liu J, *et al.* (2017) Effect of vitamin A supplementation on gut microbiota in children with autism spectrum disorders — A pilot study. *BMC Microbiol* **17**(1): 204.

133. Kato-Kataoka A, *et al.* (2016) Fermented milk containing lactobacillus casei strain Shirota preserves the diversity of the gut microbiota and relieves abdominal dysfunction in healthy medical students exposed to academic stress. *Appl Environ Microbiol* **82**: 3649–3658.
134. Kelly JR, *et al.* (2017) Lost in translation? The potential psychobiotic Lactobacillus rhamnosus (JB-1) fails to modulate stress or cognitive performance in healthy male subjects. *Brain Behav Immun* **61**: 50–59.
135. Vrieze A, *et al.* (2012) Transfer of intestinal microbiota from lean donors increases insulin sensitivity in individuals with metabolic syndrome. *Gastroenterology* **143**: 913–916.
136. Kang DW, *et al.* (2017) Microbiota transfer therapy alters gut ecosystem and improves gastrointestinal and autism symptoms: An open-label study. *Microbiome* **5**(1): 10.
137. Tomova A, *et al.* (2015) Gastrointestinal microbiota in children with autism in Slovakia. *Physiol Behav* **138**: 179–187.
138. Santocchi E, *et al.* (2016) Gut to brain interaction in Autism Spectrum disorders: A randomized controlled trial on the role of probiotics on clinical, biochemical and neurophysiological parameters. *BMC Psychiatry* **16**: 183.
139. Hashim H, *et al.* (2014) Eradication of helicobacter pylori infection improves levodopa action, clinical symptoms and quality of life in patients with Parkinson's disease. *PLoS One* **9**: e112330.
140. Sun MF, *et al.* (2018) Neuroprotective effects of fecal microbiota transplantation on MPTP-induced Parkinson's disease mice: Gut microbiota, glial reaction and TLR4/TNF-α signaling pathway. *Brain Behav Immun* **70**: 48–60.
141. Reich SG, Savitt JM. (2019) Parkinson's disease. *Med Clin North Am* **103**(2): 337–350.
142. Kalia LV, Lang AE. (2015) Parkinson's disease. *Lancet* **386**(9996): 896–912.
143. Marras C, Tanner CM. (2004) Epidemiology of Parkinson's disease, in *Movement Disorders: Neurologic Principles and Practice*. New York, NY, The McGraw-Hill Companies, p. 177.
144. Raza C, Anjum R, Shakeel NUA. (2019) Parkinson's disease: Mechanisms, translational models and management strategies. *Life Sci* **226**: 77–90.

145. Opara J, Małecki A, Małecka E, Socha T. (2017) Motor assessment in Parkinson's disease. *Ann Agric Environ Med* **24**(3): 411–415.
146. Sampson T, *et al.* (2016) Gut microbiota regulate motor deficits and neuroinflammation in a model of Parkinson's disease. *Cell* **167**: 1469–1480.
147. Fasano A, Visanji NP, Liu LW, Lang AE, Pfeiffer RF. (2015) Gastrointestinal dysfunction in Parkinson's disease. *Lancet Neurol* **14**(6): 625–639.
148. Pfeiffer RF. (2018) Gastrointestinal dysfunction in Parkinson's disease. *Curr Treat Options Neurol* **20**(12): 54.
149. Verbaan D, Marinus J, Visser M, van Rooden SM, Stiggelbout AM, van Hilten JJ. (2007) Patient-reported autonomic symptoms in Parkinson's disease. *Neurology* **69**: 333.
150. Dardiotis E, *et al.* (2018) Pylori and Parkinson's disease: Meta-analyses including clinical severity. *Clin Neurol Neurosurg* **175**: 16–24.
151. Huang HK, Wang JH, Lei WY, Chen CL, Chang CY, Liou LS. (2018) Helicobacter pylori infection is associated with an increased risk of Parkinson's disease: A population-based retrospective cohort study. *Parkinsonism Relat Disord* **47**: 26–31.
152. Shen X, Yang H, Wu Y, Zhang D, Jiang H. (2017) Association of Helicobacter pylori infection with Parkinson's diseases: A meta-analysis. *Helicobacter* **22**: 12398.
153. Quigly. (2017).
154. Dobbs SM, *et al.* (2010) Differential effect of Helicobacter pylori eradication on time-trends in brady/hypokinesia and rigidity in idiopathic parkinsonism. *Helicobacter* **15**: 279–294.
155. Scheperjans F, *et al.* (2015) Gut microbiota are related to Parkinson's disease and clinical phenotype. *Mov Disord* **30**: 350–358.
156. Keshavarzian A, *et al.* (2015) Colonic bacterial composition in Parkinson's disease. *Mov Disord* **30**(10): 1351–1360.
157. Hopfner F, *et al.* (2017) Gut microbiota in Parkinson disease in a northern German cohort. *Brain Res* **1667**: 41–45.
158. Zhao H, *et al.* (2018) Potential biomarkers of Parkinson's disease revealed by plasma metabolic profiling. *J Chromatogr B Analyt Technol Biomed Life Sci* **1081–1082**: 101–108.
159. Pournaras DJ, le Roux CW. (2013) Are bile acids the new gut hormones? Lessons from weight loss surgery models. *Endocrinology* **154**: 2255–2256.

160. Molinaro A, Wahlström A, Marschall HU. (2018) Role of bile acids in metabolic control. *Trends Endocrinol Metab* **29**(1): 31–41.

161. Shah V, *et al.* (2020) Evaluating the efficacy of Taurodeoxycholic acid in providing otoprotection using an *in vitro* model of electrode insertion trauma. *Front Mol. Neurosci* **13**: 113.

162. Vang S, Longley K, Steer CJ, Low WC. (2014) The unexpected uses of Urso- and tauroursodeoxycholic acid in the treatment of non-liver diseases. *Glob Adv Health Med* **3**(3): 58–69.

163. Yu Z, Yang J, Xiang D, Li G, Liu D, Zhang C. (2020) Circadian rhythms and bile acid homeostasis: A comprehensive review. *Chronobiol Int* **4**: 1–11.

164. Lord C, Elsabbagh M, Baird G, Veenstra-Vanderweele J. (2018) Autism spectrum disorder. *Lancet* **392**(10146): 508–520.

165. Lai MC, Lombardo MV, Baron-Cohen S. (2014) Autism. *Lancet* **383**(9920): 896–910.

166. Wiśniowiecka-Kowalnik B, Nowakowska BA. (2019) Genetics and epigenetics of autism spectrum disorder-current evidence in the field. *J Appl Genet* **60**(1): 37–47.

167. Bölte S, Girdler S, Marschik PB. (2019) The contribution of environmental exposure to the etiology of autism spectrum disorder. *Cell Mol Life Sci* **76**(7): 1275–1297.

168. Woodbury-Smith M, Scherer SW. (2018) Progress in the genetics of autism spectrum disorder. *Dev Med Child Neurol* **60**(5): 445–451.

169. Ramaswami G, Geschwind DH. (2018) Genetics of autism spectrum disorder. *Handb Clin Neurol* **147**: 321–329.

170. Chaste P, Leboyer M. (2012) Autism risk factors: Genes, environment, and gene-environment interactions. *Dialogues Clin Neurosci* **14**(3): 281–292.

171. Iglesias-Vázquez L, Van Ginkel Riba G, Arija V, Canals J. (2020) Composition of Gut Microbiota in children with autism spectrum disorder: A systematic review and meta-analysis. *Nutrients* **12**(3): 792.

172. De Angelis M, *et al.* (2013) Fecal microbiota and metabolome of children with autism and pervasive developmental disorder not otherwise specified. *PLoS One* **8**: e76993.

173. Parracho HM, Bingham MO, Gibson GR, McCartney AL. (2005) Differences between the gut microflora of children with autistic spectrum disorders and that of healthy children. *J Med Microbiol* **54**(Pt 10): 987–991.

174. Bergman EN. (1990) Energy contributions of volatile fatty acids from the gastrointestinal tract in various species. *Physiol Rev* **70**: 567–590.
175. Marchesi JR, *et al.* (2016) The gut microbiota and host health: A new clinical frontier. *Gut* **65**(2): 330–339.
176. Kaluzna-Czaplinska J, Blaszczyk S. (2012) The level of arabinitol in autistic children after probiotic therapy. *Nutrition* **28**: 124–126.
177. Parracho HMRT, Gibson GR, Knott F, Bosscher D, Kleerebezem M, McCartney AL. (2010) A double blind, placebo-controlled, crossover-designed probiotic feeding study in children diagnosed with autistic spectrum disorders. *Int J Probiotics Prebiotics* **5**: 69–74.
178. Vendrik KEW, *et al.* (2020) Fecal microbiota transplantation in neurological disorders. *Front Cell Infect Microbiol* **10**: 98.
179. Adams JB, Borody TJ, Kang DW, Khoruts A, Krajmalnik-Brown R, Sadowsky MJ. (2019) Microbiota transplant therapy and autism: Lessons for the clinic. *Expert Rev Gastroenterol Hepatol* **13**(11): 1033–1037.
180. Rhee SH, Pothoulakis C, Mayer EA. (2009) Principles and clinical implications of the brain-gut-enteric microbiota axis. *Nat Rev Gastroenterol Hepatol* **6**: 306–314.

Chapter 12

Zebrafish as a Model Organism to Understand the Gut–Brain Axis

Anthony Dao*, Henson Destine*, Crystal Li, Mausam Patel and Rahul Mittal

Abstract

The gut–brain axis (GBA) is a complex communication network that bridges the emotional and cognitive neurological centers with peripheral intestinal functions. Zebrafish (*Danio rerio*) is an emerging animal model for understanding gut–brain–microbiota interplay and its related factors such as enteric nervous system (ENS) development, behavioral processes, immunological responses, and toxicological effects. The zebrafish is a model organism for several reasons, such as genetic and physiological homology to humans, genetic tractability, rapid development of organ systems, a fully sequenced genome, as well as the ability for researchers to observe and quantify complex behaviors within zebrafish. The bidirectional correlations between gut microbiota and stress-related behavior along with neurological development

* Hearing Research and Communication Disorders Laboratory, Department of Otolaryngology, University of Miami-Miller School of Medicine, Miami, FL, USA.

* These authors contributed equally to this work.

Corresponding Author: Rahul Mittal, PhD, Hearing Research and Communication Disorders Laboratory, Department of Otolaryngology, University of Miami-Miller School of Medicine, 1600 NW 10th Avenue, Miami, FL 33136, USA, Tel: 305-243-7117, E-mail: r.mittal11@med.miami.edu.

have been discovered. In addition, studies have suggested that microbiota of zebrafish impacts central nervous system signaling. Immunologically, zebrafish is an appropriate model since numerous immune genes are conserved between them and mammals. Zebrafish have also gained popularity as a model organism to appraise the function and possibility of certain human ENS disease genes, such as those responsible for Hirschsprung's disease, through genome-wide association studies.

INTRODUCTION

The gut–brain axis (GBA) is a bidirectional complex communication network involving underlying neuro-immuno-endocrine mediators that bridge the emotional and cognitive brain centers with peripheral intestinal functions.[1] Consisting of the central nervous system (CNS), the autonomic nervous system (ANS), the enteric nervous system (ENS), and the hypothalamic–pituitary–adrenal (HPA) axis, the GBA serves to coordinate gut functions such as immune activation, intestinal permeability, enteric reflex, and enteroendocrine signaling.[1] This communication network is inherently tied to and further enhanced by the enteric microbiota composition naturally present in the gastrointestinal (GI) tract. Activities of the intestinal functional effector cells are jointly influenced by both gut microbiota and the brain's neural as well as hormonal lines of communication.[1] Consequently, this establishes a system where the gut–brain interactions can regulate and in turn be regulated by gut microbiota and GI functions. The GBA concept, more accurately termed the gut–microbiota–brain axis, involves the numerous molecular cross plays between commensal microorganisms and the brain.[2–4] Gut bacteria such as *Staphylococcus aureus* can activate sensory neurons to stimulate nociception, while other microbes such as *Lactobacillus* can activate intestinal afferent neurons. The neural pathways on enteric neurons can be modulated directly as in the case with toll-like receptors.[4] Host enteric neurotransmitters can dictate the function of microbiota, thereby increasing inflammation or modulating certain gut functions such as motility and mucus production.[4] Zebrafish is an emerging model to understand gut–brain–microbiota interplay and its related factors such as ENS development, behavioral processes, immunological responses, and toxicological effects.

ZEBRAFISH

The zebrafish, *Danio rerio*, is a model organism for research in many fields and is especially important for understanding the GBA as it offers several advantages including transparent larvae, genetic and physiological homology to humans, external fertilization, rapid development of organ systems, genetic tractability, and a fully sequenced genome.[4] The different regions of the brain such as forebrain, midbrain, and hindbrain structures are distinguishable within 24 hours post-fertilization (Fig. 12.1). The most important component is the ability for researchers to observe and quantify complex behaviors within zebrafish such as aggression, memory, and sleep habits.[5] It is possible to observe the complex behavior in zebrafish such as anxiety-like behavior, which can be recorded in a multitude of ways: shoaling behavior, water column preferences, light versus dark environments, and locomotion (Fig. 12.2).[5] In comparison to rodents (mammals), zebrafish systems have parallels between brain morphology, brain neurochemistry, and pharmacological profiles.[6] Within the 25 pairs of chromosomes and approximately 26,000 protein-coding genes of zebrafish, there is a significant amount of homology with humans: 91% similarity in amino acid sequence, 70% homology with human genes, and over 80% homology of disease-related genes.[6,7] Specifically, zebrafish have a complex serotonergic system that has all the necessary genetic makeup necessary for 5-hydroxytryptamine (5-HT) (serotonin) signaling, rate limitation, metabolism, and synthesis. The microbiomes of humans and zebrafish are also distinctly similar and share bacterial phylum that are associated with positive and negative biological effects (Fig. 12.1).

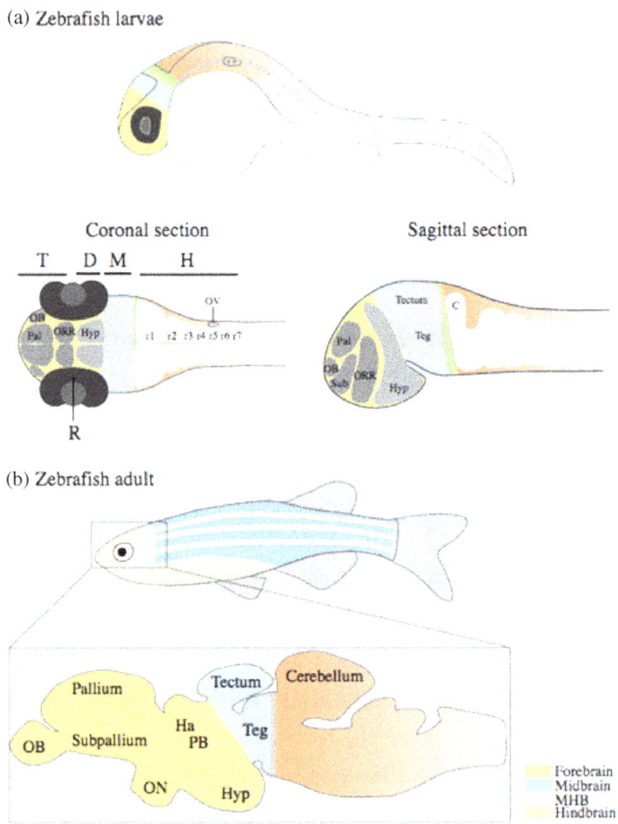

Fig. 12.1: Development of the zebrafish brain. (a) Schematic representation of the embryonic brain (30 hpf), showing the forebrain (in yellow), midbrain (in blue), MHB (in green), and hindbrain (in orange). Coronal and sagittal section schemes show brain structures primordia. The forebrain is subdivided into the telencephalon (in darker gray) and the diencephalon (containing the hypothalamus, lighter grey). (b) Simplified representation of the adult brain and main domains. Drawings not to scale. C: cerebellum; D: diencephalon; M: midbrain; MHB: midbrain–hindbrain boundary; H: hindbrain; Ha: habenula; Hyp: hypothalamus; OB: olfactory bulb; ON: optic nerve; ORR: optic recess region; OV: otic vesicle; Pal: pallium; PB: pineal body; R: retina; r1–r7: rhombomeres 1 to 7; Sub: subpallium; T: telencephalon; and Teg: tegmentum. (Reprinted from Vaz *et al.*[8] This is an open access article distributed under the Creative Commons Attribution License, which permits unrestricted use, distribution, and reproduction in any medium, provided the original work is properly cited.)

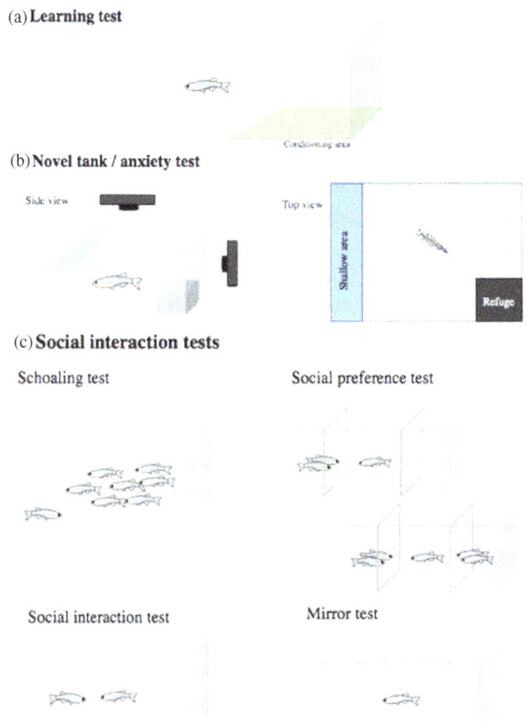

Fig. 12.2: Common behavior tests for adult zebrafish. (a) Learning test or conditioned place preference (CPP) test, consists of testing the learning capacity by repeatedly exposing the specimen to a reward in a specific location (conditioning area, in green). (b) The novel tank test is used to assess anxiety. The animal is placed in a new arena containing refuge areas (represented by a black box) or risk areas (shallow area), and the swimming pattern is analyzed. Bold fish typically explore the open and shallow areas, while socially impaired fish prefer refuge areas. (c) Social interaction tests. These can be subdivided into the shoaling test, assessing a fish behavior when in the presence of a group of conspecific fish; the social preference test, by separating the subject fish from conspecific or non-conspecific fish using a physical barrier and recording response; the social interaction test, by placing usually two fish in a tank and assessing behavior; or the mirror test. These tests are useful to test for socialization or aggressive behavior. Several social interaction tests can be performed in adult zebrafish: (1) shoaling test, by assessing the behavior of individual fish within a group; (2) social preference test, a test that studies the interest and interaction of fish that are in the same tank but separated by a transparent barrier; (3) social interaction test, similar to the social preference test, but without any physical barrier between the two animals; and (4) mirror test, where the reaction of an individual fish placed in a tank with a mirror is recorded (Fig. 12.2c). (Reprinted from Vaz *et al.*[8] This is an open access article distributed under the Creative Commons Attribution License which permits unrestricted use, distribution, and reproduction in any medium, provided the original work is properly cited.)

However, despite these advantages, there are a few limitations to the use of zebrafish as a model. Physiologically, the zebrafish gut has distinct differences compared to mammals: their gut comes from an anlage and not a primitive gut tube, and they have intestinal folds as opposed to intestinal villi.[9] The laboratory setting itself is a possible limitation as well. The gut microbiota of domesticated zebrafish in different laboratories differ in composition due to varied environmental conditions and location.[10] However, the exact impact of these variations on gut microbiota is unknown. Moreover, comparisons between domesticated and wild zebrafish have revealed a non-statistical significance between the gut microbiota suggesting these the variations may not be a confounding factor.[10]

BEHAVIORAL STUDIES AND HUMAN CORRELATION

Zebrafish have been used as model to understand the effect of probiotic supplementation on behaviors such as locomotion, shoaling, shuttling, and anxiety. A study fed zebrafish with *Lactobacillus rhamnosus* IMC 501, a widely used probiotic.[7] The researchers found a multitude of statistically significant changes in the zebrafish after being fed *L. rhamnosus* IMC 501 for 28 days. These changes include increased shoaling behavior and increased *bdnf* (brain-derived neurotrophic factor) expression levels in the brain as well as increase in Firmicutes and decrease in Proteobacteria in the gut.[7] Shoaling behavior is considered a complex social behavior that benefits zebrafish in many ways such as foraging, preventing predation, and reproductive purposes. These results suggest that *L. rhamnosus* IMC 501 influences behavior in zebrafish. In addition, increased expression of *bdnf* was observed that is responsible for serotonin signaling metabolism via mutualistic influence in the brain. These findings are in agreement with the qualitative behavioral changes induced by *L. rhamnosus* IMC 501.[7] The increase in serotonin production/activity in the serotonergic system is comparable with a very common treatment in individuals with depression and generalized anxiety using selective serotonin reuptake inhibitors (SSRIs).[7] Firmicutes and Proteobacteria are present in both human and zebrafish microbiota. A decrease in Firmicutes has been associated with depression due to the lack of short-chain fatty acids produced by these microbes leading to physiological changes.[11] The increase in Firmicutes in the gut of zebrafish administered with *L. rhamnosus* IMC 501, therefore, also has the potential to develop treatment modalities for neurological disorders. On the other hand, predominance of Proteobacteria has been associated with a variety of physiological disorders, both intestinal and extraintestinal, ranging from inflammatory bowel disease to asthma. Therefore, decrease in Proteobacteria after supplementation with *L. rhamnosus* IMC 501 may have clinical implications for designing therapeutic strategies for neurological disorders.[7,12]

Another study determined the effect of *Lactobacillus plantarum* supplementation in zebrafish. It was observed that *L. plantarum* altered gut microbiota structure, which led to decreased anxiety-related behavior, up-regulation in the serotonergic pathway, and protection against stress-induced dysbiosis.[13] A common anxiety-related behavior in zebrafish is diving to the bottom of a tank. In contrast, time spent near the top of the water's surface has been associated with relaxed behavior. Zebrafish supplemented with *L. plantarum* significantly spent more time at the upper portion of the tank suggesting relaxed behavior.[13] This behavioral change was accompanied with up-regulation of genes that encode for serotonin transportation, *slc6a4a* and *slc6a4b,* although *slc6a4b* increase was not statistically significant.[13] Zebrafish supplemented with *L. plantarum* were also seen to be resistant to stress-induced dysbiosis. This was evaluated through a chronic unpredictable stress (CUS) test, which showed a clear correlation between stress and gut microbiota. Zebrafish not treated with *L. plantarum* displayed major shifts, most notably a statistically significant decrease in the core phylum, Fusobacteria.[13] Furthermore, there was down-regulation of vitamin biosynthesis in the untreated zebrafish.[13] However, zebrafish treated with

L. plantarum maintained a steady gut microbiota through a CUS test along with no changes in vitamin biosynthesis, suggesting that *L. plantarum* facilitates in mitigation of stress-induced dysbiosis.

THE GUT CONNECTION: THE ENS

The ENS is a division of the ANS that serves as the intrinsic nervous system of the GI tract. The ENS is composed of complete feedback loops that recognize the physiological condition of the GI tract, and controls essential intestinal functions such as fluid transport, blood flow, general homeostasis, and nutrient handling.[14] The zebrafish ENS is well developed with five distinct enteric neuronal subpopulations organized by the secretion of specific neurotransmitters and enzymes.[15] The zebrafish ENS develops from neural crest cells that migrated caudally and eventually colonized the intestine.[16] Contrary to mammals, the zebrafish ENS lacks a submucosal plexus and instead is made up of myenteric neurons arranged in a single layer in small groups that express both excitatory and inhibitory neurotransmitters.[16] Additionally, zebrafish and mammals possess other morphological differences such as the lack of a stomach and genes encoding for explicit gastric functions. However, aspects such as epithelial cell types, enterocytes, architectural organization, and molecular mechanisms regulating the development of the ENS are highly conserved between zebrafish and mammals.[15] ENS development in zebrafish are regulated through conserved transcription signaling factors and influenced by surrounding tissue such as the developing gut and mesenchyme.[16] Zebrafish are well suited for investigating ENS contributions to the composition of the gut microbiota. The development of the ENS, the absolute bacterial abundance, and disease phenotypes such as the accumulation of neutrophils can all be monitored and used to accurately model what takes place in mammals.[16] The functional similarities have allowed human ENS disorders, such as Hirschsprung's disease and Bardet–Biedl syndrome, to be accurately modeled in zebrafish and functionally translated. As a result, zebrafish can be used to gauge system-level functional readouts that illustrate properties of the gut microbiota.[16] Their high fertility also begets large sample sizes that consequently increases experimental ability with phenotypic variations and both natural or induced microbiota variation.[17] Recently, zebrafish have gained extreme popularity as a model organism to appraise the function and possibility of certain human ENS disease genes, such as those responsible for Hirschsprung's disease through genome-wide association studies.[17] Zebrafish are also appropriate for conducting large-scale genetic screens to identify novel regulators that are involved in ENS development, novel ENS diseases, or gene candidates for existing ENS diseases. Expanding the usefulness of zebrafish models, these organisms can be developed in the absence of commensal bacteria or with distinct bacterial populations contributing to studies where isolation or absence of commensal bacteria populations are needed.[17]

ZEBRAFISH AS A MODEL FOR IMMUNE DISORDERS

Immunologically, zebrafish serves as an appropriate experimental model due to the conservation of various immune genes between them and mammals. Innate and adaptive immunity development, as well as immune cell-type homologues to higher vertebrates accentuate the practicality of using zebrafish models. The immunological interactions such as processes involved in intestinal inflammatory diseases and impacted by the GBA can be studied using zebrafish models.[17] Zebrafish microbiota, although consisting of a different dominant phylum, induces a host response parallel to mammalian microbiota. This composition and parallel response can be used to demonstrate disease susceptibility as in the case of intestinal inflammation. Zebrafish heterophilic-neutrophilic granulocytes (HNGs) share the same function and chemical components as the mammalian neutrophilic cytoplasmic granules.[4] The activity of these HNGs and other leukocytes in the zebrafish intestine are heavily influenced by commensal microbiota. For example, commensal microbiota colonization of zebrafish intestine increases the amount of HNGs present, while also influencing the migration of these cells.[4] Commensal microbiota also impacts HNG homeostasis to heighten recruitment of these cells to specific sites of injury. It has been shown that production of serum amyloid A, directly caused by intestinal colonization in zebrafish, is a host factor for microbial stimulation of HNG migratory behaviors.[4] The presentation of pathogens to the intestinal milieu in zebrafish has proven to unbalance the ENS, alter microbiome composition and induce innate immune responses along with potential increase of reactive oxygen species and defensin-associated mRNA.[4] Zebrafish models of chemically induced intestinal inflammation appropriately mirror responses seen in higher vertebrates as far as microbial involvement, changes to intestinal tissue, and responses to anti-inflammatory drugs.[17]

TOXICOLOGICAL EFFECTS

Neurotoxicological effects of environmental chemicals can be altered by gut microbiota through the microbiota–gut–brain axis since the microbiota can affect the toxicokinetics and toxicodynamics of xenobiotic exposures.[18] Zebrafish are primed to highlight the mechanisms used by gut microbiota to adjust the neurotoxicity of environmental chemicals. Unlike other animal models, zebrafish embryos develop external to the mother allowing the embryo itself to be exposed to xenobiotic agents.[18] The neurodevelopment of zebrafish is well conserved in comparison to humans, thereby displaying similar neurotoxicological phenotypes such as loss or expansion of brain ventricles, truncation of the telencephalon, and neuronal necrosis.[18] Host toxicity, as in the form of neurodevelopmental toxicity, can occur via dysbiosis caused by changes to the structure of gut microbiota through toxicodynamic interactions. Interactions with xenobiotic agents such as pesticides, metals, microplastics, and antibiotics that can then be biotransformed, through chemical-specific profiles, by microbial enzymes.[18] To verify a link between microbiota and neurobehavioral development in zebrafish, experiments were conducted that exposed zebrafish to a broad spectrum of antibiotics. Results displayed hyperactivity among the zebrafish in line with mammalian results showcasing hyperactivity and increased anxiety-like behavior.[18]

BEHAVIORAL PROCESSES

Zebrafish display several neurological behavioral phenotypes having similarity to humans such as anxiety, learning, fear, sociability, and psychosis.[19] The evolutionarily highly conserved phenotype between zebrafish and humans is anxiety-related behavior and HPA axis-based physiological responses to stress.[19] The HPA is a primary stress efferent axis used to coordinate adaptive responses to the wide range of stressors that the host may encounter; The HPA axis, a component of the limbic system dealing with memory and emotional responses, can be activated by environmental stress leading to cortisol — a major stress hormone in the body — being released from the adrenal glands.[19] The bidirectional correlations between gut microbiota and stress-related behavior along with neurological development have been discovered, as the microbiota of zebrafish impact CNS signaling.[20] Gut microbiota have been found to alter essential cells in the brain that respond to neuroinflammation, neurogenesis and neurotransmission, thereby playing a critical role in normal CNS functioning and behavior.[4] Consequently, complications of the microbiota play a part in the onset of psychiatric and neurodegenerative diseases.[20] Studies have shown that the structure of the intestinal microbiota can be altered by relevant social stress, and these alterations can in turn affect the development of neurological disorders and behavior. Further investigations have also suggested that stress-associated inflammatory cytokine levels, anxiety-related behavior, and symptoms of stress have all been decreased following ingestion of probiotic *Lactobacillus* strains in mammals and zebrafish.[7]

ZEBRAFISH MODELS AND NEUROLOGICAL DISORDERS

Zebrafish models have been used to understand the pathophysiology of neurological disorders involving the GBA. These areas of research include autism spectrum disorders (ASDs), neurodegenerative diseases, anesthesia, and the genetic factors that influence anesthesia, "Sickness behavior," and germ-free (GF) development.

Autism Spectrum Disorders

There is an established connection between ASD and GI issues but this connection is not well understood. A study investigated the gut of zebrafish *SHANK3* mutant models, which served as a model of Phelan–McDermid syndrome.[21] Phelan–McDermid syndrome is a monogenic form of ASD that causes a mutation in one copy of the *SHANK3* — either *shank3a* or *shank3b* in zebrafish.[21] On the molecular level, *SHANK3* is important for a plethora of functions including synaptic scaffolding protein in the CNS, expression in enterocytes and nitrergic neurons of the ENS, early developmental stages, and interaction with the Wnt signaling pathway.[21] Most importantly, the established connection between ASD and GI distress was due to limited studies suggesting that *SHANK3* is important for host/symbiont interactions, zinc metabolism, and intestinal barrier function.[21] To create frameshift mutations in *shank3a* and *shank3b*, site-specific CRISPR-Cas9 sgRNA injections were used. The two different mutants created were shank3abΔC$^{+/-}$ and shank3abΔC$^{-/-}$. These two mutants yielded a multitude of results including reduced peristalsis frequency in the digestive tract (DT) in both mutants, increased DT transit time in shank3abΔC$^{+/-}$, decreased 5-HT+ cells in both mutants, and more.[21] Interestingly, it was observed that the hypomotility issues could be partially "rescued" via an injection of human *SHANK3* mRNA.[21] Human *SHANK3* mRNA was shown to rescue DT transit from the intestinal bulb to the upper intestine, but the limitation was that overall intestinal motility was not improved within mutants.[21] It was observed that injection was given to zebrafish at the one-cell larval stage and is only transiently expressed for 48 hours.[21] Because of this short expression time, the researchers hypothesized that early *shank3a-* and *shank3b*-dependent processes such as hindbrain morphogenesis could be saved but late processes such as ENS migration and enteroendocrine cell differentiation could not be rescued.[21] Overall, the results of this study suggest a robust connection between the brain and gut microbiome in relation to GI distress and Phelan–McDermid syndrome. A niche that was not explored within this study was behavior and social communication. Due to the zebrafish's ability to exhibit complex behaviors, it is possible to study the behavioral and social communication deficits that are hallmarks of ASD along with considerations of the gut microbiome.

Hereditary Neurodegenerative Diseases

Hereditary neurodegenerative disorders share common underlying cellular mechanisms discovered through zebrafish modeling of these rare diseases. In disorders such as Charcot–Marie–Tooth (CMT) disease, spinal muscle atrophy, Hereditary Spastic Paraplegia (HSP), Cerebellar Ataxia (CATX), and amyotrophic lateral sclerosis (ALS), axon tract degeneration is a key clinical feature.[22] Zebrafish models can be used to dissect molecular mechanisms due to the early development of their peripheral, motor, and sensory neurons.[22] For example, with regard to axonal transport processes and *in vivo* axon degeneration, a subset of genes directly tied to neurodegenerative disorders, zebrafish are an ideal model to study due to their optical transparency and the availability of transgenic lines.[22] Studies have identified *Kinesin Family Member 5A* (*KIF5A*) mutations in both CMT type 2 (CMT2) and HSP patients; replicating this mutation in zebrafish results in decreased touch response, defective sensory neuronal maintenance, and affected transport and distribution of mitochondria in neurons.[22] Similarly, the dominant mutations in *Atlastin GTPase 1 (ATL1)*, which causes early onset HSP, was replicated in zebrafish using morpholino knockdown; this mutation resulted in decreased mobility in larval fish along with disrupted axon tracts of spinal motor neurons.[22] Other examples of different cellular level mechanisms causing neurodegenerative disorders were also studied. Examples such as mitochondrial dysfunction leading to axonal degeneration through the *Mitofusin* 2 mutation and cerebellar Purkinje neuron atrophy through *sorting nexin 14* mutations or *KCN3* mutations.[22] These studies all revealed the possibility of gaining insight into hereditary neurodegenerative diseases by studying the zebrafish model.

Usage in Understanding Anesthesia

Zebrafish have been used in studies to further comprehend the genetic possibilities that create risk with anesthesia. Zebrafish were elected as the specimen of choice to observe the genetic complications associated with anesthesia because: the sensory evoked locomotor behaviors of zebrafish embryos allow sedation to be evaluated, the transparency of the zebrafish larvae allow neuronal activity to be tracked, the conserved ion channels, neurotransmitters, and receptors present in zebrafish are popular anesthetic targets, and zebrafish present with flexible genetic approaches and behavioral screening methods.[23] Studies conducted have used genome engineering techniques to create double-stranded breaks with CRISPRs/Cas9 approaches followed by non-homologous end joining repair to replicate zebrafish models of human inherited disorders associated with anesthetic risk.[23] Novel genetic variants that affect responses to anesthesia, are additionally identified using unbiased mutant lines. These lines are then used for further studies in specific phenotypes by researchers.[23] Through use of a high-throughput drug screening of behavioral phenotypes, conserved behaviors, such as the photomotor response and the vibration response, can be quickly evaluated to quantify the effects of anesthetics through fast screening of both mutations and drugs.[23] Results of the study revealed that when compared to other animals, zebrafish larvae respond with analogous tendencies and patterns when exposed to propofol and other anesthetics.[23] These conclusions strengthen the support for the continued use of the zebrafish in future efforts to not only understand anesthesia, but also the genetic factors that influence responses to anesthesia.

"Sickness Behavior"

"Sickness behavior" is a well-recognized concept that encompasses symptoms such as immobility, abnormally large amounts of sleep, as well as reduced food and water consumption.[24] However, sickness behavior is also observed in zebrafish includes additional symptoms such as reduced social interaction, locomotion and reproduction. Sickness behavior in zebrafish is induced via inoculation of formalin-inactivated *Aeromonas hydrophila*.[25] *Aeromonas hydrophila* is a gram-negative pathogen that has been utilized to investigate the link between the immune system and behavioral change (nervous system). After the inoculation of formalin-inactivated *A. hydrophila*, there was an increased expression of pro-inflammatory cytokine genes such as tumor necrosis factor alpha (TNF-α), interleukin (IL)-1β, and IL-6 in the brain tissues that suggested the activation of immune responses in zebrafish.[25] On par with these findings, the locomotor activity, social preference and exploratory behavior toward a new object were reduced in *A. hydrophila* inoculated zebrafish compared to the control fish. However, the molecular mechanisms behind how the immune system induces behavioral alterations is still not very well understood. Therefore, gut microbiota manipulation experiments should be done in the context of evaluating its connection to sickness behavior.

GF Development

Though gut microbiota has been shown to play a significant role within zebrafish, it remains to be determined what happens when they do not have a gut microbiota (i.e., GF). The effects of completely eliminating normal gut microflora has yet to be extensively studied in zebrafish to validate the findings observed in other animal models such as mice. GF mice exhibit increased motor activity, decreased anxiety-like behavior, elevated levels of noradrenaline (NA), dopamine (DA), and 5-HT turnover in the striatum as well as changes in expression of synaptic plasticity-related genes.[26] Another significant finding was that GF mice exposed to gut microbiota post-natally had reduced expression of postsynaptic density protein 95 (PSD-95) and synaptophysin, which is also observed in mice having normal gut microbiota.[26] Based on this evidence, the researchers made the connection that the gut microbiota is specifically responsible for nervous system signaling mechanisms indicated with the significant changes in motor activity, behavior, gene expression, and neurotransmitter levels. GF zebrafish have yet to be explored. However, similar studies to Heijtz *et al.*[26] should reveal significant changes in nervous system processes as well. Another potential avenue of exploration is the differences/influence in gut flora passed down in live birth versus laying of eggs. Specifically, in zebrafish, a more in-depth analysis of behavioral and social changes can be examined in order to possibly discover wider implications into human health and psychiatric disorders.

CONCLUSIONS

The GBA is a bidirectional communication network, consisting of the CNS, ANS, ENS, and HPA axis, which bridge the emotional and cognitive brain centers with peripheral intestinal functions. The zebrafish emerged as the appropriate animal model for understanding the GBA interplay and its related factors as it offers several advantages. These include genetic and physiological homology to humans, external fertilization, rapid development of organ systems, genetic tractability, a fully sequenced genome as well as the ability for researchers to observe and quantify complex behaviors. Studies conducted into the behavioral processes, human correlation, ENS, immune disorders, and toxicological effects have strengthened the zebrafish as the model animal for further understanding of the GBA. Future directions utilizing the zebrafish involve exploration into the zebrafish link to autism, insight into neurodegenerative diseases, understanding the genetic factors correlated to responses to anesthesia, "sickness behavior," and the repercussions of developing GF zebrafish. In addition, emerging gene editing technologies such as CRISPR/Cas9 will facilitate in deciphering the role of genes involved in bidirectional communication between the GBA using zebrafish models.

REFERENCES

1. Carabotti M, Scirocco A, Maselli MA, Severi C. (2015) The gut-brain axis: Interactions between enteric microbiota, central and enteric nervous systems. *Ann Gastroenterol* **28**(2): 203–209.
2. Gonzalez-Santana A, Diaz Heijtz R. (2020) Bacterial Peptidoglycans from Microbiota in Neurodevelopment and Behavior. *Trends Mol Med*. **26**(8): 729–743.
3. Du D, Tang W, Zhou C, Sun X, Wei Z, Zhong J, Huang Z. (2021) Fecal Microbiota Transplantation Is a Promising Method to Restore Gut Microbiota Dysbiosis and Relieve Neurological Deficits after Traumatic Brain Injury. *Oxid Med Cell Longev*. 2021 Feb 10;2021:5816837.
4. de Abreu MS, *et al*. (2019) Modeling gut-brain interactions in zebrafish. *Brain Res Bull* **148**: 55–62, doi:10.1016/j.brainresbull.2019.03.003
5. Norton W, Bally-Cuif L. (2010). Adult zebrafish as a model organism for behavioural genetics. *BMC Neurosci* **11**: 90. doi:10.1186/1471-2202-11-90.

6. Kalueff AV, Stewart AM, Gerlai R. (2014) Zebrafish as an emerging model for studying complex brain disorders. *Trends Pharmacol Sci* **35**(2): 63–75, doi:10.1016/j.tips.2013.12.002.
7. Borrelli L, *et al.* (2016) Probiotic modulation of the microbiota-gut-brain axis and behaviour in zebrafish. *Sci Rep* **6**: 1–9, doi:10.1038/srep 30046.
8. Vaz R, Hofmeister W, Lindstrand A. (2019) Zebrafish models of neurodevelopmental disorders: Limitations and benefits of current tools and techniques. *Int J Mol Sci.* **20**(6): 1296.
9. Wallace KN, Pack M. (2003) Unique and conserved aspects of gut development in zebrafish. *Dev Biol* **255**(1): 12–29, doi:10.1016/S0012-1606(02)00034-9.
10. Roeselers G, *et al.* (2011) Evidence for a core gut microbiota in the zebrafish. *ISME J* **5**(10): 1595–1608, doi:10.1038/ismej.2011.38.
11. Huang Y, *et al.* (2018) Possible association of firmicutes in the gut microbiota of patients with major depressive disorder. *Neuropsychiatr Dis Treat* **14**: 3329–3337, doi:10.2147/NDT.S188340.
12. Rizzatti G, Ianiro G, Gasbarrini A. Antibiotic and Modulation of Microbiota: A New Paradigm? J Clin Gastroenterol. 2018 Nov/Dec;52 Suppl 1, Proceedings from the 9th Probiotics, Prebiotics and New Foods, Nutraceuticals and Botanicals for Nutrition & Human and Microbiota Health Meeting, held in Rome, Italy from September 10 to 12, 2017:S74–S77.
13. Davis DJ, *et al.* (2016) Lactobacillus plantarum attenuates anxiety-related behavior and protects against stress-induced dysbiosis in adult zebrafish. *Sci Rep* **6**: 1–11, doi:10.1038/srep33726.
14. Ganz J, Melancon E, Eisen JS. (2016), Zebrafish as a model for understanding enteric nervous system interactions in the developing intestinal tract. *Methods Cell Biol* **134**: 139–164, doi:10.1016/bs.mcb.2016.02.003.
15. Zhao X, Pack M. (2017) Modeling intestinal disorders using zebrafish. *Methods Cell Biol* **138**: 241–270, doi:10.1016/bs.mcb.2016.11.006.
16. Ganz J. (2018) Gut feelings: Studying enteric nervous system development, function, and disease in the zebrafish model system. *Dev Dyn* **247**(2): 268–278, doi:10.1002/dvdy.24597.
17. Brugman S. (2016) The zebrafish as a model to study intestinal inflammation. *Dev Comp Immunol* **64**: 82–92. doi:10.1016/j.dci.2016.02.020.

18. Bertotto LB, Catron TR, Tal T. (2020) Exploring interactions between xenobiotics, microbiota, and neurotoxicity in zebrafish. *NeuroToxicology* **76**: 235–244, doi:10.1016/j.neuro.2019.11.008.
19. Davis DJ, Bryda EC, Gillespie CH, Ericsson AC. (2016) Microbial modulation of behavior and stress responses in zebrafish larvae. *Behav Brain Res* **311**: 219–227, doi:10.1016/j.bbr.2016.05.040.
20. Richendrfer H, Pelkowski SD, Colwill RM, Creton R. (2012) On the edge: Pharmacological evidence for anxiety-related behavior in zebrafish larvae. *Behav Brain Res* **228**(1): 99–106. doi:10.1016/j.bbr.2011.11.041.
21. James DM, *et al.* (2019) Intestinal dysmotility in a zebrafish (Danio rerio) shank3a;shank3b mutant model of autism. *Mol Autism* **10**(1): 1–15, doi:10.1186/s13229-018-0250-4.
22. Kozol RA, Abrams AJ, James DM, Buglo E, Yan Q, Dallman JE. (2016) Function over form: Modeling groups of inherited neurological conditions in zebrafish. *Front Mol Neurosci* **9**: 55, doi:10.3389/fnmol.2016.00055
23. Bedell V, *et al.* (2018) Zebrafish: A pharmacogenetic model for anesthesia. *Methods Enzymol* **602**: 189–209, doi:10.1016/bs.mie.2018.02.004.
24. Clark SM, *et al.* (2015) Dissociation between sickness behavior and emotionality during lipopolysaccharide challenge in lymphocyte deficient Rag2-/- mice. *Behav Brain Res* **278**: 74–82, doi:10.1016/j.bbr.2014.09.030.
25. Kirsten K, Soares SM, Koakoski G, Carlos Kreutz L, Barcellos LJG. Characterization of sickness behavior in zebrafish. *Brain Behav Immun*. 2018 Oct;73:596-602.
26. Heijtz RD, *et al.* (2011) Normal gut microbiota modulates brain development and behavior. *Proc Natl Acad Sci USA* **108**(7): 3047–3052, doi:10.1073/pnas.1010529108.

Chapter 13

Chronic Stress and the Gut–Brain Connection: An Occupational Therapist's Perspective

Nelson Gonzalez

Abstract

Prolonged stress can influence the physiologic functioning of immune cells and induce chronic inflammation. The inflammation that results from stress can also alter the gut microbiome and significantly impact the gut–brain axis. Occupational therapist (OT) provides a holistic approach to alleviate some of the negative effects of chronic stress by improving one's ability to adapt and coordinate self-care, work, education, and leisure activities. OTs help to reduce the effects of stress and foster well-being by encouraging exercise, improving sleep quality, and promoting a healthy diet. Overall, an appropriate amount of exercise has been shown to stimulate gut bacteria to produce metabolites that are necessary to many of the body's organ systems. Some particular forms of therapeutic exercise, such as stimulation of the vagus nerve through singing, have demonstrated a positive effect on gut functioning. Sleep quality has been shown to correlate with microbial diversity in the gastrointestinal tract, while diet plays

Director of Habilitation and Rehabilitation Services for Therapies 4 Kids and T4K Adults in South Florida, FL, USA.

Corresponding Author: Nelson Gonzalez, OTR/L, RFS, CHHC, CCT, CPMT, CLT, Executive Director for Rehab Without Walls NeuroSolutions, 805 N. Whittington Parkway, Louisville, Kentucky, 40222.

an integral role in maintaining homeostasis of the microbiome. In order to manage diseases caused by chronic stress, occupational therapy may provide targeted interventions aimed at improving the overall functioning of the gut–brain axis.

INTRODUCTION

The role of an occupational therapist (OT) is to examine an individual's routines and environment in order to identify barriers to health and make meaningful changes that promote well-being. OTs often work with patients who suffer from chronic stress and inflammatory conditions associated with dysfunction of the gut microbiome. Chronic stress is pervasive in modern society and results in continuous activation of the sympathetic nervous system. Prolonged sympathetic activation leads to up-regulation of the immune system and results in a pro-inflammatory state marked by increased levels of cytokines.[1,2] There is compelling evidence that exposure to chronic stress alters the structural composition of the intestinal microbiome leading to metabolic, immunologic, and behavioral dysfunction. Occupation therapy addresses dysbiosis in the gut microbiome with the use of interventions that target exercise, sleep, and diet.

Exercise and the Gut–Brain Axis

OTs design stress reduction programs and coordinate behavioral strategies to address dysfunction within the gut–brain axis. One of the most important interventions OT provides is individualized exercise programs. Aerobic exercise has been shown to have a positive impact on the GI tract and brain functioning and can be used as a treatment modality to promote homeostasis of the gut microbiota Fig. 13.1.[3,4] Several studies have demonstrated that individuals who exercised at a higher frequency have greater diversity of gut bacteria, specifically, *Faecalibacterium prausnitzii* and species from the genus *Oscillospira*, Lachnospira, and *Coprococcus* and

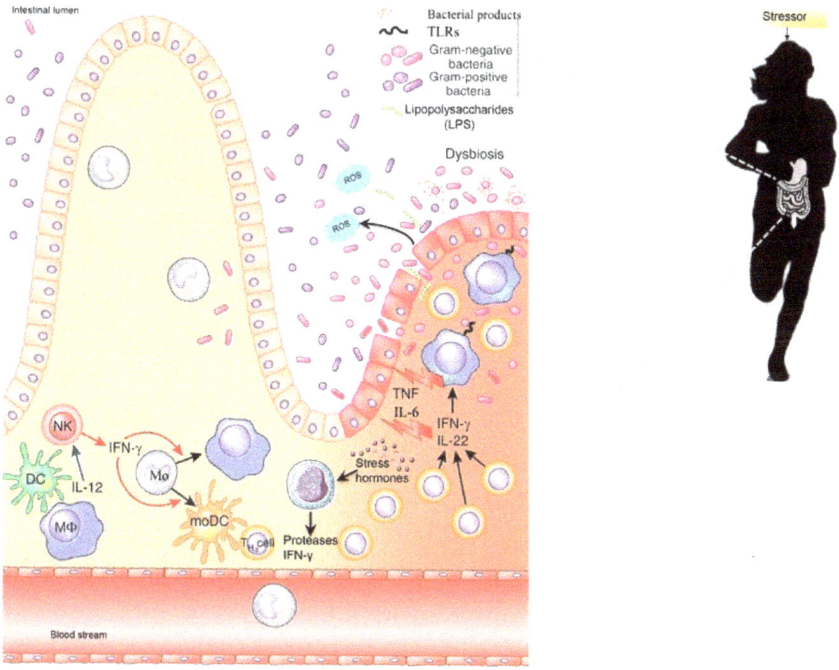

Fig. 13.1: Depiction of gastrointestinal disruption during high-intensity exercise. During intense exercise, re-distribution of blood flow away from the intestines together with thermal damage to the intestinal mucosa can cause intestinal barrier disruption, followed by inflammation. Additionally, intense exercise over a prolonged period of time increase stress hormones and lipopolysaccharides (LPS) translocation in the gastrointestinal tract, which triggers immune responses that often results in increased pro-inflammatory cytokines and intestinal permeability. Furthermore, gastrointestinal tract responds to stress by releasing hormones such as gamma-aminobutyric acid (GABA), neuropeptide Y, and dopamine that have been implicated in GI disease, anxiety, and depression. Conversely, the microbiota's production of butyrate and propionate can increase transepithelial resistance, which improves intestinal barrier function and decreases inflammation. (From Clark and Mach.[2] This article is distributed under the terms of the Creative Commons Attribution 4.0 International License (http://creativecommons.org/licenses/by/4.0/), which permits unrestricted use, distribution, and reproduction in any medium, provided we give appropriate credit to the original author(s) and the source, provide a link to the Creative Commons license, and indicate if changes were made.)

that in general, professional athletes exhibit higher diversity of gut microorganisms compared to non-athletes. One study looking at rugby players found that in a group of professional rugby players, 22 phyla were detected in their gut microbiome, while only 11 phyla were identified in age-matched low BMI control and 9 phyla in an age-matched high BMI control. The diversity of phyla among the rugby players correlated with both dietary protein intake and creatine kinase, suggesting that both diet and exercise promote diversity in the gut microbiome.[5]

D-β-hydroxybutyrate (DBHB) is responsible for the production of brain-derived neurotrophic factor (BDNF), which plays a role in synaptic transmission, memory, and cognitive function.[6] Synthesis of DBHB in the liver is stimulated by exercise and thus results in increasing levels of BDNF. Analysis of post-mortem hippocampal tissue from suicide victims showed reduced levels of BDNF and other neurotrophic-growth factors.[7]

OTs are important to implementing safe and beneficial exercise programs that do not lead to further stress and inflammation. Research has shown that disruption to the gut microbiome can result when an individual goes beyond 60% of their maximum oxygen uptake (VO_2 max) during exercise or if one rigorously exercises for more than 90 minutes.[2] One study found that mice who were forced to do treadmill running (8–12 m/min at a 5% grade for 40 min, 5 times per week) had exacerbated colitis symptoms when compared to mice that voluntary exercised on a wheel. Researchers hypothesize that the forced exercise may have been perceived as a psychological stressor by the mice, since the voluntary wheel running was a conscious decision. Other adverse effects of excessive exercise include abdominal pain, diarrhea, heartburn, and gastrointestinal bleeding.[8] Exercising at mild to moderate intensity levels has shown to be protective against health complications including constipation, diverticular disease, cholelithiasis, and colon cancer. OTs recommend the appropriate dosage of exercise for individuals who are susceptible to increased levels of stress and inflammation.

Modulation of the Gut–Brain Axis Through the Vagus Nerve

The vagus nerve plays an important role in the enteric nervous system and serves as a direct connection between the gut and central nervous system.[9] Stimulation of the vagus nerve can influence metabolic responses in the organs it innervates including those responsible for digestion, absorption, and metabolism of nutrients.[10] Decreased vagus nerve activity has been associated with conditions such as irritable bowel disease, irritable bowel syndrome, and depression and is attributed to dysfunction of gastric motility and gastric emptying.[11] Occupational therapy can utilize meditative techniques that influence actions of the vagus nerve and enteric nervous system. Taking deep breaths that cause the diaphragm to move, stimulates the vagus nerve to detect blood pressure and will cause a compensatory decrease in heart rate if blood pressure is high.[12] During meditation, individuals are encouraged to take deep breaths leading to an increase in vagal tone and a resulting state of calm. The vagus nerve is also stimulated by exposure to cold and results in increased parasympathetic activation.[13] Application of cold temperatures at the lateral neck area with a washcloth or ice pack is a technique that can be used to reduce heart rate.[13] Humming, singing, and gargling stimulate the vagus nerve by activating the vocal cords and the muscles in the posterior aspect of the throat.[14] Activity of the vagus nerve is negatively correlated with stress, morbidity, and mortality, and is thus implemented as a therapeutic strategy among OTs to promote well-being and increased healing.

Sleep and the Gut–Brain Axis

Sleep is a physiological state that is intrinsically linked to chronic inflammation and the gut–brain axis (Fig. 13.2). In general, short sleep duration and poor sleep quality have been associated with cognitive dysfunction, mood instability, and memory issues.[15] Lack of quality sleep has also been implicated in the pathogenesis of

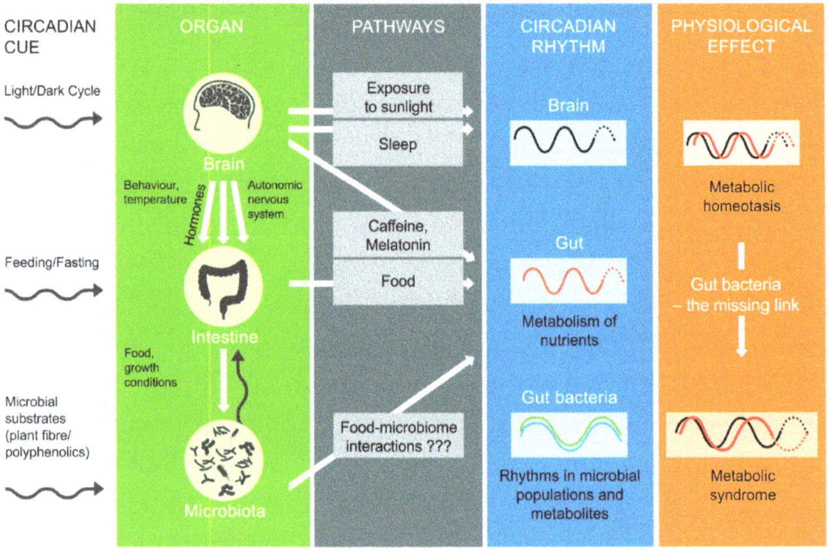

Fig. 13.2: Circadian rhythm, gut microbiota, and metabolism. Intestinal bacteria show diurnal fluctuations in terms of their abundance and functions. Host-controlled behaviors such as sleep, diet, and light exposure may potentially affect intestinal rhythms of metabolism. New evidence indicates that gut microbiota influences circadian rhythms, with consequent impact on the metabolic homeostasis of the host. (From Parkar *et al.*[20] This is an open access article distributed under the Creative Commons Attribution License, which permits unrestricted use, distribution, and reproduction in any medium, provided the original work is properly cited.)

cancer, type 2 diabetes, and Alzheimer's disease.[16] Inflammatory cytokines represent a potential critical interface between sleep physiology and gut microbiome composition. One human study found that microbial diversity of the gut was positively correlated with sleep efficiency and total sleep time, while it was negatively correlated with the sleep fragmentation.[16] OTs work with patients to create healthy sleep habits and a home environment that promote sleep. For example, OTs may recommend maintaining room temperatures between 60 to 67 degrees Fahrenheit (approximately 15.5 to 19 degrees Celsius) to experience restful sleep.[17] There are strong links between thermal environment and thermoregulation that affect the stages of sleep.[18] OTs also recommend reducing screen time such as not using the phone or computer before sleep. These devices overpower the rise of melatonin, hinders rapid eye movement (REM) sleep, and diminishes the ability to be alert the following day.[19] Along with creating a sleep hygiene program, an OT can create individualized nutritional strategies to modify the timing of feeding, fasting, quantity, and quality of the food being consumed as it relates to sleep quality.

Diet and the Gut–Brain Axis

Numerous epidemiologic studies have demonstrated the connection between diet and chronic inflammation as well as the effect of diet on the gut microbiota. One of the roles of an OT is to educate individuals on the impact nutrition plays on overall functioning. OTs advocate for well-balanced meals full of fiber, vitamins and minerals in order to support the gut microbiome and in turn improve metabolic functioning. For example, the human genome encodes a limited number of glycoside hydrolases and no polysaccharide lyases, which are necessary for digestion and absorption of glycans; a vital nutrient found in plant and animal sources.[21] These glycans traverse the small intestines in their undigested form until they encounter the abundant microbiome in the large intestine. In contrast to humans, the microbiota genome encodes thousands of enzymes that digest glycans. Dysbiosis can lead to impaired production of these vital enzymes that metabolize glycans and in turn lead to malabsorption. Furthermore, the breakdown of glycans yields glucose, which is fermented by other bacterial species to produce important metabolites such as acetate, propionate, formate, butyrate, lactate and succinate.[22] In particular, propionate and butyrate are important regulators of the immune response as they can influence the production of immune cells and inflammatory cytokines.[23] The regulatory role of these metabolites is especially important for the cells that make up the blood–brain barrier (BBB) and protect the central nervous system from systemic inflammation. A nutritious diet is key to maintaining and promoting homeostasis of the gut microbiome and must be addressed by OTs in order to augment healing.

Gut microorganisms are also important to the production and distribution of metabolites such as serotonin and gamma-aminobutyric acid (GABA), which are critical to neurologic functioning.[24] Dysfunction of the gut microbiota can lead to decreased production of GABA and serotonin resulting in behavioral, emotional and cognitive dysfunction.[25] GABA regulates neuronal activity by inhibiting the release of neurotransmitters and synchronizing cortical activity.[26] Muscle pain, headaches, memory problems, and anxiety have all been associated with decreased levels of GABA and are common conditions that occupational therapy helps to treat.[27] Serotonin is mostly produced on the surface of gut mucosa and is important for the regulation of the sleep–wake cycle, mood, appetite and digestion. Decreased appetite, irritability, and insomnia are all associated with a deficiency in serotonin.[7] Occupational therapy works with patients to create a daily meal schedule for proper nutrition, as well as confront any barriers to meal planning and shopping for healthy food. The time a person eats and the kinds of food a person consumes also affect the gut microbiota, which then affects the circadian clock.[20] Therefore, eating the right foods at the right time can improve a person's sleep, while eating poor nutritious foods at the wrong times will lead to deprived sleep quality.

CONCLUSIONS AND FUTURE DIRECTIONS

Chronic stress and anxiety are detrimental conditions that affect the gut–brain axis. Addressing chronic stress and anxiety requires a multi-disciplinary team of clinicians from varied professions. Pharmacological therapies, psychological therapies, physical activity programs, good sleep hygiene, and consuming a healthy diet collectively have the capability of improving disease resulting from dysfunction of the gut–brain axis (Fig. 13.3).[28] Occupational

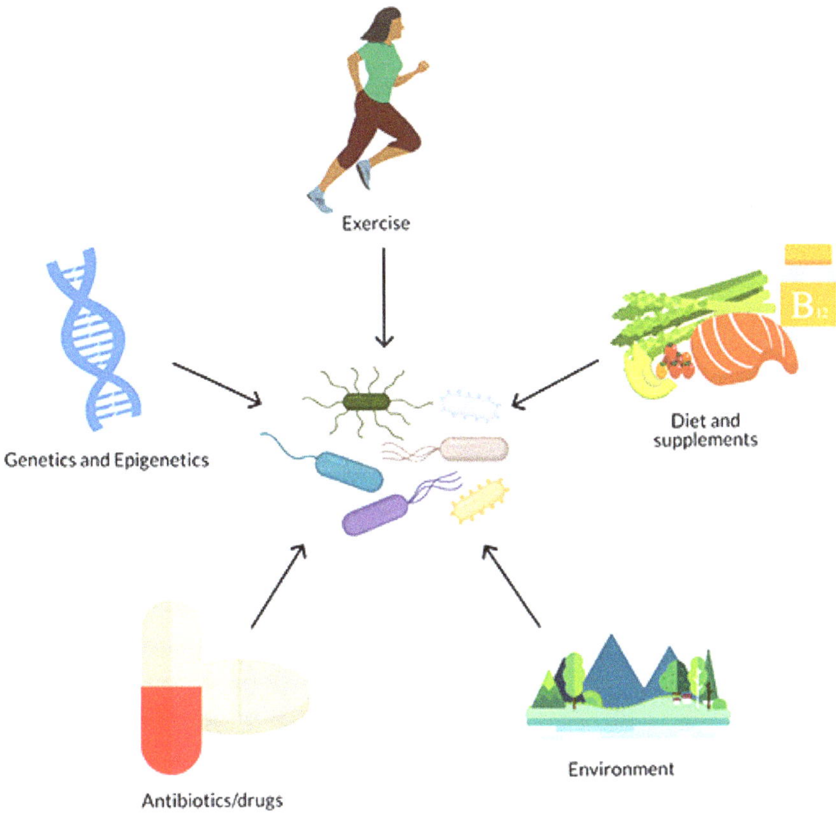

Fig. 13.3: The gut microbiome is influenced by numerous biological and lifestyle factors such as exercise, diet, genetics, antibiotics, and the environment. (From Hughes.[29] This is an open access article distributed under the terms of the Creative Commons Attribution License (CC BY). The use, distribution, or reproduction in other forums is permitted, provided the original author(s) and the copyright owner(s) are credited and that the original publication in this journal is cited, in accordance with accepted academic practice.)

therapy not only provides a holistic approach to treatment, but a whole-person approach that includes improving clients' internal and external environments. Due to the array of benefits occupational therapy provides in mitigating physical and psycho-social dysfunction, physicians should not hesitate to refer clients who are experiencing poor sleep, gut problems, anxiety, and chronic stress.

ACKNOWLEDGMENT

We would like to thank Jenna Bergman and Camron Davies for providing their input and feedback on this chapter.

REFERENCES

1. Won E, Kim Y. (2016) Stress, the autonomic nervous system, and the immune-kynurenine pathway in the etiology of depression. *Curr Neuropharmacol* **14**(7): 665–673, doi:10.2174/1570159x14666151208113006. https://pubmed.ncbi.nlm.nih.gov/27640517, https://www.ncbi.nlm.nih.gov/pmc/articles/PMC5050399/.
2. Clark A, Mach N. (2016) Exercise-induced stress behavior, gut-microbiota-brain axis and diet: A systematic review for athletes. *J Int Soc Sports Nutr* **13**(1): 43–21, doi:10.1186/s12970-016-0155-6. https://www.ncbi.nlm.nih.gov/pubmed/27924137.
3. Dalton A, Mermier C, Zuhl M. (2019) Exercise influence on the microbiome-gut-brain axis. *Gut Microbes* **10**(5): 555–568, doi:10.1080/19490976.2018.1562268. http://www.tandfonline.com/doi/abs/10.1080/19490976.2018.1562268.
4. Monda V, *et al.* (2017) Exercise modifies the gut microbiota with positive health effects. *Oxid Med Cell Longev* **2017**: 3831972–8, doi:10.1155/2017/3831972. https://dx.doi.org/10.1155/2017/3831972.
5. Clarke SF, *et al.* (2014) Exercise and associated dietary extremes impact on gut microbial diversity. *Gut* **63**(12): 1913–1920, doi:10.1136/gutjnl-2013-306541. http://dx.doi.org/10.1136/gutjnl-2013-306541.
6. Sleiman SF, *et al.* (2016) Exercise promotes the expression of brain derived neurotrophic factor (BDNF) through the action of the ketone body β-hydroxybutyrate. *eLife* **5**: 15092, doi:10.7554/eLife.15092. https://www.ncbi.nlm.nih.gov/pubmed/27253067.
7. Duman RS, Monteggia LM. (2006) A neurotrophic model for stress-related mood disorders. *Biol Psychiatry* **59**(12): 1116–1127, doi:10.1016/j.biopsych.2006.02.013. http://dx.doi.org/10.1016/j.biopsych.2006.02.013.
8. de Oliveira E, Burini R. (2009) The impact of physical exercise on the gastrointestinal tract. *Curr Opin Clin Nutr Metab Care* **12**(5):533–538, doi:10.1097/MCO.0b013e32832e6776. https://www.ncbi.nlm.nih.gov/pubmed/19535976.

9. Galland L. (2014) The gut microbiome and the brain. *J Med Food* **17**(12): 1261–1272, doi:10.1089/jmf.2014.7000. https://www.liebertpub.com/doi/abs/10.1089/jmf.2014.7000.
10. Kiecolt-Glaser JK. (2010) Stress, food, and inflammation: Psychoneuroimmunology and nutrition at the cutting edge. *Psychosom Med* **72**(4): 365–369, doi:10.1097/PSY.0b013e3181dbf489. https://www.ncbi.nlm.nih.gov/pubmed/20410248.
11. Bonaz B, Bazin T, Pellissier S. (2018) The vagus nerve at the interface of the microbiota-gut-brain axis. *Front Neurosc* **12**: 49, doi:10.3389/fnins.2018.00049. https://www.ncbi.nlm.nih.gov/pubmed/29467611.
12. Bonaz B, Picq C, Sinniger V, Mayol JF, Clarençon D. (2013) Vagus nerve stimulation: From epilepsy to the cholinergic anti-inflammatory pathway. *Neurogastroenterol Motil* **25**(3): 208–221, doi:10.1111/nmo.12076. https://onlinelibrary.wiley.com/doi/abs/10.1111/nmo.12076.
13. Jungmann M, Vencatachellum S, Van Ryckeghem D, Vögele C. (2018) Effects of cold stimulation on cardiac-vagal activation in healthy participants: Randomized controlled trial. *JMIR Form Res* **2**(2): e10257, doi:10.2196/10257. https://www.narcis.nl/publication/RecordID/oai:cris.maastrichtuniversity.nl:publications%2Fb251f69f-82b8-4bde-9759-845ac17d63b6.
14. Grape C, Sandgren M, Hansson L, Ericson M, Theorell T. (2003) Does singing promote well-being?: An empirical study of professional and amateur singers during a singing lesson. *Integr Physiol Behav Sci* **38**(1): 65. https://www.ncbi.nlm.nih.gov/pubmed/12814197.
15. Gobin CM, Banks JB, Fins AI, Tartar JL. (2015) Poor sleep quality is associated with a negative cognitive bias and decreased sustained attention. *J Sleep Res* **24**(5): 535–542, doi:10.1111/jsr.12302. https://onlinelibrary.wiley.com/doi/abs/10.1111/jsr.12302.
16. Smith RP, *et al.* (2019) Gut microbiome diversity is associated with sleep physiology in humans. *PLoS One* **14**(10): e0222394, doi:10.1371/journal.pone.0222394. https://www.ncbi.nlm.nih.gov/pubmed/31589627.
17. Onen SH, Onen F, Bailly D, Parquet P. (1994) Prévention et traitement des dyssomnies par une hygiène du sommeil [Prevention and treatment of sleep disorders through regulation] of sleeping habits]. *Presse Med* **23**(10): 485–489.

18. Okamoto-Mizuno K, Mizuno K. (2012) Effects of thermal environment on sleep and circadian rhythm. *J Physiol Anthropol* **31**(1): 14, doi:10.1186/1880-6805-31-14. https://www.ncbi.nlm.nih.gov/pubmed/22738673.
19. Chang A-M, Aeschbach D, Duffy JF, Czeisler CA. (2015) Evening use of light-emitting eReaders negatively affects sleep, circadian timing, and next-morning alertness. *Proc Natl Acad Sci USA* **112**(4): 1232–1237, doi:10.1073/pnas.1418490112. https://www.jstor.org/stable/26454261.
20. Parkar SG, Kalsbeek A, Cheeseman JF. (2019) Potential role for the gut microbiota in modulating host circadian rhythms and metabolic health. *Microorganisms (Basel)* **7**(2): 41, doi:10.3390/microorganisms7020041. https://www.narcis.nl/publication/RecordID/oai:pure.knaw.nl:publications%2F2de1d0b2-27c6-4240-ab42-391cb9d35add.
21. Flint HJ, Scott KP, Duncan SH, Louis P, Forano E. (2012) Microbial degradation of complex carbohydrates in the gut. *Gut Microbes* **3**(4): 289–306, doi:10.4161/gmic.19897. https://pubmed.ncbi.nlm.nih.gov/22572875 https://www.ncbi.nlm.nih.gov/pmc/articles/PMC3463488/.
22. Zmora N, Suez J, Elinav E. (2019) You are what you eat: Diet, health and the gut microbiota. *Nat Rev Gastroenterol Hepatol* **16**(1): 35–56, doi:10.1038/s41575-018-0061-2. https://www.ncbi.nlm.nih.gov/pubmed/30262901.
23. Cryan JF, *et al.* (2019) The microbiota-gut-brain axis. *Physiol Rev* **99**(4): 1877–2013, doi:10.1152/physrev.00018.2018 [doi].
24. Evrensel A, Ceylan ME. (2015) The gut-brain axis: The missing link in depression. *Clin Psychopharmacol Neurosci* **13**(3): 239–244, doi:10.9758/cpn.2015.13.3.239. https://www.ncbi.nlm.nih.gov/pubmed/26598580.
25. Frankiensztajn LM, Elliott E, Koren O. (2020) The microbiota and the hypothalamus-pituitary-adrenocortical (HPA) axis, implications for anxiety and stress disorders. *Curr Opin Neurobiol* **62**: 76–82. doi:10.1016/j.conb.2019.12.003.
26. Gelman DM, Marín O. (2010) Generation of interneuron diversity in the mouse cerebral cortex. *Eur J Neurosc* **31**(12): 2136–2141, doi:10.1111/j.1460-9568.2010.07267.x. https://onlinelibrary.wiley.com/doi/abs/10.1111/j.1460-9568.2010.07267.x.
27. Allen DL, McCall GE, Loh AS, Madden MC, Mehan RS. (2010) Acute daily psychological stress causes increased atrophic gene expression and myostatin-dependent muscle atrophy. *Am J Physiol Regul Integr Comp*

Physiol **299**(3): 889–898, doi:10.1152/ajpregu.00296.2010. http://ajpregu.physiology.org/content/299/3/R889.

28. Briguglio M, *et al.* (2020) Healthy eating, physical activity, and sleep hygiene (HEPAS) as the winning triad for sustaining physical and mental health in patients at risk for or with neuropsychiatric disorders: Considerations for clinical practice. *Neuropsychiatr Dis Treat* **16**: 55–70, doi:10.2147/NDT.S229206. https://www.ncbi.nlm.nih.gov/pubmed/32021199.
29. Hughes RL. (2020) A review of the role of the gut microbiome in personalized sports nutrition. *Front Nutr* **6**: 191.

Chapter 14

Migraine and the Gut Microbiome

Adwight Risbud[1,*], Mehdi Abouzari[1,*] and Hamid R. Djalilian[1,2]

Abstract

The pathophysiology of migraine is both exceedingly complex and not well understood at present. Among the various proposed mechanisms, the gut-brain axis has increasingly gained support as a key modulator in the disorder. In a bidirectional axis involving the gastrointestinal and nervous systems, the gut microbiome has been shown to play a particularly important function in maintaining proper regulation of the axis. Changes to the microbiome are thought to result in many of the symptomatic manifestations observed in migraine patients. With a growing understanding of the gut microbiome, various treatment strategies have been explored, utilizing the principles of dysbiosis and gut microbiota diversity. In this chapter, we explore the potentially crucial role of the gut microbiome in the pathophysiology of migraine in greater detail and conclude with a discussion of the current evidence on migraine therapy aimed at maintaining the microbiota through dietary and nutritional modifications.

[1] Department of Otolaryngology — Head and Neck Surgery, University of California, Irvine, USA.

[2] Department of Biomedical Engineering, University of California, Irvine, USA.

*Contributed equally to this work.

Corresponding Author: Hamid R. Djalilian, MD, Division of Neurotology and Skull Base Surgery, Department of Otolaryngology — Head and Neck Surgery, University of California, Irvine, 19182 Jamboree Road, Otolaryngology-5386, Irvine, CA 92697, Phone: (714) 456-5753, Fax: (714) 456-5747, E-mail: hdjalili@hs.uci.edu.

INTRODUCTION

Migraine is a widespread and highly burdensome condition with an estimated global prevalence of 14%.[1] It is among the leading causes of years lived with disability and is a disorder that is, unfortunately, both underdiagnosed and undertreated.[2] The International Headache Society classifies migraine as a primary headache disorder involving recurrent attacks that may be associated with visual, sensory, or other nervous system symptoms. Based on the specific criteria fulfilled, migraine can be further categorized into various subtypes, such as *migraine with* or *without aura*, *chronic migraine*, *vestibular migraine*, and *retinal migraine*.[3] Migraine is also one of the leading causes of vertigo.[4]

While many theories exist with regard to the underlying mechanisms of a migraine, the pathophysiology of this disorder remains largely undefined despite its high prevalence and burden on society. One major factor implicated in migraine pathophysiology is the *gut-brain axis*, a term used to indicate the interconnection between the gastrointestinal and central nervous systems.[5] In this bidirectional axis, regulation of these two systems occurs through a wide spectrum of factors, including neuronal, hormonal, and inflammatory mediators, as well as the composition of the gut microbiota.[6] Dysregulation in any of these factors is thought to result in disruption of the axis and contributes to migraine and a host of other neurological conditions, including multiple sclerosis, Parkinson's disease, and Alzheimer's disease.[7]

Communication between the gut microbiome and nervous system is facilitated through both direct and indirect signaling pathways. While the former occurs via direct connection with the vagus nerve and its stimulating end terminals, the latter requires the release and activation of hormones, cytokines, and microbiota-derived neurotransmitters.[6] Given the involvement of the gut microbiota with these chemical mediators, it is possible that the role of the gut microbiota in migraine pathophysiology stems from its larger influence on inflammatory pathways and immune function.[8] The key mediators in this axis include neuropeptides such as calcitonin gene-related peptide (CGRP), neuropeptide Y (NPY), and vasoactive intestinal peptide (VIP).[9] In addition to modulating many neural processes involved in migraine pathophysiology, these substances also exert a direct effect as antimicrobial agents on certain gut bacteria and also function to inhibit a number of digestive enzymes.[6] The central nervous system simultaneously functions to alter the gut microbiome through the release of molecules and modulation by both the sympathetic and parasympathetic divisions of the autonomic nervous system.[10] These dynamic changes in the gut microbiota are thought to determine many of the manifestations of a migraine and form the basis of research on this special connection between the gut microflora and brain.[10]

DYSBIOSIS AND MIGRAINE

To illustrate the effects of imbalances in the gut microbiome, termed *dysbiosis*, one recent study utilized a germ-free mouse model in which mice were deprived of gut microflora and were compared to those colonized with commensal bacteria.[11] In the microbiota-deficient mice, a decrease in the stress responses that normally induce migraine-like pain was observed. In the same study, antibiotics and probiotics were also administered to determine the effect of dysbiosis, yielding similar results with regard to visceromotor responses and nociception in animals with perturbed gut flora. The administration of antibiotics was found to prolong the chronicity of nitroglycerin-induced migraine-like pain in the mouse model through up-regulation of the tumor necrosis factor α (TNFα), a proinflammatory cytokine involved in chronic pain signaling, and enhancement of CGRP, an important biomarker in migraine.[9,12] Following the administration of probiotics and recovery of the gut microbiota in these models, the prolongation of migraine-like pain and TNFα up-regulation was completely reversed. This study highlights not only a critical role of gut microflora in nociception and host adaptation to the environment but also provides a molecular basis for the potential therapeutic value of probiotic therapy for the treatment of migraine. The recent literature on probiotic supplementation will be discussed later in this chapter.

Changes in the gut microbiome can also be caused by physical and psychological stressors, which act on the hypothalamic-pituitary-adrenal (HPA) axis to release cortisol from the adrenal glands and alter the microbiota.[7] In turn, changes in gut integrity and composition can lead to a release of pro-inflammatory cytokines such as TNFα and IL-1β, which further stimulate the HPA axis and have been shown to mediate migraine pain.[13] A schematic representation of the proposed association between the gut-brain axis, microbiome, and migraine is shown in Fig. 14.1. The importance of maintaining a diverse microflora composition has been demonstrated in a recent metagenome-wide association study that examined shotgun-sequenced stool samples from elderly women suffering from migraine and healthy control subjects.[14] Significant differences were found in the composition between these two groups, with migraine patients exhibiting much lower levels of diversity in their gut microbiota as well as lower counts of metabolically beneficial bacteria, including short-chain fatty acid (SCFA)-producing species. Equally alarming in the migraine

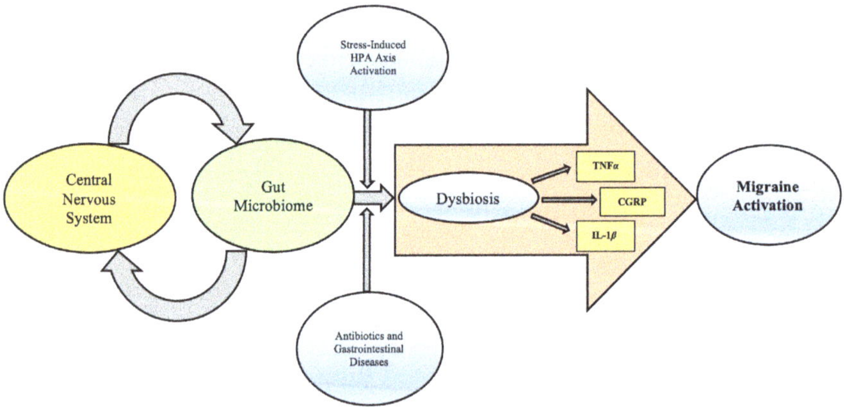

Fig. 14.1: Schematic overview of the proposed association between the gut microbiome and initiation of migraine pain.

patients of this study was the presence of deleterious organisms such as the *Clostridium* species, which were discovered at higher rates than in healthy subjects.

Another harmful species of bacteria that has been well studied in its association with migraine is *Helicobacter pylori* (*H. pylori*). Results of a meta-analysis showed that compared to about one-third of healthy control subjects, nearly 45% of migraine patients were colonized with *H. pylori*, with additional studies demonstrating symptomatic relief in migraine patients following the eradication of the bacteria.[15,16] The association between migraine and *H. pylori* infection is believed to result from the release of inflammatory cytokines and immune cells, including other interleukins (IL-10 and IL-12), CGRP, and vasoactive peptides such as serotonin and VIP, many of which have also been observed at increased levels during migraine attacks.[17]

In discussing the role of the gut microbiota, it is important to note the close association between migraine and gastrointestinal (GI) disorders, as evidenced by shared symptomologies and patient comorbidities. Migraine patients very commonly experience symptoms of abdominal discomfort, diarrhea, constipation, and dyspepsia, and are more likely to be diagnosed with GI conditions than the general population. In particular, many patients with irritable bowel syndrome (IBS) and celiac disease have been found to suffer concomitantly from migraine, likely due in part to the effect that these conditions have on altering the gut microflora composition. In one study that surveyed patients with IBS, celiac disease, and gluten sensitivity, the prevalence of chronic headaches and migraine ranged from 21% to 56% in these patients, compared to only 14% to those in controls.[18]

TREATMENT STRATEGIES: DIET AND NUTRITION

Nutrition plays a key role in the maintenance of the gut microbiome, both directly through effects on the gut mucosa, as well as through neuroprotection and improved immune function. The importance of SCFAs on both gut and brain immunity has also been well established.[19] These nutrients, which are produced by gut bacteria, are sensitive to both dietary factors as well as antibiotic agents that may disrupt the diversity and composition of the host flora. Dysbiosis and extreme dietary shifts can be detrimental to the absorption of nutrients beneficial to brain functioning and further exacerbate the GI symptoms associated with migraine attacks by impeding intestinal motility and gut barrier integrity.[20]

Although the influence of the immune system on migraine pathophysiology itself remains strongly debated, there is emerging evidence showing increased levels of immune cells such lymphocytes (CD3, CD4, CD8, and CD19) in patients suffering from chronic migraine compared to patients with episodic or no migraine.[21] The impact of diet on other markers of immunity and inflammation, including antibodies to food allergens and associated cytokines, has also been studied in randomized controlled trials. One study demonstrated both a decrease in IgG antibodies and inflammatory cytokines following the elimination of various food allergens from the diet, as well as a corresponding decrease in the frequency, duration, and maximum severity of migraine attacks.[22]

In general, the avoidance or reduction of food triggers has becoming increasingly recommended by physicians and other providers for migraine patients. For reasons not completely understood, the worsening of dizziness, headaches, and migraine attacks followed the consumption of caffeinated beverages, alcohol, certain dairy products, and processed meats, among numerous other foods.[23,24] Interestingly, many of these dietary sensitivities have been observed in other vestibular conditions such as Meniere's disease, which has recently been hypothesized to exist on the same disease spectrum as migraine.[25] Future studies may help to explain the potential inflammatory or dysbiotic effects of these specific diet-related triggers and further demonstrate the possible role the gut microbiota plays in migraine and migraine-related conditions. While the elimination of specific foods has been implicated in the improvement of migraine symptoms, researchers have also sought to investigate the benefits of dietary supplementation.[26] Based on the evidence previously discussed regarding gut microbiota composition and dysbiosis, it is no surprise that probiotic supplementation has been widely used, both empirically and in clinical trials for the treatment of migraine.

Over the past several years, different probiotic regimens have been evaluated for various migraine subtypes. In one study assessing the effects of a probiotic mixture in patients with either chronic or episodic migraine, patients with both subtypes reported decreased severity and frequency of migraine attacks, as well as a reduction in abortive medication usage following a 10-week trial of the 14-strain mixture, as compared to the placebo group.[27] In another recent trial investigating the efficacy of a *Lactobacillus casei* Shirota probiotic in patients with vestibular migraine, patients in the treatment group experienced an improvement in the number of vertigo attacks, in addition to anxiety and depressive symptoms over a 4-month study period.[28] Unfortunately, while the promising results of these studies have been supported by many similar trials, other studies have failed to confirm the reported benefit of probiotic supplementation on migraine symptoms or its effect on gut barrier integrity and inflammation.[29] A recent systematic review on 70 randomized controlled trials investigating the use of probiotics for migraine prophylaxis revealed considerable heterogeneity in the findings and lack of consistency in the study methods.[30]

Regarding essential nutrients, the effect of low-fat diets has been a popular focus of many studies, particularly given the importance of SCFAs. Omega-3 fatty acids promote the proliferation of SCFA-producing bacterial species including *Lactobacillus*, thus playing a critical part in the neuroprotective role of the gut microbiota as previously discussed.[31] Multiple reports have demonstrated a reduction in both pain and frequency of migraine headaches in patients following 3-month trials of high omega-3/low omega-6 diets.[32,33] Although the outcomes observed in these studies have been attributed to the vasodilatory and serotonin-releasing properties of the diets rather than their effect on SCFA production in migraine patients, the results nonetheless show the promise of low-fat diets in migraine prophylaxis.[34]

Vitamin D is another example of a nutrient that has been shown to affect the composition of gut microbiota. In one study, healthy individuals who were given vitamin D3 supplementation demonstrated a reduction in the counts of *H. pylori* and other bacterial species, including *Pseudomonas*, *Escherichia*, and *Shigella,* in addition to increased bacterial richness after 8 weeks of treatment.[35] Moreover, it has been reported that the presence of migraine is associated with vitamin D deficiency and that supplementation may help reduce both the frequency and intensity of migraine attacks.[36] Other vitamins and micronutrients commonly prescribed due to their reported efficacy in migraine prevention and severity include riboflavin, magnesium, coenzyme Q-10, and herbal preparations, such as feverfew. Unlike vitamin D, however, the effects of these nutrients on the gut microbiota composition and function are yet to be studied.[37]

Lastly, a possible link between migraine and obesity has been explored, with studies showing a higher incidence of episodic and chronic migraine in obese individuals as well as increased levels of migraine biomarkers such as CGRP and TNFα.[38,39] In animal models, it was demonstrated that obesity could exert a wide range of effects on the gut microbiota, including changes in the microflora composition, decreased gut permeability, and structural integrity.[40] Preliminary data from human studies has suggested the benefit of weight reduction in both adults and adolescents in reducing migraine intensity, frequency, and attack duration, through either lifestyle modification or bariatric surgery.[41,42]

CONCLUSIONS

In the past decade, significant advances have been made in uncovering the complex mechanisms of migraine and other neurological disorders. From this investigation, the gut microbiome has emerged as an integral component to the pathophysiology, and we have gained a better understanding of the association between the gut microbiome and migraine, which involves a multitude of physical and chemical mediators, and exists in a delicate balance. To date, researchers have proposed various approaches to treating migraine by utilizing the current knowledge of gut microbiota diversity and dysbiosis. Despite numerous strategies, however, the evidence on the true benefit of the treatment methods remains limited by inadequate study designs and confounding variables, among other investigational challenges. Standardization of research protocols and reporting in the future can help mitigate conflicting results and effectively characterize these potential therapeutic options. As we continue to explore this critical association between migraine and the gut microbiome, the pathophysiology of migraine will be further elucidated and pave the way for more effective treatments targeted to the intricate processes underlying this debilitating neurovascular disorder.

REFERENCES

1. Stovner LJ, *et al.* (2018) Global, regional, and national burden of migraine and tension-type headache, 1990–2016: A systematic analysis for the Global Burden of Disease Study 2016. *Lancet Neurol* **17**(11): 954–976, https://doi.org/10.1016/S1474-4422(18)30322-3.
2. Foster SA, *et al.* (2020) Economic burden and risk factors of migraine disease progression in the US: A retrospective analysis of a commercial payer database. *J Med Econ* **12**: 1–9, doi:10.1080/13696998.2020.1814790.
3. Headache Classification Committee of the International Headache Society (IHS). (2018) The International Classification of Headache Disorders, 3rd edition. *Cephalalgia* **33**(9): 629–808.
4. Dieterich M, Obermann M, Celebisoy N. (2016) Vestibular migraine: The most frequent entity of episodic vertigo. *J Neurol* **263**(Suppl 1): S82–S89, doi:10.1007/s00415-015-7905-2.
5. Dodick DW. (2018) A phase-by-phase review of migraine pathophysiology. *Headache* **58**(Suppl 1): 4–16, doi:10.1111/head.13300.

6. Holzer P, Farzi A. (2014) Neuropeptides and the microbiota-gut-brain axis. *Adv Exp Med Biol* **817:** 195–219.
7. Mayer EA, Tillisch K, Gupta A. (2015) Gut/brain axis and the microbiota. *J Clin Invest* **125**(3): 926–938, doi:10.1172/JC176304.
8. Arzani M, *et al.* (2020) Gut brain axis and migraine headache: A comprehensive review. *J Headache Pain* **21**(1): 15, https://doi.org/10/1186/s10194-020-1078-9.
9. Pérez-Pereda S, *et al.* (2020) Serum CGRP, VIP, and PACAP usefulness in migraine: A case control study in chronic migraine patients in real clinical practice. *Mol Biol Rep* **47**: 7125–7138, https://doi.org/10.1007/s11033-020-05781-0.
10. Hindiyeh N, Aurora SK. (2015) What the gut can teach us about migraine. *Curr Pain Headache Rep* **19**(7): 33, doi:10.1007/s11916-015-0501-4.
11. Tang Y, Liu S, Shu H, Yanagisawa L, Tao F. (2020) Gut microbiota dysbiosis enhances migraine-like pain via TNFα upregulation. *Mol Neurobiol* **57**(1): 461–468, doi:10.1007/s12035019-01721-7.
12. Leung L, Cahill CM. (2010) TNF-alpha and neuropathic pain — A review. *J Neuroinflammation* **7**: 27, doi:10.1186/1742-2094-7-27.
13. Ramachandran R. (2018) Neurogenic inflammation and its role in migraine. *Semin Immunopathol* **40**(3): 301–314, doi:10.1007/s00281-018-0676-y.
14. Chen J, Wang Q, Wang A, Lin Z. (2020) Structural and functional characterization of the gut microbiota in elderly women with migraine. *Front Cell Infect Microbiol* **9**: 470, https://doi.org/10.3389/fcimb.2019.00470.
15. Su J, Zhou XY, Zhang GX. (2014) Association between helicobacter pylori infection and migraine: A meta-analysis. *World J Gastroenterol* **20**(40): 14965–14972, doi:10.3748/wjg.v20.i40.14965.
16. Savi L, Ribaldone DG, Fagoonee S, Pellicano R. (2013) Is helicobacter pylori the infectious trigger for headache? A review. *Infect Disord Drug Targets* **13**(5): 313–317, doi:10.2174/1871526513666131201125021.
17. Munno I, Marinaro M, Bassi A, Cassiano MA, Causarano V, Centonze V. (2001) Immunological aspects in migraine: Increase of IL-10 plasma levels during attack. *Headache* **41**(8): 764–767, doi:10.1046/j.1526-4610.2001.01140.x.
18. Dimitrova AK, *et al.* (2013) Prevalence of migraine in patients with celiac disease and inflammatory bowel disease. *Headache* **53**(2): 344–355, doi:10.1111/j.15264610.2012.02260.x.

19. Sandhu KV, Sherwin E, Schellekens H, Stanton C, Dinan TG, Cryan JF. (2017) Feeding the microbiota gut-brain axis: diet, microbiome, and neuropsychiatry. *Transl Res* **179**: 223–244, doi:10.1016/j.trsl.2016.10.002.
20. van Hemert S, *et al.* (2014) Migraine associated with gastrointestinal disorders: Review of the literature and clinical implications. *Front Neurol* **5**: 241, doi:10.3389/fneur.2014.00241.
21. Grazzi L, Corsini E, Ciusani E, Usai S, Vasco C, Bussone G. (2014) Evaluation of immune parameters in chronic migraine with medication overuse. *Neurol Sci* **35**(Suppl 1): 171–173, doi:10.1007/s10072-014-1763-z.
22. Aydinlar EI, *et al.* (2013) IgG-based elimination diet in migraine plus irritable bowel syndrome. *Headache* **53**(3): 514–525, doi:10.1111/j.1526-4610.2012.02296.x.
23. Rockett FC, de Oliveira VR, Castro K, Chaves ML, Perla Ada S, Perry ID. (2012) Dietary aspects of migraine trigger factors. *Nutr Rev* **70**(6): 337–356, doi:10.1111/j.17534887.2012.00468.x.
24. Finocchi C, Sivori G. (2012) Food as trigger and aggravating factor of migraine. *Neurol Sci* **33**(Suppl 1): S77–S80, doi:10.1007/s10072-012-1046-5.
25. Sarna B, Abouzari M, Lin HW, Djalilian HR. (2020) A hypothetical proposal for association between migraine and Meniere's disease. *Med Hypotheses* **134**: 109430, doi:10.106/j.mehy.2019.109430.
26. Hindiyeh NA, Zhang N, Farrar M, Banerjee P, Lombard L, Aurora SK. (2020) The role of diet and nutrition in migraine triggers and treatment: A systematic literature review. *Headache* **60**(7): 1300–1316, doi:10.111/head.13836.
27. Martami F, *et al.* (2019) The effects of a multispecies probiotic supplement on inflammatory markers and episodic and chronic migraine characteristics: A randomized double-blind controlled trial. *Cephalalgia* **39**(7): 841–853, doi:10.1177/0333102418820102.
28. Qi X, Fan G, Jia H. (2020) The probiotic *Lactobacillus casei* Shirota attenuates symptoms of vestibular migraine: A randomised placebo-controlled double-blind clinical trial. *Benef Microbes* **11**(5): 469–476, doi10.3920/BM2020.0058.
29. de Roos NM, van Hemert S, Rovers JMP, Smits MG, Witteman BJM. (2017) The effects of a multispecies probiotic on migraine and markers of intestinal permeability-results of a randomized placebo-controlled study. *Eur J Clin Nutr* **71**(12): 1455–1462, doi:10.1038/ejcn.2017.57.

30. Naghibi MM, Day R, Stone S, Harper A. (2019) Probiotics for the prophylaxis of migraine: A systematic review of randomized placebo controlled trials. *J Clin Med* **8**(9): 1441, doi:10.3390/jcm8091441.
31. Hildebrandt MA, *et al.* (2009) High-fat diet determines the composition of the murine gut microbiome independently of obesity. *Gastroenterology* **137**(5): 1716–1724.e1-2, doi:10.1053/j.gastro.2009.08.042.
32. Bic Z, Blix GG, Hopp HP, Leslie FM, Schell MJ. (1999) The influence of a low-fat diet on incidence and severity of migraine headaches. *J Womens Health Gend Based Med* **8**(5): 623-630, doi:10.1089/jwh.1.1999.8.623.
33. Bunner AE, Agarwal U, Gonzales JF, Valente F, Barnard ND. (2014) Nutrition intervention for migraine: A randomized crossover trial. *J Headache Pain* **15**(1): 69, doi:10.1186/11292377-15-69.
34. Jahromi SR, Ghorbani Z, Martelletti P, Lampl C, Togha M. (2019) Association of diet and headache. *J Headache Pain* **20**(1): 106, https://doi.org/10.1186/s10194-019-1057-1.
35. Bashir M, *et al.* (2016) Effects of high doses of vitamin D3 on mucosa-associated gut microbiome vary between regions of the human gastrointestinal tract. *Eur J Nutr* **55**(4): 1479–1489, doi:10.1007/s00394-015-0966-2.
36. Ghorbani Z, *et al.* (2019) Vitamin D in migraine headache: A comprehensive review on literature. *Neurol Sci* **40**(12): 2459–2477, doi:10.1007/s10072-019-04021-z.
37. D'Onofrio F, Raimo S, Spitaleri D, Casucci G, Bussone G. (2017) Usefulness of nutraceuticals in migraine prophylaxis. *Neurol Sci* **38**(Suppl 1): 117–120, doi:10.1007/s10072-017-29011.
38. Bond DS, Roth J, Nash JM, Wing RR. (2011) Migraine and obesity: Epidemiology, possible mechanisms and the potential role of weight loss treatment. *Obes Rev* **12**(5): e362–e371, https://doi.org/10/111/j.1526-4610.20212.02260.
39. Laino D, *et al.* (2016) Headache, migraine and obesity: An overview on plausible links. *J Biol Regul Homeost Agents* **30**(2): 333–338.
40. Nagpal R, Newman TM, Wang S, Jain S, Lovato JF, Yadav H. (2018) Obesity-linked gut microbiome dysbiosis associated with derangements in gut permeability and intestinal cellular homeostasis independent of diet. *J Diabetes Res*, https://doi.org/10.1155/2018/3462092.

41. Jahromi SR, *et al.* (2018) Bariatric surgery promising in migraine control: A controlled trial on weight loss and its effect on migraine headache. *Obes Surg* **28**(1): 87–96, doi:10.1007/s11695-017-2793-4.
42. Verrotti A, *et al.* (2013) Impact of a weight loss program on migraine in obese adolescents. *Eur J Neurol* **20**(2): 394–397, https://doi.org/10.111/j.1468-1331.2012.03771

Chapter 15

Interactions of the Gut Nervous System with Bacteria

Chalystha Yie Qin Lee[1], Suzanne Hosie[1], Ashley E. Franks[2] and Elisa L. Hill-Yardin[1,*]

Abstract

The gut–brain axis refers to the bidirectional communication between the brain and the enteric nervous system (ENS) in the gastrointestinal (GI) tract. Tightly regulated interactions between these two components are crucial for maintaining physiological gut homeostasis. The ENS also sends neurochemical signals to modulate immune function within the GI tract. The presence of the gut microbiome adds another dimension of complexity to this relationship. Trillions of microbes that collectively make up the microbiome interact with neurons, both in the brain and the ENS, to affect behavior and gut function in health and disease. Due to its close proximity, the ENS is a critical component of the interface between the nervous system of the host and bacteria. Recent microbiome studies point toward the occurrence of dysbiosis in many neurological disorders. To understand the changes in gut–brain axis

[1] School of Health and Biomedical Sciences, RMIT University, Bundoora, VIC, Australia.
[2] School of Life Sciences, La Trobe University, Bundoora, VIC, Australia.
Corresponding Author: Elisa Hill-Yardin, PhD, School of Health and Biomedical Sciences, RMIT University, Building 223, Bundoora West Campus, 225–245 Clements Drive, Bundoora, VIC 3083, Australia, Tel +613-9925-7609, E-mail: elisa.hill@rmit.edu.au.

function in these disorders, pre-clinical animal models are an essential tool to elucidate the complex interactions between host genetics and influences from the external environment. Exciting advances in this field are focused on deciphering mechanisms of the gut–brain axis to identify potential targets for personalized medicine and improve health.

INTRODUCTION

The gut-brain axis is the tightly-regulated bidirectional communication between the brain and the enteric nervous system (ENS), and plays a major role in maintaining physiological gut homeostasis. This function is further enriched by trillions of microbes found in the gut including bacteria, fungi, and archaea; commonly referred to as the microbiota. Neurons, both in the brain and the gut, interact with the microbiome to affect mood and behavior although the precise biological mechanisms involved are unclear. Microbiota–gut–brain axis communication occurs via neuroimmune and neuroendocrine signaling, the vagus nerve, and through the actions of microbial metabolites.[1,2] The ENS is a critical component of the interface between the nervous system of the host and bacteria. Recent microbiome research suggests that dysbiosis is present in many neurological diseases including autism, Parkinson's disease (PD), Alzheimer's disease, and multiple sclerosis.[2–8] To further elucidate the pathogenesis of these disorders including changes in gut–brain axis function, pre-clinical animal models are a crucial tool in assessing interactions between the host physiology, genetics and the environment.[9] It is anticipated that unraveling interactions of the gut–brain axis in health and neurological disorders including a specific focus on the gut–microbiota interface, will identify potential targets for personalized medicine and improve quality of life for patients.

The Gut–Brain Axis in Health and Disease

The Enteric Nervous System

The ENS is often referred to as "the brain in our gut" and contains a complex network of approximately 500 million neurons — equivalent to the neuronal population of the spinal cord.[10] The intrinsic network of neurons comprising the ENS includes two neuronal plexuses located between the smooth muscle layers of the gastrointestinal (GI) tract: the myenteric plexus (MP) and submucosal plexus (SMP). The MP and SMP predominantly regulate gut motility, and permeability and secretion, respectively.[10–12] The ENS also interacts with the immune system including via bidirectional signaling to the gut-associated lymphoid tissue (GALT) within the GI tract. Interactions between intestinal immune cells and enteric neurons are integral to the proper functioning of the sensory nervous system. For example, immune cell activation in response to infection provides feedback to the neural sensory circuitry via efferent neuroimmune inflammatory reflex pathways to formulate a defensive physiological response.[13] Neuroimmune interactions are also influenced by the microbial community within the GI tract. Accordingly, shifts in microbial ecology and gut–brain axis communication are associated with diseases ranging from inflammatory bowel disease (IBD) and obesity to neurodegenerative and neurodevelopmental disorders such as Parkinson's disease and autism[14–16] (Figure 15.1).

Neuroimmune Interactions in the GI Tract

The main functions of the GI tract are digestion and absorption of nutrients. On the flip side, these features also expose the gut to external toxins and pathogens entering the bloodstream via the mucosal barrier. The presence of gut microbiota in the intestinal lumen adds an additional layer of complexity to this physiological system. Hence, the GI tract requires a complex filtering system to screen ingested materials while ensuring that nutrients are absorbed and pathogens are excluded without causing excessive inflammatory responses to beneficial bacteria in the gut.

Immune function within the GI tract is modulated by signals from the ENS under physiological homeostasis.[13] Multiple neuronal subtypes in the GI tract express receptors to cytokine messengers from the immune system and immune cells also express the cellular machinery enabling responses to neurotransmitters. Abnormalities in development (e.g., due to external stimuli or gene mutations) of the GI tract can alter the organization of the ENS and, in turn, responses to inflammatory insult. For example, mice exhibiting altered numbers of enteric neurons show differing responses to inflammatory stimuli such as via the trinitrobenzene sulfonic acid (TNBS) and dextran sulfate sodium (DSS) model. Interestingly, in these models, the increased inflammatory response is proportional to the neuronal density within the GI tract,[17] further highlighting the importance of interactions of the ENS with immune function.

Muscularis macrophages are a distinct population of intestinal macrophages that reside within the intestinal muscularis adjacent to enteric neurons.[18] At resting state, the muscularis macrophage population secrete bone morphogenetic protein 2 (BMP2), a growth factor that acts on the ENS to regulate peristalsis.[18–20] In turn, enteric neurons produce macrophage colony-stimulating factor (M-CSF), also known as colony-stimulating factor 1 (CSF1) to influence muscularis macrophage development. Furthermore, commensal bacteria within the GI tract modulate expression of both BMP2 and CSF1, as evidenced by decreases in BMP2 (expressed by muscularis macrophages) and CSF1 (expressed by enteric neurons) observed in antibiotic-treated mice. Interestingly, depletion of these proteins was restored by supplementation of the endotoxin lipopolysaccharide (LPS) or by fecal microbiota transfer from control mice.[18] These findings indicate that the gut microbiota play an important role in the dynamic bidirectional neuroimmune crosstalk between muscularis macrophages and the nervous system of the GI tract.

A large component of the body's immune system includes GALT comprising aggregates of immune cells containing multiple B cell follicles, T cell interfollicular regions, and a dendritic cell (DC) subepithelial dome.[21] Peyer's patches, cecal patches, and colonic patches are macroscopically visible components of the GALT, present in the small intestine, cecum, and large intestine, respectively. The GALT is characterized by overlying subepithelial regions in close contact with the lumen of the GI tract that contain specialized microfold cells (M cells). M cells are responsible for the uptake and presentation of antigens to an underlying DC to initiate an adaptive immune response. Despite general similarities in GALT found in the cecum and Peyer's patches of the small intestine, Peyer's patches exhibit a larger proportion of mature M cells compared to the cecal patch.[22] Given that the cecum is proposed to modulate microbial communities within the GI tract,[23] the comparatively sparse M cell activity in the cecal patch may serve to prevent excessive immune responses to adjacent gut bacteria.

Neuroinflammation and the Gut–Brain Axis

Inflammation in both the brain and gut is implicated in a growing repertoire of neurological diseases. Neuroinflammation is indicated by sustained activity and proliferation of immune cells that is a result of infection, injury, or disease.[24,25] Astrocytes and microglia are major glial subtypes that contribute to immune responses within the central nervous system (CNS) including via producing pro-inflammatory chemokines and cytokines. Constant and prolonged activation of these cells can result in an aberrant immune response, affecting neural signaling, and, subsequently, altering cognitive function.[26,27] Within the ENS, there are at least four glial subtypes (I–IV) based on cellular morphology and location in the

GI tract, however, their functional roles are not well defined.[28] Major markers of these glial populations include glial fibrillary acidic protein (GFAP), S100-beta and Sox10, but different glial subtypes do not correlate with differences in the expression of these proteins.[28] The major marker of microglia in the brain is ionized calcium-binding adaptor molecule 1 (Iba-1), which labels muscularis macrophages in the GI tract.[29] While these neuroimmune cell subtypes are substantially well-characterized in the CNS, more fundamental research is required to thoroughly characterize these populations of cells in the ENS.

Unsurprisingly, glial cell reactivity and dysfunction likely contribute, both directly and indirectly, in the pathobiology of neurodevelopmental disorders, such as autism spectrum disorder (ASD; autism).[30–32] Autism is a neurodevelopmental disorder characterized by impaired social interactions, and restrictive and repetitive behavior.[33] Extensive microglial and astrocyte immunoreactivity and abnormal histology such as a patchy loss of neurons in the cerebellum were reported in various regions of the brain in post-mortem analysis of tissues from autism patients compared to neurotypical controls.[34] These findings were further supported by an increase in microglial density and altered spatial organization suggesting heightened interactivity with neurons in autism.[35]

Children with neurodevelopmental disorders such as autism often experience GI problems that are more severe and frequent than in neurotypical children.[36–38] Given that GI symptoms correlate with the severity of autism and with increased irritability, anxiety and social withdrawal,[39] it would be of interest to compare large-scale datasets including microbial profiles in this context. That said, elevated levels of inflammatory cytokines in serum samples from autistic children compared to neurotypical controls have been reported[40–43] and are similarly associated with the severity of autism-related symptoms.[44,45] In further support of altered gut–brain axis interactions in neurodevelopmental disorders, cytokine levels indicative of inflammation are correlated with GI symptomology and several key autism-associated gut microbial populations (e.g., Clostridiales).[6] How interactions of the ENS with the immune system and bacteria are altered in these patient groups, however, is only beginning to be understood.

The Interface Between the ENS and Bacteria

The Mucus Barrier

Mucus provides a supportive environment for a large proportion of commensal microbes and is an important component of the mucosal barrier that protects the host from pathogens within the lumen of the GI tract. There are a number of biological pathways by which the ENS may influence mucus structure and function.[3] For example, mucus production is regulated by molecular pathways involved in developmental processes and nervous system activity such as the Spdef and Wnt/Notch signaling pathways. Altered levels of inflammatory cytokines such as interleukin-4 (IL-4), IFN-γ, and tumor necrosis factor alpha (TNF-α) also affect mucus properties in mice.[46] The mucus layer additionally plays an important role maintaining gut–brain axis function as part of the first line of defence against infiltration of microorganisms, digestive enzymes and acids, digested food particles, microbial by-products and food-associated toxins. As well as being a physical barrier to bacteria, mucus houses a large component of the beneficial and pathogenic microbes adherent to the wall of the GI tract. Altered mucus properties may, therefore, also impact microbial populations and underly dysbiosis as well as increased mucosal permeability reported in patients with neurological disorders including autism, PD, Alzheimer's disease, and multiple sclerosis.[3,47]

It was recently revealed that enteric neural interactions with the immune system regulate gut permeability via important molecular pathways.[48,49] These findings show that submucosal enteric neurons drive mucosal barrier function by release of vasoactive intestinal peptide (VIP) that in turn stimulates Group 3 innate lymphoid cells (ILC3s) to increase release of the cytokine IL-22.

In human epithelial cell lines, IL-22 promotes GI permeability via the paracellular pathway by increasing expression of Claudin-2, which forms pores at tight junctions on the basolateral membrane of epithelial cells.[50] These elegant studies identify neuroimmune pathways that, if modified in neurological disease states, may result in impaired mucosal barrier function.

Neural Regulation of Gut Physiology

Although the GI tract receives extrinsic regulation from the CNS *in vivo*, GI motility and secretion can function independently of extrinsic neuronal inputs in an *ex vivo* setting.[10,51] Major neuronal subpopulations of the ENS include primary afferent neurons, interneurons, and motor neurons involved in sensory–motor regulation. Primary afferent neurons in the MP and SMP respond to chemical and mechanical stimuli from the lumen.[11] These neurons synapse with enteric interneurons that regulate the peristaltic reflex to propel luminal content along the GI tract.[52] In addition, enteric motor neurons act on a range of effector cells including smooth muscle cells, pacemaker cells, mucosal glands, and immune cells to govern GI function. The spatiotemporal coordination of gut contractions generates the motor patterns observed along the GI tract. These rhythmic contractions can also occur in response to local stimuli including gut distension, which activate local neuronal pathways. Migrating motor complexes in the colon, for example, are a subtype of neurally driven, regular, and spontaneous contractions that span significant portions of the large bowel.[51,53–56] Neuronal pathways regulating the neuronal circuitry underlying secretion of water, acids, bicarbonate, hormones, neuropeptides, enzymes and mucus in the GI tract minimally comprise a two-neuron reflex: an afferent neuron and a secretomotor neuron that extends into the lamina propria from the submucosal plexus. However, these secretory processes may comprise a more complex mechanism involving a broader group

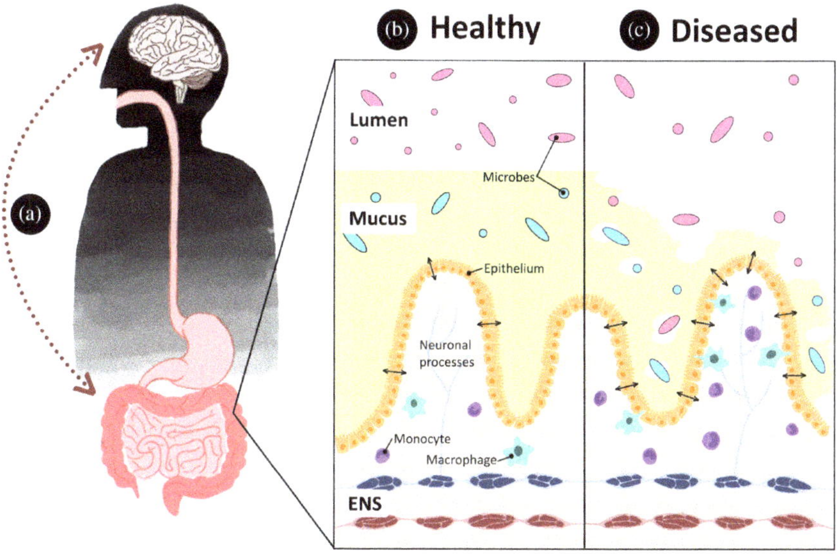

Fig. 15.1: The gut–brain axis in health and disease. (a) Bidirectional communication between the brain and gut allows the brain to be influenced by gut function and vice versa. This occurs via various pathways including the involvement of microbial metabolites and neuroimmune interactions. (b) Under physiological homeostatic conditions, the enteric nervous system (ENS) functions to modulate gut motility and permeability as well as secretion of water and electrolytes. Microbes within the lumen and mucus layer further enrich this complex interaction between gut and brain. (c) In neurological diseases, the intestinal barrier integrity may be compromised due to various factors including microbial dysbiosis, changes in the structure of the mucus layer, mucosal permeability, and inflammation (represented by increased numbers of monocytes and macrophages within the lamina propria adjacent to the epithelium). Arrows indicate mucosal epithelial paracellular and transcellular permeability.

of enteric neurons, including those in the MP.[57] The activity of these circuits is triggered by sensory epithelial enteroendocrine cells embedded in the mucosal epithelium that directly sense and communicate information regarding the luminal content.[12,52,58] Further research is needed to fully define the fundamental enteric circuitry involved in gut functions in different species, and how the activity of these neuronal populations intersect with the structure and function of gut microbial communities in health and disease.

Interactions of the Nervous System with the Microbiome

Bacteria Excite Submucosal Neurons

To understand whether microbes might directly impact neurons in the GI tract, Mao *et al.* utilized whole cell patch clamp recordings of the soma of sensory enteric neurons. In this experimental paradigm, bacterial strains were added directly to the adjacent mucosal surface (innervated by these neurons) and changes in neuronal activity recorded using the whole cell patch clamp technique. Addition of either *Lactobacillus rhamnosus* (JB-1) or *Bacteroides fragilis* to the intestinal mucosa preparation elicited a sensory response from myenteric primary afferent neurons.[59] Moreover, the administration of *Lactobacillus reuteri* DSM-17938 (DSM) and its secreted vesicles reduces motility (including a decrease in the velocity and frequency of propagating contractile clusters) in jejunum preparations but increases motility in the colon in mice.[60] In rats, it was reported that ingestion of *L. reuteri* enhances tonic inhibition of colon contractile activity via its action on intermediate conductance calcium-activated potassium channels (IK_{Ca}).[61] Overall, these reports show that bacterial species influence the activity of individual neurons and the GI tract to impact motility in rodents.

Innervation of GALT

Whether impairments in ENS activity and connectivity, for example, in the context of neurological disorders, affect GALT function is unclear. Peyer's patches are heavily innervated by the ENS, a network that extends to all GALT compartments,[62] but to better understand changes in this neuronal network in models of neurological disease, further research is required. In both humans and mice, the size and density of GALT differs from the small intestine to the cecum/appendix and colon.[21] Peyer's patches are scarcely found in the duodenum and increase in size and density from the jejunum to the ileum. Cecal patches are macroscopic structures found in the blind end of the cecum (in mice) or around the ileocecal valve (in humans). Colonic patches, as the name suggests, are found in the colon and rectum. How ENS innervation of these organized immune structures changes along the length of the GI tract and whether this neuronal circuitry is affected in neurological disorders remains understudied. As would be anticipated, gut microbes create an additional dimension to the relationship between the GALT and the ENS. The richness and heterogeneity of the gut microbial population in healthy individuals varies significantly along the GI tract,[63] which gives rise to further questions such as whether the diverse profile of the microbial community in specific gut regions is shifted in neurological disorders.

Our recent findings in mice expressing an autism-associated mutation in the nervous system identified an increased density of macrophages within the cecal patch, suggesting neuroimmune changes. These macrophages also showed altered morphology including reduced cellular volume and increased sphericity, suggesting an overall increase in the reactive state of these cells and potentially, an aberrant immune response in mutant mice.[64] Based on these observations, and the presence of comorbid GI dysfunction that may be related to inflammatory responses in a broad range of brain disorders (e.g., including PD and Alzheimer's disease), further investigations in the clinical setting and in pre-clinical models could identify common mechanisms and potential targets for therapies modulating the immune system within the GI tract.

The cecum is a pouch-like sac at the juncture of the ileum and colon in mice and typically contains one or more aggregates of lymphoid tissue, termed the cecal patch, that are morphologically and functionally similar to Peyer's patches located predominantly in the ileum. Both the cecal and Peyer's patch tissue are primary sites for the development of immunoglobulin A (IgA)-secreting plasma cells important in mucosal immune function. Following secretion, IgA is transported into the intestinal lumen and contributes to the maintenance of gut homeostasis by discriminating between pathogenic and commensal bacteria as well as modulating gut microbial activity.[65,66] Furthermore, IgA deficiency is associated with autoimmune disorders such as celiac disease and can result in defective immune defences against intestinal pathogens.[67] Masahata *et al.*[23] reported that in mice with the cecal patch surgically removed, a decrease in IgA-secreting plasma cells is observed in the colon and the composition of fecal microbiota is altered. While cecal and Peyer's patches possess similar immunological characteristics, IgA-secreting cells derived from the cecal patch express higher levels of CCR10 (C-C chemokine receptor type 10), which drives migration of IgA-secreting plasma cells to both the small and large intestine.[23] In contrast, IgA-secreting plasma cells originating from Peyer's patches are preferentially recruited to the small intestine.

The cecum is considered the equivalent of the human appendix in rodents and is proposed to play a homeostatic role in regulating neuroimmune interactions in the gut. This region of the GI tract shapes microbial communities via immune-mediated processes.[68] Due to its narrow structure in humans, biofilm lining and lumen content that is rich in microorganisms, the appendix has been proposed as a "safe house" for gut commensal flora. In the case of purging (e.g., diarrhea) in response to disease or infection, it is hypothesized that commensal microbes can be reintroduced to repopulate the gut, thus re-establishing GI homeostasis.[69] Nutrient-rich biofilms on the surface of the intestinal epithelial lining facilitate the growth and maintenance of the gut microbial community. The adherent nature of these biofilms additionally allow bacteria to live for an extended period in a flowing environment without being expelled from the gut.[65] The ENS also drives regular motility patterns in the cecum that likely serve to mix or "churn" digesta and microbial content for maximal digestion.[70] Alterations in the ENS will likely cause cecal dysmotility or impaired permeability, and subsequently influence biofilm structure as well as microbial composition. Furthermore, these changes are predicted to have downstream effects on immune responses. In animal models of neurodevelopmental disorders, it has already been demonstrated that microbial dysbiosis is common, however, the precise interactions between alterations in neurotransmission and inflammation are under study.[53,71,72]

Gut Dysbiosis

Dysbiosis in Neurological Diseases

A broad body of evidence now suggests that patients with neurological disorders such as autism, PD, Alzheimer's disease, and multiple sclerosis commonly experience GI dysfunction and have dysbiosis.[3] These findings implicate dysfunctions in the microbiota–gut–brain axis in disease pathophysiology and a burgeoning area of research is focused on identifying abnormalities in cell signaling pathways involved in neuroimmune function.

In addition to impairments in social interactions alongside restrictive and repetitive behavior, GI dysfunction is a major comorbidity for autism patients[36,73,74] and includes symptoms such as abdominal pain, diarrhea, constipation and bloating. Altered levels of mucosa-associated bacterial species are also reported in autism patients with GI dysfunction, with *Akkermansia muciniphila*, *Dorea*, *Blautia*, *Sutterella* and *Neisseria* having decreased abundance, while mucosa-associated Clostridiales (*Lachnospiraceae* and *Ruminococcaceae*), *Burkholderia*, *Ruminococcaceae*, *Lachnospiraceae* and *Sutterella* have increased abundance.[6,75–78] Autism is now understood to have strong genetic underpinnings, however, unraveling interactions between genetics, in particular with respect to nervous system impairments and microbial communities could yield clinically relevant information. Changes in microbial signatures that correlate with GI symptoms highlight that the gut microbiome may provide a potential therapeutic target for modulating GI dysfunction in the future. Given that in 2016, 1 in 54 children were diagnosed with autism in the United States,[79] understanding how the gut–brain axis influences well-being in autism is of high importance.

PD is the second most common neurodegenerative disease observed in people over 60 years of age[80] and is increasingly correlated with GI disorders prior to the onset of characteristic motor symptoms such as tremor and coordination of complex movement. Although the pathophysiology of PD remains unclear, cellular accumulation of α-synuclein appears to cause neuronal death[81,82] and Parkinson's patients with colonic inflammation also showed α-synuclein deposits in colon tissue.[83] Mucosal biopsy samples of PD patients showed increased abundance of *A. muciniphila*, and *Ralstonia*, and a decrease in the abundance of *Faecalibacterium* (*Blautia*, *Coprococcus*, *Roseburia*) and *Prevotella*.[7,84–86] Very few microbial analyses have been undertaken in pre-clinical models of PD and the findings from microbial studies in patient samples to date are inconsistent, which may reflect differences in sample processing and analytical approaches. These findings suggest that further research is required to gain insights into potential gut–brain axis mechanisms that may, in the future, be modulated to improve quality of life and slow symptom progression.

Alzheimer's disease is an increasingly prevalent neurodegenerative disease characterized by progressive cognitive decline and this patient group is also commonly reported to experience comorbid GI dysfunction. Patients with Alzheimer's disease who also had symptoms indicative of irritable bowel syndrome (IBS) showed dysbiosis that included an increased abundance of mucolytic bacteria including *A. muciniphila* and *Prevotella denticola*.[8] Similarly, stool samples of Alzheimer patients showed an increase in the abundance of *Escherichia/Shigella* (pro-inflammatory taxa) and a decrease in abundance of *Eubacterium rectale* (anti-inflammatory taxa).[87] In keeping with altered gut–brain axis function, microbial dysbiosis in Alzheimer's disease has been implicated in increasing gut permeability, which may influence systemic inflammation and impairment of the blood–brain barrier.[88,89] Consequently, strategies involving probiotic supplementation in patients with Alzheimer's disease are underway with potential positive outcomes.[90] In addition, however, larger clinical trials across broader patient populations are required to establish potential benefits associated with microbiota-based therapeutics considering the multi-factorial nature of Alzheimer's disease.

Multiple sclerosis involves an aberrant immune system that causes demyelination in the CNS. Several studies in patients with multiple sclerosis have found increased abundance of mucosal bacteria including *A. muciniphila*, *Methanobrevibacter* and *Acinetobacter calcoaceticus* and decreased abundance of *Butyricimonas*, *Faecalibacterium* and *Parabacteroides distasonis*.[4,91–93] Such alterations in the mucosal microbiome potentially favor the growth of pathogenic bacteria that alter the composition of the mucus layer and, therefore, may exacerbate core symptoms of these disorders.[94,95] Together, these findings suggest a potential role of the gut microbiota in neurological disorders. Further research focused on unraveling mechanisms underpinning these changes may provide biological targets and diversify disease treatment and management options.

Strategies in Gut–Brain Axis Research

Pre-Clinical Models of Neurological Disease

Given the substantial evidence for interactions between the host nervous system and the microbiome in health and disease, it is critical that appropriate pre-clinical models and informative assays are applied to understand underlying biological mechanisms. A major advantage of studying animal models is the ability to control environmental and genetic effects as well as to identify specific molecular pathways altered in pathological processes. Although animal models are not able to fully replicate clinical conditions, the degree of relevance of animal models of disease can be assessed based on face, construct and predictive validity. Specifically, an animal model that displays endophenotypes relevant to the disease of interest (i.e., face validity) as well as demonstrating that the mode of generation of the model resembling the disease pathology (i.e., relevant gene mutation or environmental insult) can serve as a valuable tool for assessing mechanistic insights into disease progression. In addition, animal models that display predictive validity, that is, treatment with medication commonly prescribed to this patient group also reverses endophenotypes relevant to the disorder are also particularly valuable. Multiple rodent models of neurological disease including PD, Alzheimer's disease, epilepsy, multiple sclerosis, as well as autism have enabled the dissection of neuronal pathways involved in pathology,[53,96–103] however, how altered nervous system activity affects interactions with the microbial community in these models is not well defined.

Assessing Gut–Brain Axis Function in Animal Models

Techniques that measure behavior, brain and gut neurophysiology, neurochemistry, histology and whole intestinal transit, as well as microbial assays are needed to comprehensively characterize changes in the microbiota–gut–brain axis in pre-clinical models of disease.

Behavior

Multiple well-established behavioral assays are available to assess for disease-relevant endophenotypes in rodents. Assays assessing motor function and gait, for example, are commonly utilized to determine if PD-relevant traits are present. Behavioral tests identifying memory dysfunction such as the Morris water maze, the Barnes maze, and touch screen technologies testing subtle cognitive changes[104] are also used in establishing the face validity of models of Alzheimer's disease, whereas a range of social interaction and anxiety-relevant tests are readily available to assess for autism-like traits in mouse models[105] including changes in social interaction such as abnormal levels of aggression.[106,107] The use of these behavioral tests in rodent models is important when assaying effects of modulating components of the gut–brain axis pathway in the context of neurological disorders.

GI transit and motility

A growing number of neurological disorders are associated with GI dysfunction and hence several assays for these traits are now of interest in the field of gut–brain axis research. Whole animal studies to assess GI transit can be utilized via gavage administration of dye and assessment of transit times by measuring the dye front in the region of interest within the GI tract following the terminal end point of the experiment.[18,53,108] Additionally, transit in live animals can be assessed using X-ray imaging technology following ingesting radioactive solutions such as barium sulfate via gavage and sequential images compared to identify transit speeds in different gut regions of the individual animal.[109,110] To determine the influence of the ENS on gut motility, state-of-the-art video-imaging systems enable detection of changes in *ex vivo* gut contractile patterning[51] at baseline and in response to perturbation of the nervous system using pharmacological tools or indeed responses to different compositions of microbiota added to the lumen of the GI tract in these organ bath experimental setups. Additional assays for changes in the contractile characteristics of the smooth muscle within the gut include the assessment of contractile strength and frequency by muscle strip preparations and response to stretch (for both longitudinal and circular muscle preparations) and pharmacological stimulation.[111–113]

GI secretion and membrane permeability

In addition to organ bath motility assessment, the functional integrity of the mucosal barrier is central to the proper functioning of the gut–brain axis in health and disease. A fundamental physiological measurement of the mucosal barrier integrity can be carried out using an Ussing chamber, whereby the permeability of the mucosal tissue from a small segment of the GI tract can be assessed by measurement of the flow of small dye particles (i.e., FITC) over time.[47] Similarly, these experimental approaches can detect the transepithelial resistance (TER) to clarify if the integrity of the tissue is significantly compromised in response to certain stimuli or as a result of a disease state. A further comparative measure of mucosal function utilizing the Ussing chamber apparatus includes recording the short-circuit current to identify changes in neuronal-driven secretory function.[114,115]

Enteric neuron structure and function

At the cellular level, the use of multiple immunofluorescent labels to identify cell types within the ENS and the mucosa are well established,[10] and multiple cell classes are classified on this basis as well as their morphological and connectivity traits. The most commonly utilized immunofluorescence labels within the ENS include antisera to the pan-neuronal marker, Hu (also known as Hu/CD[116]), nitric oxide synthase (NOS), cholinergic markers including choline acetyltransferase (ChAT) and vesicular acetylcholine transporter (vAChT), serotonin and serotonergic receptor subtypes, neuropeptides such as VIP, neuropeptide Y (NPY), peptide YY (PYY), somatostatin (SST), cholecystokinin (CCK), and the calcium-binding proteins, calbindin and calretinin.

To assess cellular function, electrophysiological techniques are also invaluable in understanding the roles of various neuronal subpopulations within the ENS. To date most neurophysiological recordings have utilized sharp electrode recordings,[117–119] which enable electrical activity profiles (including synaptic and action potential characteristics) to be obtained. In addition, this approach enables the filling of the cell with biocytin included in the recording pipette and subsequent morphological analyses. More recent research has applied the whole cell patch clamp technique in ENS preparations,[59,120–123] which enables detailed analyses of specific currents at the cell membrane.[123] This approach also enables cell filling and subsequent morphological analysis and, importantly opens avenues for simultaneous analysis of cellular gene expression using techniques established in the CNS.[124–126] An important example of the use of patch clamp electrophysiology in this field is that of Mao *et al.*, who successfully recorded from intestinal primary afferent neurons within the MP and showed increased activity with the application of *L. rhamnosus* or *B. fragilis* to the mucosal surface of the same tissue preparation.[59] The application of powerful techniques for functional and gene expression analyses as described will significantly advance our understanding of microbial influences on nervous system function.

Microbial analyses and engineering approaches

Undertaking the appropriate analysis of microbial populations is now recognized as a central component in understanding the role of the gut–brain axis in the maintenance of physiological homeostasis and disease. To date, most studies in animal models have simply assessed for the presence or absence of certain species, however, this is now recognized to be an overly simplistic way of assessing change in microbial populations. Microbiome analysis assays measuring abundance and diversity of species are often conducted using various molecular techniques. Automated Ribosomal Inter-Spacer Analysis (ARISA) and deep sequencing provide patterning of the microbiome through the generation of Operational Taxonomic Units.[9] Functional approaches to provide a holistic view of the microbiome's capability can be conducted directly through community level physiological profiling (CLPP) or predictively by Phylogenetic Investigation of Communities by Unobserved States (PICRUSt). Metagenomics, shotgun sequencing, and the incorporation of meta-omics are also necessary for thoroughly understanding the structure and function of microbial communities in control conditions and in models of disease. The microbiome exists as a complex ecosystem and the overall microbial function and outputs are critical to understanding the role of this extensive population.[127] Importantly, the presence of a specific microbial species may not necessarily change community-level function in ecosystems such as the gut microbiome. Many community-level functions, such as the production of metabolites, are redundant between species and may not produce different functional outcomes due to environmental changes. Unfortunately, few studies report the use of both structural and functional approaches for identifying changes in microbial profiles. Nevertheless, active research in the fields of synthetic biology,[128] nanotechnology,[129,130] and bacteriophage and polyclonal antisera platforms may deliver tools to engineer microbes in the disease context. These approaches should assist to produce desired outcomes when fundamental research identifies relevant drivers to detrimental or productive change in microbiome composition. Research into this field is anticipated to identify potential therapeutic targets to improve quality of life in multiple neurological disorders via modulating microbial–nervous system interactions.

Treatment of Neurological Diseases via the Gut–Brain Axis

The impact of the microbiota–gut–brain axis in neurological disorders is evident. Despite this, current intervention and treatment strategies for these disorders are often focussed solely on restoring brain function. Mounting evidence demonstrates that these routinely prescribed medications for neurological disease may in fact impact the ENS–bacteria interface and alter microbiome profiles. This includes pharmaceutical treatments such as carbidopa/levodopa for PD[131] and glatiramer acetate for multiple sclerosis.[92] The effects of medications such as Risperidone, Bumetanide, and Aripiprazole for autism, and Acetylcholinesterase inhibitors (AChEI) and memantine (N-methyl-d-aspartate [NMDA] receptor antagonist) for Alzheimer's disease on the ENS and the gut microbiota in patients remain unknown.

In considering the dynamic communication between the gut and the brain, future approaches for developing novel therapies for patients will likely involve the gut–brain axis. Several strategies to modify the gut–brain axis include prebiotics, probiotics, and fecal microbiota transplantation (FMT). Prebiotics are defined by the International Scientific Association for Probiotics and Prebiotics as "a substrate that is selectively utilized by host microorganisms conferring a health benefit," whereas probiotics are "live microorganisms that, when administered in adequate amounts, confer a health benefit on the host."[132] Several human studies investigating the beneficial effects of various prebiotic supplements in healthy and diseased (type 2 diabetes, IBS, and autism) populations demonstrate that learning and cognition is enhanced, whereas stress, anxiety, and depression levels are diminished.[2] On the other hand, probiotics also showed promising results as a form of microbiota-targeted intervention. Various cocktails of bacterial strains have demonstrated efficacy in improving clinical symptoms for multiple neurological disorders including Alzheimer's disease, schizophrenia, and depression.[2] FMT is the administration of a fecal solution from a donor to the recipient's GI tract[133] and is widely accepted as the most effective microbiota intervention for recurrent *Clostridium difficile* infections. The potential effects of FMT on symptoms of neurological disorders such as autism and PD, however, remain inconclusive due to contrasting results and limited human studies despite some positive outcomes.[134] Overall, much work remains to be done to investigate the clinical impact of targeting the microbiota–gut–brain axis on the progression of neurological disease.

CONCLUSIONS

Interactions between the gut microbiome and the nervous system affect mood and behavior in health and disease. The largest and most prominent interface between these two systems is located within the GI tract, which also contains the majority of the body's immune system. Bidirectional communication between the gut and the brain occurs through various mechanisms such as the actions of microbial metabolites and neuroimmune interactions. Dysfunction of the microbiota–gut–brain axis appears to impact symptoms and potentially the progression of many neurological diseases, ranging from autism to Alzheimer's disease. Novel intervention strategies targeting the microbiota are emerging with the aim to modulate microbial profiles to improve intestinal barrier integrity and subsequently clinical outcomes in health and disease.

REFERENCES

1. Cryan JF, Dinan TG. (2012) Mind-altering microorganisms: The impact of the gut microbiota on brain and behaviour. *Nat Rev Neurosci* **13**(10): 701–712, doi:10.1038/nrn3346.
2. Cryan JF, *et al.* (2019) The microbiota-gut-brain axis. *Physiol Rev* **99**(4): 1877–2013, doi:10.1152/physrev.00018.2018.
3. Herath M, Hosie S, Bornstein JC, Franks AE, Hill-Yardin EL. (2020) The role of the gastrointestinal mucus system in intestinal homeostasis: Implications for neurological disorders. *Front Cell Infect Microbiol* **10**: 248, doi:10.3389/fcimb.2020.00248.
4. Jangi S, *et al.* (2016) Alterations of the human gut microbiome in multiple sclerosis. *Nat Commun* **7**(1): 12015, doi:10.1038/ncomms12015.
5. Kelly JR, Minuto C, Cryan JF, Clarke G, Dinan TG. (2017) Cross talk: The microbiota and neurodevelopmental disorders. *Front Neurosci* **11**: 490, doi:10.3389/fnins.2017.00490.
6. Luna RA, *et al.* (2017) Distinct microbiome-neuroimmune signatures correlate with functional abdominal pain in children with autism spectrum disorder. *Cell Mol Gastroenterol Hepatol* **3**(2): 218–230, doi:10.1016/j.jcmgh.2016.11.008.
7. Petrov VA, *et al.* (2017) Analysis of gut microbiota in patients with Parkinson's disease. *Bull Exp Biol Med* **162**(6): 734–737, doi:10.1007/s10517-017-3700-7.
8. Zhuang Z-Q, *et al.* (2018) Gut microbiota is altered in patients with Alzheimer's disease. *J Alzheimer's Dis* **63**: 1337–1346, doi:10.3233/JAD-180176.
9. Nithianantharajah J, Balasuriya GK, Franks AE, Hill-Yardin EL. (2017) Using animal models to study the role of the gut–brain axis in autism. *Curr Dev Disord Rep* **4**(2): 28–36, doi:10.1007/s40474-017-0111-4.
10. Furness JB. (2012) The enteric nervous system and neurogastroenterology. *Nat Rev Gastroenterol Hepatol* **9**(5): 286–294, doi:10.1038/nrgastro.2012.32.

11. Costa M, Brookes SJH, Hennig GW. (2000) Anatomy and physiology of the enteric nervous system. *Gut* **47**(4): 415–419, doi:10.1136/gut.47.suppl_4.iv15.
12. Fung C, Vanden Berghe P. (2020) Functional circuits and signal processing in the enteric nervous system. *Cell Mol Life Sci* **77**: 4505–4522, doi:10.1007/s00018-020-03543-6.
13. Margolis KG, Gershon MD, Bogunovic M. (2016) Cellular organization of neuroimmune interactions in the gastrointestinal tract. *Trends Immunol* **37**(7): 487–501, doi:10.1016/j.it.2016.05.003.
14. Dinan TG, Cryan JF. (2017) The microbiome-gut-brain axis in health and disease. *Gastroenterology Clin North Am* **46**(1): 77–89, doi:10.1016/j.gtc.2016.09.007.
15. Lee CYQ, Franks AE, Hill-Yardin EL. (2020) Autism-associated synaptic mutations impact the gut-brain axis in mice. *Brain Behav Immun* **88**: 275–282, doi:10.1016/j.bbi.2020.05.072.
16. Xu M, Xu X, Li J, Li F. (2019) Association between gut microbiota and autism spectrum disorder: A systematic review and meta-analysis. *Front Psychiatry* **10**: 473, doi:10.3389/fpsyt.2019.00473.
17. Margolis KG, *et al.* (2011) Enteric neuronal density contributes to the severity of intestinal inflammation. *Gastroenterology* **141**(2): 588–598.e2, doi:10.1053/j.gastro.2011.04.047.
18. Muller PA, *et al.* (2014) Crosstalk between muscularis macrophages and enteric neurons regulates gastrointestinal motility. *Cell* **158**(2): 300–313, doi:10.1016/j.cell.2014.04.050.
19. De Schepper S, *et al.* (2018) Self-maintaining gut macrophages are essential for intestinal homeostasis. *Cell* **175**(2): 400–415.e13, doi:10.1016/j.cell.2018.07.048.
20. Gabanyi I, *et al.* (2016) Neuro-immune interactions drive tissue programming in intestinal macrophages. *Cell* **164**(3): 378–391, doi:10.1016/j.cell.2015.12.023.
21. Mowat AM, Agace WW. (2014) Regional specialization within the intestinal immune system. *Nat Rev Immunol* **14**(10): 667–685, doi:10.1038/nri3738.
22. Kimura S, *et al.* (2015) Visualization of the entire differentiation process of murine M cells: Suppression of their maturation in cecal patches. *Mucosal Immunol* **8**(3): 650–660, doi:10.1038/mi.2014.99.

23. Masahata K, *et al*. (2014) Generation of colonic IgA-secreting cells in the caecal patch. *Nat Commun* **5**(1): 3704, doi:10.1038/ncomms4704.
24. Matta SM, Hill-Yardin EL, Crack PJ. (2019) The influence of neuroinflammation in autism spectrum disorder. *Brain Behav Immun* **79**: 75–90, doi:10.1016/j.bbi.2019.04.037.
25. Onore C, Careaga M, Ashwood P. (2012) The role of immune dysfunction in the pathophysiology of autism. *Brain Behav Immun* **26**(3): 383–392, doi:10.1016/j.bbi.2011.08.007.
26. Sidoryk-Wegrzynowicz M, Wegrzynowicz M, Lee E, Bowman AB, Aschner M. (2011) Role of astrocytes in brain function and disease. *Toxicol Pathol* **39**(1): 115–123, doi:10.1177/0192623310385254.
27. Tay TL, Savage JC, Hui CW, Bisht K, Tremblay MÈ. (2017) Microglia across the lifespan: From origin to function in brain development, plasticity and cognition. *J Physiol* **595**(6): 1929–1945, doi:10.1113/jp272134.
28. Boesmans W, Lasrado R, Vanden Berghe P, Pachnis V. (2015) Heterogeneity and phenotypic plasticity of glial cells in the mammalian enteric nervous system. *Glia* **63**(2): 229–241, doi:10.1002/glia.22746.
29. Avetisyan M, *et al*. (2018) Muscularis macrophage development in the absence of an enteric nervous system. *Proc Natl Acad Sci* **115**(18): 4696–4701, doi:10.1073/pnas.1802490115.
30. McGann JC, Lioy DT, Mandel G. (2012) Astrocytes conspire with neurons during progression of neurological disease. *Curr Opin Neurobiol* **22**(5): 850–858, doi:10.1016/j.conb.2012.03.009.
31. Reemst K, Noctor SC, Lucassen PJ, Hol EM. (2016) The indispensable roles of microglia and astrocytes during brain development. *Front Hum Neurosci* **10**: 566, doi:10.3389/fnhum.2016.00566.
32. Salter MW, Stevens B. (2017) Microglia emerge as central players in brain disease. *Nat Med* **23**(9): 1018–1027, doi:10.1038/nm.4397.
33. American Psychiatric Association, DSM-5 Task Force. (2013) *Diagnostic and Statistical Manual of Mental Disorders: DSM-5* (5th ed.). Washington, DC, American Psychiatric Association.
34. Vargas DL, Nascimbene C, Krishnan C, Zimmerman AW, Pardo CA. (2005) Neuroglial activation and neuroinflammation in the brain of patients with autism. *Ann Neurol* **57**(1): 67–81, doi:10.1002/ana.20315.

35. Morgan JT, Chana G, Abramson I, Semendeferi K, Courchesne E, Everall IP. (2012) Abnormal microglial–neuronal spatial organization in the dorsolateral prefrontal cortex in autism. *Brain Res* **1456**: 72–81, doi:10.1016/j.brainres.2012.03.036.
36. McElhanon BO, McCracken C, Karpen S, Sharp WG. (2014) Gastrointestinal symptoms in autism spectrum disorder: A meta-analysis. *Paediatrics* **133**(5): 872–883, doi:10.1542/peds.2013-3995.
37. Schieve LA, *et al*. (2012) Concurrent medical conditions and health care use and needs among children with learning and behavioral developmental disabilities, National Health Interview Survey, 2006–2010. *Res Dev Disabil* **33**(2): 467–476, doi:10.1016/j.ridd.2011.10.008.
38. Vuong HE, Hsiao EY. (2017) Emerging roles for the gut microbiome in autism spectrum disorder. *Biol Psychiatry* **81**(5): 411–423, doi:10.1016/j.biopsych.2016.08.024.
39. Gorrindo P, Williams KC, Lee EB, Walker LS, McGrew SG, Levitt P. (2012) Gastrointestinal dysfunction in autism: Parental report, clinical evaluation, and associated factors. *Autism Res* **5**(2): 101–108, doi:10.1002/aur.237.
40. Careaga M, Rogers S, Hansen RL, Amaral DG, Van de Water J, Ashwood P. (2017) Immune endophenotypes in children with autism spectrum disorder. *Biol Psychiatry* **81**(5): 434–441, doi:10.1016/j.biopsych.2015.08.036.
41. Gesundheit B, *et al*. (2013) Immunological and autoimmune considerations of autism spectrum disorders. *J Autoimmun* **44**: 1–7, doi:10.1016/j.jaut.2013.05.005.
42. Inga Jácome MC, *et al*. (2016) Peripheral inflammatory markers contributing to comorbidities in autism. *Behav Sci* **6**(4): 29, doi:10.3390/bs6040029.
43. Zerbo O, *et al*. (2014) Neonatal cytokines and chemokines and risk of autism spectrum disorder: The early markers for autism (EMA) study: A case-control study. *J Neuroinflammation* **11**(1): 113, doi:10.1186/1742-2094-11-113.
44. Ashwood P, Krakowiak P, Hertz-Picciotto I, Hansen R, Pessah I, Van de Water J. (2011) Elevated plasma cytokines in autism spectrum disorders provide evidence of immune dysfunction and are associated with impaired behavioral outcome. *Brain Behav Immun* **25**(1): 40–45, doi:10.1016/j.bbi.2010.08.003.

45. Masi A, *et al.* (2017) Cytokine levels and associations with symptom severity in male and female children with autism spectrum disorder. *Mol Autism* **8**: 63, doi:10.1186/s13229-017-0176-2.
46. Sharba S, *et al.* (2019) Interleukin 4 induces rapid mucin transport, increases mucus thickness and quality and decreases colitis and Citrobacter rodentium in contact with epithelial cells. *Virulence* **10**(1): 97–117, doi:10.1080/21505594.2019.1573050.
47. Herath M, Franks AE, Bornstein JC, Hill-Yardin EL. (2020) Altered mucosal structure and barrier function in the neuroligin-3R451C mouse model of autism. *Neurogastroenterol Motil* **32**(3).
48. Seillet C, *et al.* (2020) The neuropeptide VIP confers anticipatory mucosal immunity by regulating ILC3 activity. *Nat Immunol* **21**(2): 168–177, doi:10.1038/s41590-019-0567-y.
49. Talbot J, Hahn P, Kroehling L, Nguyen H, Li D, Littman DR. (2020) Feeding-dependent VIP neuron–ILC3 circuit regulates the intestinal barrier. *Nature* **579**(7800): 575–580, doi:10.1038/s41586-020-2039-9.
50. Wang Y, Mumm JB, Herbst R, Kolbeck R, Wang Y. (2017) IL-22 increases permeability of intestinal epithelial tight junctions by enhancing claudin-2 expression. *J Immunol* **199**(9): 3316–3325, doi:10.4049/jimmunol.1700152.
51. Swaminathan M, *et al.* (2016) Video imaging and spatiotemporal maps to analyse gastrointestinal motility in mice. *J Vis Exp* **2016**(108): 53828–53828. doi:10.3791/53828.
52. Furness JB, Callaghan BP, Rivera LR, Cho HJ. (2014) The enteric nervous system and gastrointestinal innervation: Integrated local and central control, in *Microbial Endocrinology: The Microbiota-Gut-Brain Axis in Health and Disease* (pp. 39–71). New York, NY, Springer New York.
53. Hosie S, *et al.* (2019) Gastrointestinal dysfunction in patients and mice expressing the autism-associated R451C mutation in neuroligin-3. *Autism Res* **12**(7): 1043–1056, doi:10.1002/aur.2127.
54. Leembruggen AJL, *et al.* (2019) Colonic dilation and altered ex vivo gastrointestinal motility in the neuroligin-3 knockout mouse. *Autism Res* **13**(5): 691–701, doi:10.1002/aur.2109.
55. Roberts RR, Murphy JF, Young HM, Bornstein JC. (2007) Development of colonic motility in the neonatal mouse-studies using spatiotemporal maps. *Am J Physiol Gastrointest Liver Physiol* **292**(3): G930–G938, doi:10.1152/ajpgi.00444.2006

56. Spencer NJ, Hu H. (2020) Enteric nervous system: Sensory transduction, neural circuits and gastrointestinal motility. *Nat Rev Gastroenterol Hepatol* **17**(6): 338–351, doi:10.1038/s41575-020-0271-2.
57. Mazzuoli G, Schemann M. (2012) Mechanosensitive enteric neurons in the myenteric plexus of the mouse intestine. *PLoS One* **7**(7): e39887, doi:10.1371/journal.pone.0039887.
58. Kaelberer MM, Rupprecht LE, Liu WW, Weng P, Bohórquez DV. (2020) Neuropod Cells: Emerging biology of the gut-brain sensory transduction. *Annu Rev Neurosci* **43**(1): 337–353, doi:10.1146/annurev-neuro-091619-022657.
59. Mao YK, Kasper DL, Wang B, Forsythe P, Bienenstock J, Kunze WA. (2013) Bacteroides fragilis polysaccharide A is necessary and sufficient for acute activation of intestinal sensory neurons. *Nat Commun* **4**(1): 1465, doi:10.1038/ncomms2478.
60. West CL, *et al.* (2020) Microvesicles from Lactobacillus reuteri (DSM-17938) completely reproduce modulation of gut motility by bacteria in mice. *PLoS One* **15**(1): e0225481, doi:10.1371/journal.pone.0225481.
61. Wang B, *et al.* (2010) Lactobacillus reuteri ingestion and IK_{Ca} channel blockade have similar effects on rat colon motility and myenteric neurones. *Neurogastroenterol Motil* **22**(1): 98–107, e33, doi:10.1111/j.1365-2982.2009.01384.x.
62. Al-Shalan HAM, Hu D, Nicholls PK, Greene WK, Ma B. (2019) Innervation and nerve-immune cell contacts in mouse Peyer's patches. *Histol Histopathol* **35**(4): 371–383, doi:10.14670/hh-18-158.
63. Vasapolli R, *et al.* (2019) Analysis of transcriptionally active bacteria throughout the gastrointestinal tract of healthy individuals. *Gastroenterology* **157**(4): 1081–1092.e3, doi:10.1053/j.gastro.2019.05.068.
64. Sharna SS, Balasuriya GK, Hosie S, Nithianantharajah J, Franks AE, Hill-Yardin EL. (2020) Altered caecal neuroimmune interactions in the Neuroligin-3^{R451C} mouse model of Autism. *Front Cell Neurosci* **14**: 85, doi:10.3389/fncel.2020.00085.
65. Everett ML, Palestrant D, Miller SE, Bollinger RR, Parker W. (2004) Immune exclusion and immune inclusion: A new model of host-bacterial interactions in the gut. *Clin Appl Immunol Rev* **4**(5): 321–332, doi:10.1016/j.cair.2004.03.001.

66. Strugnell RA, Wijburg OLC. (2010) The role of secretory antibodies in infection immunity. *Nat Rev Microbiol* **8**(9): 656–667, doi:10.1038/nrmicro2384.

67. Latiff AHA, Kerr MA. (2007) The clinical significance of immunoglobulin A deficiency. *Ann Clin Biochem* **44**(2): 131–139, doi:10.1258/000456307780117993.

68. Bollinger RR, Barbas AS, Bush EL, LinSS, Parker W. (2007) Biofilms in the large bowel suggest an apparent function of the human vermiform appendix. *J Theor Biol* **249**(4): 826–831, doi:10.1016/j.jtbi.2007.08.032.

69. Laurin M, Everett ML, Parker W. (2011) The cecal appendix: One more immune component with a function disturbed by post-industrial culture. *Anat Rec* **294**(4): 567–579, doi:10.1002/ar.21357.

70. Hulls C, Lentle RG, de Loubens C, Janssen PWM, Chambers P, Stafford KJ. (2012) Spatiotemporal mapping of ex vivo motility in the caecum of the rabbit. *J Comp Physiol B* **182**(2): 287–297, doi:10.1007/s00360-011-0610-2.

71. Hsiao EY, *et al.* (2013) Microbiota modulate behavioral and physiological abnormalities associated with neurodevelopmental disorders. *Cell* **155**(7): 1451–1463, doi:10.1016/j.cell.2013.11.024.

72. Sgritta M, *et al.* (2019) Mechanisms underlying microbial-mediated changes in social behavior in mouse models of autism spectrum disorder. *Neuron* **101**(2): 246–259.e6, doi:10.1016/j.neuron.2018.11.018.

73. Chaidez V, Hansen RL, Hertz-Picciotto I. (2014) Gastrointestinal problems in children with autism, developmental delays or typical development. *J Autism Dev Disord* **44**(5): 1117–1127, doi:10.1007/s10803-013-1973-x.

74. Kohane IS, *et al.* (2012) The co-morbidity burden of children and young adults with autism spectrum disorders. *PLoS One* **7**(4): e33224, doi:10.1371/journal.pone.0033224.

75. Kushak RI, Winter HS, Buie TM, Cox SB, Phillips CD, Ward NL. (2017) Analysis of the duodenal microbiome in autistic individuals: Association with carbohydrate digestion. *J Pediatr Gastroenterol Nutr* **64**(5): e110–e116, doi:10.1097/mpg.0000000000001458.

76. Wang L, *et al.* (2011) Low relative abundances of the mucolytic bacterium akkermansia muciniphila and Bifidobacterium spp. in feces of children

with autism. *Appl Environ Microbiol* **77**(18): 6718–6721, doi:10.1128/AEM.05212-11.

77. Williams BL, *et al.* (2011) Impaired carbohydrate digestion and transport and mucosal dysbiosis in the intestines of children with autism and gastrointestinal disturbances. *PLoS One* **6**(9): e24585, doi:10.1371/journal.pone.0024585
78. Williams BL, Hornig M, Parekh T, Lipkin WI. (2012) Application of novel PCR-based methods for detection, quantitation, and phylogenetic characterization of Sutterella species in intestinal biopsy samples from children with autism and gastrointestinal disturbances. *MBio* **3**(1): e00261-11, doi:10.1128/mBio.00261-11.
79. Maenner MJ, *et al.* (2020) Prevalence of autism spectrum disorder among children aged 8 years — Autism and developmental disabilities monitoring network, 11 sites, United States, 2016. *MMWR Surveill Summ* **69**(4): 1–12, doi:10.15585/mmwr.ss6904a1.
80. de Lau LML, Breteler MMB. (2006) Epidemiology of Parkinson's disease. *Lancet Neurol* **5**(6): 525–535, doi:10.1016/S1474-4422(06)70471-9.
81. Braak H, *et al.* (2003) Staging of brain pathology related to sporadic Parkinson's disease. *Neurobiol Aging* **24**(2): 197–211, doi:10.1016/S0197-4580(02)00065-9.
82. Kirik D, *et al.* (2002) Parkinson-like neurodegeneration induced by targeted overexpression of α-synuclein in the nigrostriatal system. *J Neurosci* **22**(7): 2780–2791, doi:10.1523/JNEUROSCI.22-07-02780.2002.
83. Holmqvist S, *et al.* (2014) Direct evidence of Parkinson pathology spread from the gastrointestinal tract to the brain in rats. *Acta Neuropathol* **128**(6): 805–820, doi:10.1007/s00401-014-1343-6.
84. Heintz-Buschart A, *et al.* (2018) The nasal and gut microbiome in Parkinson's disease and idiopathic rapid eye movement sleep behavior disorder. *Mov Disord* **33**(1): 88–98, doi:10.1002/mds.27105.
85. Keshavarzian A, *et al.* (2015) Colonic bacterial composition in Parkinson's disease. *Mov Disord* **30**(10): 1351–1360, doi:10.1002/mds.26307.
86. Scheperjans F, *et al.* (2015) Gut microbiota are related to Parkinson's disease and clinical phenotype. *Mov Disord* **30**(3): 350–358, doi:10.1002/mds.26069.

87. Cattaneo A, *et al.* (2017) Association of brain amyloidosis with pro-inflammatory gut bacterial taxa and peripheral inflammation markers in cognitively impaired elderly. *Neurobiol Aging* **49**: 60–68, doi:10.1016/j.neurobiolaging.2016.08.019.
88. Kowalski K, Mulak A. (2019) Brain-gut-microbiota axis in Alzheimer's disease. *J Neurogastroenterol Motil* **25**(1): 48–60, doi:10.5056/jnm18087.
89. Vogt NM, *et al.* (2017) Gut microbiome alterations in Alzheimer's disease. *Sci Rep* **7**(1):13537, doi:10.1038/s41598-017-13601-y.
90. Akbari E, *et al.* (2016) Effect of probiotic supplementation on cognitive function and metabolic status in Alzheimer's disease: A randomized, double-blind and controlled trial. *Front Aging Neurosci* **8**: 256, doi:10.3389/fnagi.2016.00256.
91. Berer K, *et al.* (2017) Gut microbiota from multiple sclerosis patients enables spontaneous autoimmune encephalomyelitis in mice. *Proc Natl Acad Sci* **114**(40): 10719–10724, doi:10.1073/pnas.1711233114.
92. Cantarel BL, *et al.* (2015) Gut Microbiota in multiple sclerosis. *Poss Influence Immunomodulators* **63**(5): 729–734, doi:10.1097/JIM.0000000000000192.
93. Cekanaviciute E, *et al.* (2017) Gut bacteria from multiple sclerosis patients modulate human T cells and exacerbate symptoms in mouse models. *Proc Natl Acad Sci* **114**(40): 10713–10718, doi:10.1073/pnas.1711235114.
94. Buscarinu MC, *et al.* (2019) The contribution of dut barrier changes to multiple sclerosis pathophysiology. *Front Immunol* **10**: 1916, doi:10.3389/fimmu.2019.01916.
95. Camara-Lemarroy CR, Metz L, Meddings JB, Sharkey KA, Wee Yong V. (2018) The intestinal barrier in multiple sclerosis: Implications for pathophysiology and therapeutics. *Brain* **141**(7): 1900–1916, doi:10.1093/brain/awy131.
96. Cimdins K, *et al.* (2019) Amyloid precursor protein mediates neuronal protection from rotenone toxicity. *Mol Neurobiol* **56**(8): 5471–5482, doi:10.1007/s12035-018-1460-7.
97. Mitew S, *et al.* (2018) Pharmacogenetic stimulation of neuronal activity increases myelination in an axon-specific manner. *Nat Commun* **9**(1): 306, doi:10.1038/s41467-017-02719-2.

98. Moore Z, Mobilio F, Walker FR, Taylor JM, Crack PJ. (2020) Abrogation of type-I interferon signalling alters the microglial response to Aβ_{1-42}. *Sci Rep* **10**(1): 3153, doi:10.1038/s41598-020-59917-0.
99. Pennati A, Nylen EA, Duncan ID, Galipeau J. (2020) Regulatory B cells normalize CNS myeloid cell content in a mouse model of multiple sclerosis and promote oligodendrogenesis and remyelination. *J Neurosci* **40**: 5105–5115, doi:10.1523/JNEUROSCI.2840-19.2020.
100. Peter S, *et al.* (2016) Dysfunctional cerebellar Purkinje cells contribute to autism-like behaviour in Shank2-deficient mice. *Nat Commun* **7**(1): 12627, doi:10.1038/ncomms12627.
101. Schmeisser MJ, *et al.* (2012) Autistic-like behaviours and hyperactivity in mice lacking ProSAP1/Shank2. *Nature* **486**(7402): 256–260, doi:10.1038/nature11015.
102. Tabuchi K, *et al.* (2007) A Neuroligin-3 mutation implicated in autism increases inhibitory synaptic transmission in mice. *Science* **318**(5847): 71–76, doi:10.1126/science.1146221.
103. Wimmer VC, *et al.* (2010) Axon initial segment dysfunction in a mouse model of genetic epilepsy with febrile seizures plus. *J Clin Invest* **120**(8): 2661–2671, doi:10.1172/JCI42219.
104. Nithianantharajah J, *et al.* (2015) Bridging the translational divide: Identical cognitive touchscreen testing in mice and humans carrying mutations in a disease-relevant homologous gene. *Sci Rep* **5**(1): 14613, doi:10.1038/srep14613.
105. Silverman JL, Yang M, Lord C, Crawley JN. (2010) Behavioural phenotyping assays for mouse models of autism. *Nat Rev Neurosci* **11**(7): 490–502, doi:10.1038/nrn2851.
106. Burrows EL, *et al.* (2015) A neuroligin-3 mutation implicated in autism causes abnormal aggression and increases repetitive behavior in mice. *Mol Autism* **6**(1): 62, doi:10.1186/s13229-015-0055-7.
107. Hosie S, *et al.* (2018) Altered amygdala excitation and CB1 receptor modulation of aggressive behavior in the Neuroligin-3^{R451C} mouse model of autism. *Front Cell Neurosci* **12**: 234, doi:10.3389/fncel.2018.00234.
108. Margolis KG, *et al.* (2014) Pharmacological reduction of mucosal but not neuronal serotonin opposes inflammation in mouse intestine. *Gut* **63**(6): 928–937, doi:10.1136/gutjnl-2013-304901.

109. Girón R, Pérez-García I, Abalo R. (2016) X-ray analysis of gastrointestinal motility in conscious mice. Effects of morphine and comparison with rats. *Neurogastroenterol Motil* **28**(1): 74–84, doi:10.1111/nmo.12699.
110. Robinson AM, *et al.* (2017) Alterations of colonic function in the Winnie mouse model of spontaneous chronic colitis. *Am J Physiol Gastrointest Liver Physiol* **312**(1): G85–G102, doi:10.1152/ajpgi.00210.2016.
111. Seifi M, *et al.* (2014) Molecular and functional diversity of GABA-A receptors in the enteric nervous system of the mouse colon. *J Neurosci* **34**(31): 10361–10378, doi:10.1523/JNEUROSCI.0441-14.2014.
112. Seifi M, Swinny JD. (2016) Immunolocalization of AMPA receptor subunits within the enteric nervous system of the mouse colon and the effect of their activation on spontaneous colonic contractions. *Neurogastroenterol Motil* **28**(5): 705–720, doi:10.1111/nmo.12768.
113. Swaminathan M, Fung C, Finkelstein DI, Bornstein JC, Foong JPP. (2019) α-synuclein regulates development and function of cholinergic enteric neurons in the mouse colon. *Neuroscience* **423**: 76–85, doi:10.1016/j.neuroscience.2019.10.029.
114. Cox HM, Tough IR. (2002) Neuropeptide Y, Y1, Y2 and Y4 receptors mediate Y agonist responses in isolated human colon mucosa. *Br J Pharmacol* **135**(6): 1505–1512, doi:10.1038/sj.bjp.0704604.
115. Tough IR, Lewis CA, Fozard J, Cox HM. (2003) Dual and selective antagonism of neurokinin NK1 and NK2 receptor-mediated responses in human colon mucosa. *Naunyn-Schmiedeberg's Arch Pharmacol* **367**(2): 104–108, doi:10.1007/s00210-002-0671-6.
116. Fairman CL, Clagett-Dame M, Lennon VA, Epstein ML. (1995) Appearance of neurons in the developing chick gut. *Dev Dyn* **204**(2): 192–201, doi:10.1002/aja.1002040210.
117. Foong JPP, Parry LJ, Gwynne RM, Bornstein JC. (2010) 5-HT1A, SST1, and SST2 receptors mediate inhibitory postsynaptic potentials in the submucous plexus of the guinea pig ileum. *Am J Physiol Gastrointest Liver Physiol* **298**(3): G384–G394, doi:10.1152/ajpgi.00438.2009.
118. Gwynne RM, Bornstein JC. (2009) Electrical stimulation of the mucosa evokes slow EPSPs mediated by NK1 tachykinin receptors and by P2Y1

purinoceptors in different myenteric neurons. *Am J Physiol Gastrointest Liver Physiol* **297**(1): G179–G186, doi:10.1152/ajpgi.90700.2008.

119. Wood JD. (2016) Enteric neurobiology: Discoveries and directions, in *The Enteric Nervous System: 30 Years Later* (pp. 175–191). Cham, Springer International Publishing.
120. Hao MM, *et al.* (2011) Early emergence of neural activity in the developing mouse enteric nervous system. *J Neurosci* **31**(43): 15352–15361, doi:10.1523/JNEUROSCI.3053-11.2011.
121. Hao MM, *et al.* (2012) Early development of electrical excitability in the mouse enteric nervous system. *J Neurosci* **32**(32): 10949–10960, doi:10.1523/JNEUROSCI.1426-12.2012.
122. Nurgali K, Stebbing MJ, Furness JB. (2004) Correlation of electrophysiological and morphological characteristics of enteric neurons in the mouse colon. *J Comp Neurol* **468**(1): 112–124, doi:10.1002/cne.10948.
123. Osorio N, Delmas P. (2011) Patch clamp recording from enteric neurons in situ. *Nat Protoc* **6**(1): 15–27, doi:10.1038/nprot.2010.172.
124. Cauli B, *et al.* (2000) Classification of fusiform neocortical interneurons based on unsupervised clustering. *Proc Natl Acad Sci USA* **97**(11): 6144–6149, doi:10.1073/pnas.97.11.6144.
125. Földy C, *et al.* (2016) Single-cell RNAseq reveals cell adhesion molecule profiles in electrophysiologically defined neurons. *Proc Natl Acad Sci* **113**(35): E5222–E5231, doi:10.1073/pnas.1610155113.
126. Hill EL, *et al.* (2007) Functional CB1 receptors are broadly expressed in neocortical GABAergic and glutamatergic neurons. *J Neurophysiol* **97**(4): 2580–2589, doi:10.1152/jn.00603.2006.
127. Wood JL, Franks AE. (2018) Understanding microbiomes through trait-based ecology. *Microbiol Aust* **39**(1): 53–56, doi:10.1071/MA18014.
128. Bereza-Malcolm LT, Mann G, Franks AE. (2015) Environmental sensing of heavy metals through whole cell microbial biosensors: A synthetic biology approach. *ACS Syn Biol* **4**(5): 535–546, doi:10.1021/sb500286r.
129. Dyett BP, Yu H, Strachan J, Drummond CJ, Conn CE. (2019) Fusion dynamics of cubosome nanocarriers with model cell membranes. *Nat Commun* **10**(1): 4492, doi:10.1038/s41467-019-12508-8.

130. Kalantar-zadeh K, Ha N, Ou JZ, Berean KJ. (2017) Ingestible sensors. *ACS Sens* **2**(4): 468–483, doi:10.1021/acssensors.7b00045.
131. Hill-Burns EM, *et al.* (2017) Parkinson's disease and Parkinson's disease medications have distinct signatures of the gut microbiome. *Mov Disord* **32**(5): 739–749, doi:10.1002/mds.26942.
132. FAO/WHO. (2001) *Health and Nutritional Properties of Probiotics in Food Including Powder Milk with Live Lactic Acid Bacteria — Joint FAO/WHO Expert Consultation*. http://www.fao.org/3/a-a0512e.pdf
133. Smits LP, Bouter KEC, de Vos WM, BorodyTJ, Nieuwdorp M. (2013) Therapeutic potential of fecal microbiota transplantation. *Gastroenterology* **145**(5): 946–953, doi:10.1053/j.gastro.2013.08.058.
134. Vendrik KEW, *et al.* (2020) Fecal microbiota transplantation in neurological disorders. *Front Cell Infect Microbiol* **10**: 98, doi:10.3389/fcimb.2020.00098.

Index

Ingram Content Group UK Ltd.
Milton Keynes UK
UKHW050035040523
420996UK00005B/28

9 789811 221149